AF678563

The Law and Governance of Water Resources

NEW HORIZONS IN ENVIRONMENTAL AND ENERGY LAW

Series Editors: Kurt Deketelaere, *Professor of Law, University of Leuven, Belgium and University of Dundee, Scotland and* Zen Makuch, *Reader in Law, Barrister, Imperial College, London, UK*

Environmental law – including the pressing considerations of energy law and climate change – is an increasingly important area of legal research and practice. Given the growing interdependence of global society and the significant steps being made towards environmental protection and energy efficiency, there are few people untouched by environmental and energy lawmaking processes.

At the same time, environmental energy law is at a crossroads. The command and control methodology that evolved in the 1960s and 19970s for air, land and water protection may have reached the limit of its environmental protection achievements. New life needs to be injected into our environmental protection regimes – perhaps through the concept of sustainability in its environmental, economic and social forms. The same goes for energy policy and law, where liberalisation, environmental protection and security of supply are at the centre of attention. This important series seeks to press forward the boundaries of environmental and energy law through innovative research into environmental and energy law, doctrine and case law. Adopting a wide interpretation of environmental and energy law, it includes contributions from both leading and emerging international scholars.

Titles in this series include:

Climate Law and Developing Countries
Legal and Policy Challenges for the World Economy
Benjamin Richardson, Stepan Wood, Heather McLeod-Kilmurray and Yves le Bouthillier

The Law and Governance of Water Resources
The Challenge of Sustainability
Douglas Fisher

The Law and Governance of Water Resources

The Challenge of Sustainability

Douglas Fisher

Professor of Law, Queensland University of Technology, Australia

NEW HORIZONS IN ENVIRONMENTAL AND ENERGY LAW

Edward Elgar

Cheltenham, UK • Northampton, MA, USA

© Douglas Fisher 2009

All rights reserved. No part of this publication may be reproduced, stored in a retrieval system or transmitted in any form or by any means, electronic, mechanical or photocopying, recording, or otherwise without the prior permission of the publisher.

Published by
Edward Elgar Publishing Limited
The Lypiatts
15 Lansdown Road
Cheltenham
Glos GL50 2JA
UK

Edward Elgar Publishing, Inc.
William Pratt House
9 Dewey Court
Northampton
Massachusetts 01060
USA

Co-published by IWA Publishing, Alliance House, 12 Caxton Street, London
SW1H 0QS, UK
Tel. +44(0) 20 7654 5500, Fax +44(0) 20 7654 5555
publications@iwap.co.uk
www.iwapublishing.com
ISBN 1843392399
ISBN 9781843392392

A catalogue record for this book
is available from the British Library

Library of Congress Control Number: 2009936743

Mixed Sources
Product group from well-managed
forests and other controlled sources
www.fsc.org Cert no. SA-COC-1565
© 1996 Forest Stewardship Council
FSC

ISBN 978 1 84720 629 9

Typeset by Cambrian Typesetters, Camberley, Surrey
Printed and bound by MPG Books Group, UK

Contents

Preface

The cover of the edition of *Scientific American* of August 2008 indicated that it included an article entitled 'Running out of Water: a six-point plan to avert a global crisis'. It noted that 'the technologies and policy tools required to conserve existing freshwater and to secure more of it are known'. These needed to be implemented. Accordingly, 'governments and authorities at every level have to formulate and execute concrete plans for implementing the political, economic and technological measures that can ensure water security now and in the coming decades'. Each of these measures – it is assumed – will be implemented within an existing legal framework or within a legal framework that needs to be established to enable these measures to be implemented. It is the purpose of the following chapters to review the fundamental doctrines of the legal system which underpin and support the management of water resources at all levels including international, regional and national.

The effectiveness of the legal instruments that create or reflect fundamental legal doctrine depends ultimately upon how they are structured and upon the language used to create that structure. The focus of the following chapters is thus upon the text of these instruments, but the text of these instruments in the wider context in which they have effect. So the linguistic detail of the provision in the instrument is just as important as – if not more important than – the context in which it has effect. But both are critical in reviewing the legal arrangements for the sustainable use and development of water resources that have been emerging over recent years. There has been much reliance in the following chapters upon the text of these instruments and how they are structured. This should assist in determining the appropriate function for the law to perform in a system of sustainable water resources governance.

The texts of the principal international instruments and of a growing number of national instruments are readily available. A number of significant international instruments are less readily available. I have accordingly relied upon the text that has appeared in a number of other publications. I would accordingly like to acknowledge the availability of many of these in publications of the Food and Agriculture Organisation and of the Economic Commission for Asia and the Far East both of the United Nations. In addition I would like to acknowledge the text of a number of interstate agreements in the United States of America contained in *Dividing the Waters: the Resolution of Interstate Water Conflicts in the United States* by George William Sherk, published in 2000 by Kluwer Law International. Although the following chapters rely as much as possible upon sources which include the text of the instruments in question, it has not proved to be possible in all circumstances. I would therefore like to acknowledge in particular the information included in *Regulatory Frameworks for Water Resources Management*, which is a comparative study written by Salman M.A. Salman and Daniel D. Bradlow and published by the World Bank in Washington DC in 2006. This has enabled me to review material that would have been otherwise difficult to obtain.

One of the themes in the following chapters is the relationship between the several elements of the relevant legal system. These include on the one hand principles, policies, goals and objectives, and on the other hand rights protectable by the law and obligations enforceable by the law. The relationship between these has been become increasingly significant. In this context I would like to acknowledge the concept of 'interstitial normativity' introduced in the essay entitled 'Sustainable development and unsustainable arguments' by Vaughan Lowe and included in *International Law and Sustainable Development* edited by Alan Boyle and David Freestone and published by Oxford University Press in 1999.

Sustainable development in one form or another is increasingly emerging as a critical element of the legal arrangements underpinning the management, not only of natural resources in general, but also of water resources in particular. The following chapters may contribute to the development of some of the relevant ideas. These chapters could not have been put together without the assistance of Gail Fellows in the preparation of the manuscript and of Anne Overell in the subsequent checking and formalising of the text. My very sincere thanks to them for their assistance.

D.E. Fisher
October 2008

Introduction

To ensure that water resources are used and developed sustainably is a challenge for the global as well as the local communities with a shared interest in these resources. The way in which water resources are managed is ultimately the responsibility of everyone within the community. The responsibilities of every person depend upon the nature of each person's interest and its function within the system. Governance is a function that goes beyond management to include all of the arrangements – institutional, constitutional and legal – that contribute to how water resources are used and developed from all perspectives. The role performed by the law is critical. The law by itself cannot ensure sustainable water resources governance. But, in the absence of appropriate and effective legal arrangements, the sustainable use and development of water resources are unlikely to be achieved.

A water resources governance system performs two essential functions:

- to align the system with the hydrological cycle;
- to identify clearly and unambiguously who controls water and who is responsible for the way water resources are used and developed at each and every point in the hydrological cycle.

Traditionally it has been local custom and practice which have set these parameters. The wider public interest became increasingly acknowledged and made effective through centralised institutional arrangements. These frequently emerged through the legal system as rights of control incidental to rights of ownership – both public and private – of water resources and of water across the hydrological cycle. The emergence of nation states during the 16th and 17th centuries introduced the concept of the sovereignty of a nation state which has substantially influenced the way in which the community of interest of nation states in international water resources has been recognised and developed by international normative arrangements.

The relationship between sustainable water resources governance and the law thus introduces four complex and challenging concepts: sustainability, the hydrological cycle, governance and the law. The creation of a coherent structure that recognises and makes effective such a governance system is a challenge to be addressed by each member of the global community and by each member of the local community.

The legal arrangements for the management of water resources have ebbed and flowed as much as the water resources themselves. There has tended to be a mixture of formal as well as informal arrangements. Whatever their character, these arrangements have included rights and duties of various kinds. These rights and duties have been distributed across the range of institutions and persons involved in and affected by the management of water resources. Some of the earlier civilisations – for example in Egypt, in China or in Mesopotamia – recognised customary rights at the local level, but within a context of obligations or duties imposed and enforced by an institution protective of the public interest. The more formalised structure of rights and obligations in Roman law provided greater opportunities for individuals to protect their rights and to ensure compliance by others with their obligations. But during the later period of the Roman Empire there was to some extent a return to centralised control of water resources. Such a diversity of approaches can be seen at the beginning of the 21st century that are informal as well as formal. While earlier civilisations have probably been motivated by what is now described as the principle of sustainability, this principle is increasingly performing a more formal function within these sets of arrangements.

It is trite to observe that current legal arrangements have evolved out of past legal arrangements. Islamic law has influenced and continues to influence the way water resources are managed in many parts of the world. The structures of Roman law have influenced the way water resources are managed in many parts of Europe. Indeed Roman law influenced the development of English law during the Middle Ages and later periods. The doctrinal foundations of Roman law and of English law have influenced – as a result of European colonisation – the legal foundations for water resources management in different parts of the world.

Despite these historical and cultural differences, a number of common characteristics have emerged. The international community recognises that water resources are essentially shared resources. There is consequently a reciprocal responsibility to ensure their sustainable use and development. This has led to the principle of equitable utilisation of water resources and the correlative responsibility not to harm these resources and not to harm the environment of which they are a part. While the concept of sovereignty recognises that a state may use and develop its natural resources as it sees fit, this has increasingly been constrained by the responsibility to ensure that activities within the state are undertaken consistently with these wider responsibilities. This does not always happen in practice. Simply because there never has been and presumably never will be complete and perfect compliance with legal arrangements. Nevertheless, nation states within the international community are in their own ways responding to these challenges. In other words, a move towards greater coherence between the international legal arrangements and the national legal arrangements.

The nature of international law as such – particularly in the absence of an international legislature – means that it evolves slowly, in accordance with emerging custom and practice, influenced by specific and more detailed arrangements between and among states, and disclosing a relatively strategic and flexible approach to the normative structure of its arrangements. This is important not only for its own sake but also because such an approach is increasingly reflected in the way national arrangements are structured. These arrangements comprise a set of relatively flexible principles and directions comprising a strategic normative framework which is implemented – in the case of national arrangements – through more detailed sets of plans leading to the recognition or creation of sets of individual rights and obligations for the use and development of water resources in particular locations and for particular purposes.

This is essentially a normative structure with effect at three levels:

- strategic;
- regulatory;
- operational.

The general principles of customary international law tend to be strategic in nature. Agreements between and among states about water resources – interstate agreements – are strategic, but increasingly have become regulatory and operational. Most importantly, however, the arrangements within states are not only strategic but also regulatory and operational. The balance between these three levels depends – as one would expect – not only upon the doctrinal foundations of the legal system but also upon the political context within which the legal system operates. The structure for the management of water resources within a state is in this sense probably no different from the structure for the management of natural resources generally. In both cases there has probably always been a recognition of the public interest in how these resources are managed.

If these resources are to be used and developed sustainably – which is the current assumption – then the legal system may respond in a number of different ways. It may state strategic objectives and leave it to the public and private sector participants to perform their functions accordingly. In this case the achievement of sustainable development is not enforceable. Alternatively, the legal arrangements may be structured as a system of direct regulation which prohibits engaging in the regulated activities without a specific authorisation by a public sector agency. In this case the decision of the public sector agency is controlled – either by way of guidance or by way of direction – so as to achieve in practice sustainable development. A third alternative is for a set of rights and duties to be structured so that compliance with them will lead – or be expected to lead – to sustainable development. This set of rights and duties applies generally and automatically. It is the responsibility of public sector and private sector participants to comply with these arrangements. A failure to comply will result in appropriate legal sanctions.

In practice it is likely that a combination of these various approaches will emerge. This necessarily makes the arrangements more complicated from the legal perspective. This is because they comprise sets of public rights and public duties and sets of private rights and private duties. The exercise of these rights and the discharge of these duties operate within a strategic framework of principles and directions. Some of the elements of this system are enforceable as a matter of law while others are not. Whatever their enforceability, each impacts upon and is impacted upon by the other. It is consequently the relationship among them that is critical. Thus there is on the one hand what may be described as a macrolegal system which states the values, the principles, the goals and the objectives that comprise the normative framework of the system. And on the other hand there is a microlegal system which states the rights that are protectable and the duties that are enforceable in particular sets of circumstances, and these comprise the normative fabric of the system. A norm may be part of the framework – strategic. Or it may be part of the fabric – regulatory or operational. Their relationship has been described as 'interstitial normativity'.[1] The legal relationship between the interstices of this normative system depends upon its structure and its terminology.

International legal arrangements are often described as a combination of 'hard law' and 'soft law'. The former is potentially enforceable and the latter not. Much depends upon the status of the instrument and the way it is structured. For example, resolutions of the General Assembly of the United Nations or statements following upon international conferences do not contain rules of law in the sense of enforceable rules. Similarly, preambular statements in multilateral agreements and statements of exhortation or encouragement in such agreements do not constitute such rules of law. Nevertheless the application and interpretation of rules of law in multilateral or bilateral agreements are influenced by such statements.

These distinctions reflect to some extent the statement of a value, or perhaps even of a fundamental right, on the one hand and the realisation of this value or right through a set of regulatory or operational rules on the other hand. Much of international law can be characterised in this way. For example, the principle of equitable utilisation of water resources on the one hand and its realisation on the other hand by its operational application and its operational application in particular sets of circumstances in an interstate agreement between the relevant parties or in a determination by the International Court of Justice settling a dispute between parties. This reflects the nature of international law. The concept of 'interstitial normativity' links these elements of the system in a practical way.

It is increasingly common for the constitutions of nation states or their statutory arrangements to include statements of value or of fundamental rights relating to natural resources and the environment with implications for water resources. These statements perform various functions within the system and there is no uniformity of status. Some are part of a coherent framework, while others sit independently within the system. But whatever their function and whatever their status, increasingly one is beginning to inform the other. Even more importantly, they are part of the overall normative framework which includes sets of rights and duties cast in traditional and enforceable form: once again – an example of 'interstitial normativity'.

These international and national arrangements for the governance of water resources disclose on the one hand a framework of values, principles, goals and objectives that are on the face of it unenforceable, and on the other hand a complex set of enforceable rights and duties. The first we have described elsewhere as paralegal rules and the second as legal rules.[2] It is their relationship that is important. Paralegal rules may be described as descriptions of the fundamental *Grundnorms* which underlie the system and drive it in a particular direction, and legal rules may be described as prescriptions of the procedures and processes for decision-making by regulators and of the behaviour and conduct required of those undertaking operational activities. An example of the former is sustainability or sustainable development supported by the principles of sustainable development and an example of the latter is a duty imposed upon the users of water resources to take reasonable and practicable measures to ensure that their use of the water resource does not cause damage or injury to another user of the water resource. Accordingly:

> All of these elements are linked together in various ways as part of a legal structure. A legal structure comprises a set of rights, duties and instrumental techniques. All of these different elements are bound together by a set of rules and a set of paralegal rules. So we have a set of concepts which gives the whole system substance and two sets of rules which give it validity, coherence and possibly enforceability. It is language which binds these components of the whole system together.[3]

Integration is generally regarded as a necessity to achieve an effective system for sustainable water resources governance. Integration operates in a number of different ways: environmentally, hydrologically, institutionally, even legally. The relationship between legal and paralegal rules is essentially about what has been described as 'normative integration'.[4] This is a process to bring together all of the normative elements of the system, whether they are enforceable in the traditional sense or not. It has been suggested that this might be achieved – particularly in relation to international law – in two ways. The first is by creating a focus on interrelationships within the international legal system[5] and the second is to encourage the judiciary to use 'integration as a judicial reasoning tool'.[6] Neither is easy. But in either case the foundations need to be laid carefully before any result may ensue. The judiciary, for example, can apply and interpret constitutions and statutory provisions only in accordance with their structure and terms. If the judiciary is the ultimate guardian of the rule of law, then the legal rules and the paralegal rules need to be structured both in design and language to enable the judiciary to give effect to them so as to achieve their ultimate objectives. Clearly the legislative, executive and judicial organs of a state must be moving in the same direction. While each performs a different function, it is ultimately the responsibility not only of the legislature, the executive and the judiciary but also of everyone within the community – global, national and local – to comply with it. This is integration in the most extensive sense of the word. Thus:

> In the final analysis, integration needs to affect the entire spectrum of institutional, corporate, legal and, if I dare say, *individual* behaviour if sustainable development is to have any chance of being meaningfully realised.[7]

It is the totality of humankind that is interested in and concerned about how water resources are managed. And it is the responsibility of the totality of humankind to use and manage water resources sustainably.

It has already been suggested in this introduction that the creation of a sustainable water resources governance system is a challenge. Its incorporation within the legal system is even more of a challenge. It involves the integration of the concepts and functions related to hydrology, governance and sustainability. To be effective, the law must be aligned to address each of these at their relevant points of contact. It is normative integration in the sense discussed across international, regional, national, local and individual relationships. It applies to relationships between citizen and citizen; between citizen and state; between state and state; and among states within the international community. The normative arrangements at all of these levels differ considerably. But they impact upon each other. So it is a complex set of vertical and horizontal legal relationships. These legal relationships are continuously evolving and adapting – sometimes quite quickly.

There is a large and growing literature on all of the perspectives involved in sustainable water resources governance and the law. The approach adopted in the following chapters is consequently not a detailed and comprehensive description and review of any particular legal system – international or national – as it relates to the management of water resources. But it draws upon as many of these legal systems as possible. Nor is it an examination of any particular governance system or any particular issue associated with the management of water resources. Nor is it about the effectiveness in practical terms of any set of arrangements. It is concerned almost entirely with the fundamental doctrines, the principles, the legal rules and the paralegal rules that comprise and support the normative arrangements for the management of water resources. It is an exercise in legal doctrine, in legal structure, in legal form and ultimately in legal substance.

The following chapters seek to address six issues:

- how legal doctrines and legal structures for the management of water resources have evolved over the millennia;
- how the doctrines and structures of international legal arrangements currently address the management of water resources;
- how the doctrines and structures of national legal systems currently address the management of water resources;

- what is the nature and function of sustainable development in the context of the management of water resources;
- whether and to what extent these legal arrangements acknowledge and incorporate the sustainable development of water resources;
- how legal arrangements may be structured to achieve the sustainable development of water resources.

It is no surprise that the relationship between sustainable water resources governance and the law involves a discussion of a number of themes, some of which have already been identified. For example, the challenges of the hydrological cycle; the unique characteristics of water; the nature and function of sustainability and sustainable development; the function of governance; the fundamental doctrines of the law such as sovereignty and property; and the range of assorted public and private rights and duties all of which are set in a contextual framework of paralegal rules. Each of these themes appears, disappears, and reappears in different contexts and for different purposes. If a musical analogy may be permitted, there is a central theme – sustainable water resources governance and the law – but this is separated out into a number of linked themes. These linked themes are stated, varied, and repeated in contrapuntal and fugal fashion until – it is hoped – at the end there emerges a reconstruction of the original theme as a model for the future of a coherent and integrated system for sustainable water resources governance.

Much of the following chapters is concerned with the form, the structure and the language of a number of international and national legal instruments. In particular the structure and terminology of legal rules are in many respects remarkably different from the structure and terminology of paralegal rules. It is however their structure and their terminology which may or may not lead to coherence within the system in the sense of a set of integrated relationships between these legal and paralegal rules. Clearly there is more to legal arrangements than structure and language. But structure and language are critical if the ultimate objective is the design of a set of arrangements which is intended to achieve the sustainable development of water resources. It is in this way that the structure and language of these rules – whether legal or paralegal – indicates the function that they perform within the overall system. Are these rules norms of competence, of procedure, of conduct or behaviour, of performance, of outcome or result, or of process or methodology? The answers to these questions go to the doctrinal foundations of any legal system. These are the doctrinal foundations that give substance to the system.

Let us focus upon the fundamental point to be addressed in the following chapters. To be meaningful, legal rules are expressed in the form of language. To be enforceable, legal rules are expressed in the form of precise and unambiguous language. Precise language should itself be meaningful. But in the context of the law relating to natural resources – including water resources – the language used to express protectable rights and enforceable obligations does not necessarily conform to the degree of precision required for the creation of enforceable rights and obligations. The reason is not difficult to uncover. The objectives and perspectives of natural resources management are frequently so general and wide that they may arguably be lacking in sufficient precision to sustain enforceability.

This is not an adequate response. The discipline of natural resources law lies at the boundaries between law and policy, between independent legal duty and executively created obligation, between mandated procedures and discretionary substance, and between enforceable rules and deliberative criteria. These boundaries are increasingly becoming irrelevant. What is emerging is a normative framework comprising legal and paralegal rules which seeks to integrate all of these elements. Language is the medium through which a structure is created. It is accordingly the text and the context of the law operating coherently together which create the foundations upon which an effective structure for a sustainable water resources governance system can be built. But now is the time to turn from the architectural analogy to the substance of sustainable water resources governance.

NOTES

1. Lowe (1999, p. 31).
2. Fisher (2008).
3. Fisher (2008, p. 108).
4. French (2006, pp. 110–116).
5. French (2006, pp. 110–113).
6. French (2006, pp. 113–116).
7. French (2006, p. 116).

GENERAL REFERENCES

Fisher, D.E. (2008), 'Legal and paralegal rules for biodiversity conservation: a sequence of conceptual, linguistic and legal challenges', in Michael Jeffery et al. (eds), *Biodiversity Conservation: Law + Livelihoods* (Cambridge: Cambridge University Press), pp. 94–131.

French, Duncan (2006), 'Supporting the principle of integration in the furtherance of sustainable development: a sideways glance', *Environmental Law and Management* **18**, 103–117.

Lowe, Vaughan (1999), 'Sustainable development and unsustainable arguments', in Alan Boyle and David Freestone (eds), *International Law and Sustainable Development* (Oxford: Oxford University Press), pp. 19–37.

PART I

The challenge of sustainable water resources governance

Neither the international legal system nor national legal systems operate in a vacuum. Although the law is an intellectual discipline, it impacts upon how people behave and the consequences of their behaviour – as indeed it is designed to do. The function of the law in relation to the management of water resources must therefore be seen in the context of the hydrological cycle, how it is used by humankind, and the impacts of its use. To this there is added the formalism of legal arrangements. But legal arrangements themselves have effect within their own context. The context is the set of values which underpin the society or the community in question, and these include sustainability.

This part seeks to identify the challenges facing the community in putting together a set of effective legal arrangements for sustainable water resources governance. It does so in a number of ways:

- by reviewing the values underpinning the use of water resources over the millennia;
- by reviewing the availability of water and the impacts of its use;
- by reviewing the concept of sustainability from a range of perspectives;
- by reviewing the notion of governance in the context of sustainable development.

More particularly Chapter 1 deals with the uses and abuses of the hydrosphere. Chapter 2 looks at sustainability from a number of different perspectives including language and culture, with a specific emphasis upon the function that may be performed by sustainable development within the legal system. In does so in relation to sustainable development generally and in relation to the sustainable use and development of water resources more particularly. Chapter 3 seeks to identify the function of a governance system, and in particular a governance system for water resources use and development. Chapter 3 concludes with a synoptic review of three models of governance for sustainable development: one national, one regional and one international. Accordingly this part seeks to identify the challenges facing sustainable water resources governance from three perspectives: namely hydrology, governance and law.

1. Water resources governance in context

INTRODUCTION

The hydrosphere is an extremely complex set of natural arrangements. Many of these arrangements remain something of a mystery for human beings. Yet each human being has an interest in and in many cases a concern about what happens to the water resources that are part of this hydrosphere. This is so whether the concern is as an inhabitant of the earth, as a supplier or consumer of water, or as a strategic planner concerned with sustainable water resources governance. Civilisations have prospered and decayed over the millennia in consequence of the availability of water resources and of the ways these water resources have been managed. The cultural values of a civilisation and the ways in which these values have been incorporated in its normative, cultural, institutional and legal arrangements have significantly influenced the fate of the civilisation.

Perhaps the beginning of the 21st century has been a turning point. The evolution of different cultural values over the last 30 years of the 20th century has focused attention upon sustainability. It is now trite to observe that natural resources should be developed and used sustainably. This includes water resources. But what does this mean? How is it to be achieved? What system of governance is appropriate for the sustainable development and use of water resources? What function is there for the law to perform in achieving sustainability?

This is not to suggest that sustainability is a new concept. Natural resources have been used sustainably since time immemorial – but in different ways, in accordance with different cultural arrangements, and with differing degrees of success. Some traditional practices continue to be used to this day, while others were abandoned long ago. Much has changed in the interim: the creation of nation states; rapid and extensive industrialisation and urbanisation; a dramatic increase in population; and extended economic aspirations of both developed and developing nations. All of these changes have impacted upon the way water resources are managed. Water is a finite resource. Although it is renewable, it does not renew itself in accordance with the needs and aspirations of individuals, groups or states within the global community.

THE VALUES UNDERPINNING WATER USE

(a) A Range of Values

Water is valued for a number of diverse and potentially conflicting reasons. These may be religious or secular; economic or ecological; aesthetic or political; in effect the range of values of any particular culture. Water is a natural resource. It is part of the environment. It is removed from the environment for various purposes. It is returned to the environment, but often in different quantitative and qualitative forms. As a natural resource, water has been and continues to be subjected to competing priorities of use: for example, as an element of the means of production and as an element essential for human existence. Despite their common interest in water as a resource, civilisations have responded somewhat differently to its governance.

As soon as human beings associate with other human beings in groups of whatever size, practices and customs emerge as the basis upon which individuals and groups interact with each other. In other words, an inchoate system of law. It is no surprise that such practices and customs emerged very early in the development of communities to respond to the use of water resources. Almost inevitably the extraction and use of water involve the use of technical apparatus, albeit of a primitive kind. For example, the use of containers to extract and hold the water and the use of digging instruments to divert the water or to store it in a confined location. This led, it has been suggested, to a need for 'minimum water control in order to satisfy water demands and to ensure an equitable water distribution between different uses and users'.[1] In other words, the evolution of accepted practices, customs, norms or even inchoate laws. The nature of these arrangements reflected, as they indeed continue to reflect today, the culture of the community, the society, or the nation in question. Let us consider some examples.

(b) Philosophical and Religious Foundations of Water Use[2]

Chinese legal doctrine grew out of the theories of Confucius: the close relationship between human and natural forces or the unity of man and nature. Water in China belonged to the Emperor and was managed administratively in accordance with the procedures described in the Li-Ki, which indicated what was to be done over the months of the year. For example, in the Yueh Ling – the proceedings of government in different months – the direction for the rainy period of May was 'put the dykes and dams in good repair, clear the ditches and larger channels, and open all paths allowing no obstruction to exist'.[3] This has been described as a system of water resources governance 'in harmony with the behaviour of nature and with the seasons'.[4] Such arrangements are consistent not only with the rhythms of nature but also with the precept of Confucius that the prudent and productive management of water resources requires the exercise of political and administrative power.[5]

Significantly arrangements such as these 'in harmony with the rhythm of nature'[6] continued for many years. For example, the dredging of river channels, the building of dykes and the draining of waters contributed to the development of the city of Suzhou – once an important provincial capital during the early Imperial period. The effect was significant:

> The resultant delicate network of polders and canals was created and maintained via the combined efforts of town-dwellers and rural dwellers alike, to the mutual benefit of each. The city dwellers would benefit from the rise of this water-based 'Paradise on Earth', and the rural dwellers from the draining of surrounding swamps and development of effective irrigation systems.[7]

An example, perhaps, of the shared or common responsibility of the wider community for the governance of its water resources.

The inspiration for managing water resources in accordance with the rhythm of nature has been as much religious as philosophical, but in each case complemented by secular administration. Islamic law, for example, finding its principles in the Qur'an, has adopted an approach of shared responsibility for water resources. Water is held by the community for the benefit of all. 'Thus, everyone has the right to use clean water and to be protected from the overuse of others.' Reciprocally everyone is 'accountable for what he or she uses and for the harm caused by his or her use'.[8] There was and there remains a close relationship between rights, liabilities and the benefit of the community. Islamic law was predicated upon the essentially public character of water.[9]

The stability of early Egyptian society was very much dependent upon the waters of the river Nile. Critical was the capacity to control its floodwaters. Land and water belonged to the Pharaoh – 'a living god on earth' – supported by a sophisticated administrative structure.[10] There have been irrigation ditches for over 5,000 years. Measurements were taken of the floodwaters so that predictions for the future could be made. Many of these functions were undertaken by the administration in conjunction with the priesthood. Thus 'the worship of gods and goddesses, the religious practices, and the whole tenor of life of the people living on the banks of the rivers are inextricably linked with the rhythm of the river'.[11] Similarly, 'the hydrological and hydrogeological system in the Nile valley in Egypt was in an ecological equilibrium'.[12] Thus, while it was possible to modify to some extent the flow of the waters of the river Nile, these modifications seem to have remained entirely consistent with the forces of nature. The early Egyptians were no doubt aware of the technical limitations upon their capacity to influence the flow of rivers. Nevertheless, for religious reasons they were disposed not to interfere with the rhythm of the river to an extent that would be inconsistent with their cultural principles.

(c) Utilitarian and Economic Foundations of Water Use

The development of Roman law – so critical for later arrangements in many parts of the world – was dramatically different. While limited private rights were always recognised, much of the focus in the earlier periods was on an extensive system of public administration of public water resources managed in accordance with a regime of public law. Public administration remained important. However towards the end of the period of empire, the focus turned to a set of arrangements affording private rights in relation to water according to the civil law but subsequently finding their source in administrative concessions or grants. Thus, 'the traditional permissiveness of classical Roman law was superseded by an entirely different system, which made the administrative concession the only legitimate mode of acquisition of a right to divert water from public watercourses for irrigation and/or industrial purposes'.[13]

What emerged, therefore, was a legally based, formal and institutionalised system based upon utilitarian rather than philosophical or religious principles. This represented a cultural shift of considerable dimensions. While water may not have been used much as a source of power or energy, 'the Digest of Justinian is full of mention of uses of water for irrigation, manufacturing and consumption';[14] an acknowledgment, in other words, of the potential economic benefits associated with the use of water. Legal arrangements such as these were the catalyst for economic growth in parts of Europe during the middle ages and later.

In England the development of the common law in relation to water was influenced as much by the principles of Roman law as by the land related concepts of the feudal system. The formality of the common law has always been able to adapt to the changing circumstances in which it has had application. It has been noted that 'England's precocious economic development in the thirteenth and sixteenth centuries' and later during the industrial revolution owed much to the available water resources.[15] There was much competition for access to flowing water 'between domestic, agricultural, and manufacturing interests'.[16] These disputes were settled by the courts of common law. Thus the formalism of an adaptable common law system was able to sustain economic development for many years. But such a system of private rights and duties eventually gave way – just as Roman law did – to a set of utilitarian arrangements of an administrative nature designed more directly to secure economic growth; in essence a system of allocation by the state of rights in relation to water rather than a set of communal or individual rights finding their source in custom or rules of law.

(d) A Traditional System of Sustainable Water Use

Do the arrangements and experiences of the past provide inspiration for the future? According to Judge Weeramantry of the International Court of Justice the answer is clear: '[t]here are some principles of traditional legal systems that can be woven into the fabric of modern environmental law'.[17] One is the concept of sustainable development. The number of examples offered in support of this thesis included in particular 'the ancient irrigation-based civilisation of Sri Lanka'.[18] Over hundreds of years extending at least over two millennia rivers were dammed and reservoirs created to form a system of tanks and channels to control the movement of water and of silt. Quintessentially it involved simultaneously the development of the natural resources in question and the protection of the environment within which these resources were managed. According to Judge Weeramantry the inspiration for this approach was 'a famous principle laid down by an outstanding monarch that "not even a little water that comes from the rain is to flow into the ocean without being made useful to man"'.[19] This brings together what has been described as the philosophical, utilitarian and economic foundations of the system. In the contemporary terminology of sustainable development, it is the integration of social, economic and ecological values.

Even more importantly, however, this approach was part of the legal system. Thus, 'the felling of certain forests was prohibited, game sanctuaries were established, and royal edicts decreed that the natural resource of water was to be used to the last drop without any wastage'.[20] In Sri Lanka there emerged, therefore, a sophisticated system of water resources governance within which legal arrangements played a significant part. In particular the environment – including its water resources – was protected in accordance with a number of legal mechanisms including formal regal decrees as well as 'the practices of a sophisticated system of customary law';[21] in other words an integrated system of legal and customary arrangements involving in different ways all members and institutions within the community with a diverse range of interests and responsibilities leading to an inchoate, if not an actual, set of rights, duties and liabilities. This led Judge Weeramantry to conclude in the circumstances of the case that the Internaational Court of Justice 'must hold the balance even between the environmental considerations and the developmental considerations raised by the respective parties. The principle that enables the court to do so is the principle of sustainable development'.[22]

THE AVAILABILITY OF WATER RESOURCES FOR HUMAN USE

(a) Uses of Water

The availability of water and the uses to which it has been put have varied over the years depending upon the needs and the circumstances of each society. Although water is in theory a renewable resource, its availability at any particular location and at any particular point in time is a reflection of the vagaries of the hydrosphere – in particular those of the water cycle and of climate generally. The last two or three hundred years have witnessed increasing pressure or stress upon water resources. Demand has increased dramatically for reasons associated with industrialisation, urbanisation, demographic changes and economic growth generally. This has inspired and justified the move towards sustainability. Thus:

> For the first time, the goal of attaining an equitable and 'sustainable' society has been raised, and despite difficult and unresolved questions about the meaning of these terms, the debate has reached the entire international community. Fresh water resources play an integral role in this debate because they play an integral role in determining our overall quality of life.[23]

Water has always been used to support human existence. It has been used for irrigation for thousands of years and as a source of energy for a lesser period. Water is now used for a whole range of purposes including the essential purpose of sustaining human life. It must be recognised, however, that in many parts of the world there is not enough water to sustain life, or at least water that is drinkable without causing disease.[24] In any event irrigation remains the major use of water globally.[25] The overall statistics present a clear picture:

> Total freshwater use in 1990 was about 40 times greater than in 1700. In the twentieth century alone, water use spurted upward ninefold. Most of the increase probably derived from population growth, which in the same years was about fourfold. This means that per capita water use in the 1990s was slightly more than twice the level of 1900.[26]

Consistently with this, the population of the world has doubled over the last fifty years,[27] while at the same time the use of water across the world has increased fourfold.

(b) Sources of Water

But where does this water come from? It has been estimated – perhaps more of a guess than an estimate – that there are 1.4 billion cubic kilometres of water on earth. This water circulates in accordance with the hydrologic or water cycle. It some cases the movement of water is rapid, while in other cases water moves slowly over thousands of years. In any event it is only a very small portion of water circulating in accordance with the water cycle that is effectively available for use by human beings. About 97.5% of water on the earth is the water in the oceans. Fresh water thus comprises 2.5%. Most of this 2.5% is inaccessible to human beings. Twenty-four million cubic kilometres are in the form of ice and permanent snow and 16 million cubic kilometres are located too far below the surface of the earth to be accessible. In practical terms, therefore, human beings have access to approximately 90 thousand cubic kilometres of the freshwater in lakes and rivers.[28] This is increasingly supplemented by water available from underground aquifers.

(c) Availability of Water

While the amount of water that is available is relatively stable, its actual availability is distributed very unevenly across the continents.[29] Statistics are available which show the total annual renewable fresh water available by country (in cubic kilometres) and per capita water availability by country (in cubic metres). Figures are available for 1950 and 1995 and projections through to 2050.[30] Some comparative figures are illuminating. The following relate to per capita annual water availability in cubic metres for 1950, then 1995, and finally a medium projection for 2025: for Algeria, an arid country, 1,691; 527; and 313; for Bangladesh, with a monsoonal climate, 56,411; 19,936; and 13,096; for Brazil, a relatively wet country with a large area of land, 128,763; 43,707; and 32,087; for China, with a dynamically expanding economy, 5,047; 2,295; and 1,891; for Germany, a relatively dry country, 2,501; 2,096; and 2,114; for New Zealand, a relatively wet country with a small population, 171,384; 91,828; and 67,036; for Saudi Arabia, a dry country, 1,421; 249; and 107; for the Russian Federation, a country with a large area of land, 44,015; 30,298; and 34,233; for South Africa, a relatively dry country, 3,654; 1,206; and 698.

These figures represent the quantities of freshwater available to be used for all purposes. The mere fact that water is available in fairly large quantities does not necessary mean that it is easier to manage than if it is available in much smaller quantities. The challenges are simply different. Management is even more of a challenge because of an additional factor. Namely, about 47% of the area of the earth is made up of river basins shared by more than one country. This is particularly so in Africa and Latin America.[31] Arrangements therefore have to be made for water resources governance not only within countries but also across boundaries.

(d) Pressures on Water Availability and Use

It is no surprise – given these figures – that water resources in some countries are under stress, and in some cases under severe stress. In technical terms water stress is the ratio between human water withdrawals and water availability.[32] If it is less than 0.1 it is acceptable. But if it is 0.4 or above, it becomes critical. Thus:

> Such high values are nowadays often found in semi-arid tropical and sub-tropical regions of the Earth, while the water supply in most mid-latitude countries is so far not critical. In these regions, local deficits in drinking water of metropolitan areas are smoothed by existing water networks. The situation may worsen in the near future even in water rich countries due to an increasing water demand from different water use sectors.[33]

A number of other indicators of pressure on water resources include:

- a reduction in water supply notwithstanding measures to save water and to rationalise its use;
- deteriorating quality of water with implications for public health;
- the impacts of climate change upon aquatic ecosystems;
- increasing salinity in estuaries and the movement of sea water into aquifers;
- increasing salinisation of the soil;
- increasing salinisation of groundwater;
- more frequent catastrophic floods and droughts;
- increasing water pollution.[34]

These pressures on water resources are likely to produce an increasingly uneven availability of water across the continents, accompanied by increasing degradation of the water resources themselves and of the environment of which they are a part. The report of the fourth World Water Forum held in 2006 significantly noted:

> More than one billion people currently have no access to safe drinking water, and an estimated 2.7 billion people, or one third of the word's population, will face major water shortages by 2025.[35]

CONCLUSION

How water resources are used has been an issue since the dawn of civilisation. The relationship between water and humans has been seen over the millennia from a range of different perspectives. At the beginning of the 21st century, four 'major global problems' have been identified in relation to fresh water:

- shortage of renewable supplies;
- unequal distribution of supplies;
- problems of water quality and health;
- disastrous effects of unrestrained construction of dams and reservoirs.[36]

The explosive rate of world population growth is one factor.[37] Inadequate or misdirected efforts to harness water is another.[38] Ignorance about the functioning of basic hydrologic processes is a third factor.[39] Two of the main threats to water resources are consequently pollution and overexploitation. These lead to 'loss of habitats and biodiversity, threats to human health and limits to economic and social development'.[40] These threats strike at the very foundations of sustainable development.

Finally, there is climate change. The increase of greenhouse gas emissions into the atmosphere has been predicted to impact on water resources. Thus:

- low flows wll be more affected than high flows;
- dry climates will be more affected than humid climates.[41]

More specifically, 'human-induced climate change affects the amount and distribution of precipitation and hence the quantity of water available in watercourses and rechargeable aquifers'.[42] There is similarly a risk of increased evaporation.[43] As early as 1995, it was noted that:

> Climate change has, over the last fifteen years or so, also resulted in significant changes in the geographical pattern of rainfall within the Middle East.[44]

The challenges for the future have been stated in even more positive terms:

> Water is becoming scarcer everywhere and [this] is attributed to global atmospheric change. Unless arrested and reversed, climatic warming, together with the long range transportation of acidic and toxic pollutants, will increase demand for water, decrease water supply reliability, increase vulnerablity to droughts and floods, damage the integrity of aquatic ecosystems, and increase the potential for social conflict in many regimes of the world. This could be attributed to an increasing population pressing on limited water resources.[45]

How then, can the law address increasing demand for water, unpredictable and variable distribution or supply of water, overallocation of water, and deteriorating water quality? Perhaps through enforceable rules designed to achieve sustainable development.

NOTES

1. Caponera (1992, p. 11).
2. For more detailed discussion see Caponera (1960, pp. 245–247); Caponera (1978, pp. 96–102); Caponera (1992, pp. 14–23); de Villiers (2001, pp. 67–81); Fahlbusch (2007, pp. 77–82); Murray and Cook (2004, pp. 2–28); and Roughton (2007, pp. 111–115).
3. Book IV, Section I, Part III (10) of the Li Ki: see Legge (1968, p. 264).
4. Caponera (1960, p. 246).
5. McNeill (2000, p. 159).
6. Murray and Cook (2004, p. 25).
7. Murray and Cook (2004, p. 25).
8. Roughton (2007, p. 111).
9. Ahmad (2000, pp. 16 and 172–174).
10. Caponera (1978, p. 97) and Caponera (1992, p. 14).
11. Aswathanarayana (2001, p. 1).
12. Fahlbusch (2007, p. 78).
13. Caponera (1992, p. 42).
14. Getzler (2006, p. 13).
15. Getzler (2006, p. 1)
16. Getzler (2006, p. 1)
17. Weeramantry (1998, p. 208)
18. Weeramantry (1998, p. 208)
19. Weeramantry (1998, pp. 209 and 210).
20. Weeramantry (1998, p. 208).
21. Weeramantry (1998, p. 211).
22. Weeramantry (1998, p. 204).
23. Gleick (1993b, p. 105).
24. Gleick (1993a, p. 3).
25. Siebert and Döll (2007, p. 107).
26. McNeill (2000, pp. 120 and 121).
27. Gleick (1993b, p. 105).
28. Shiklomanov (1993, p. 13), Saeijs and van Berkel (1997, p. 3), McNeill (2000, p. 119), de Villiers (2001, pp. 30 and 31) and Lozán et al. (2007, p. 13).
29. Shiklomanov (1993, pp. 14–18) and Saeijs and van Berkel (1997, pp. 6–9).
30. de Villiers (2001, pp. 367–373).
31. de Villiers (2001, p. 81).
32. Lozán et al. (2007, p. 15).
33. Lozán et al. (2007, p. 15).
34. Lozán et al. (2007, pp. 16 and 17).
35. Beintema et al. (2006, p. 1).
36. Saeijs and van Berkel (1997, p. 5).
37. Falkenmark (1997, p. 21).
38. Gleick (1993a, p. 3).
39. Gleick (1993a, p. 3).
40. European Environment Agency (1995, p. 79).
41. Kaya (2003, p. 195).
42. Brown Weiss (2005, p. 62).
43. Tarlock (2005, p. 36).
44. Storer (1995, p. 262).
45. Grover (2006, p. 224).

GENERAL REFERENCES

Ahmad, Ali (2000), 'Islamic water law as a comparative model for maintaining water quality', *The Journal of Islamic Law and Culture* 5, 159–185.

Aswathanarayana, U. (2001), *Water Resources Management and the Environment* (Lisse: A.A. Balkema Publishers).

Beintema, Nienke, et al. (2006), *World Water Forum Bulletin* (Winnipeg: International Institute for Sustainable Development).

Brown Weiss, Edith (2005), 'Water transfers and international trade law' in Edith Brown Weiss et al. (eds), *Fresh Water and International Economic Law* (Oxford: Oxford University Press), pp. 61–89.

Caponera, D.A. (1960), 'Water law principles in the Chinese legal system', *Indian Journal of International Law* 239–275.

Caponera, D.A. (1978), 'Water laws in hydraulic civilizations', in G.L. Ullmen (ed.), *Society and History* (The Hague: Mouton Publishers), pp. 91–106.

Caponera, D.A. (1992), *Principles of Water Law and Administration* (Rotterdam: A.A. Balkema Publishers).

De Villiers, M. (2001), *Water Wars: is the World's Water Running Out?* (London: Phoenix Press).

European Environment Agency (1995), *Environment in the European Union 1995: Report for the Review of the Fifth Envirornmental Action Programme* (Luxembourg: European Communities Official Publications).

Fahlbusch, Henning (2007), 'Water and its use in early history', in José L. Lozán et al. (eds), *Global Change: Enough Water for All?* (Hamburg: Wissenschaftliche Auswertungen), pp. 77–82.

Falkenmark, M. (1997), 'Water scarcity – challenges for the future', in Edward H.P. Brans et al. (eds), *The Scarcity of Water – Emerging Legal and Policy Responses* (London: Kluwer Law Inernational), pp. 21–39.

Getzler, J. (2006), *A History of Water Rights at Common Law* (Oxford: Oxford University Press).

Gleick, Peter H. (1993a), 'An introduction to global fresh water issues', in Peter H. Gleick (ed.), *Water in Crisis: a Guide to the World's Fresh Water Resources* (Oxford: Oxford University Press), pp. 3–12.

Gleick, Peter H. (1993b), 'Water in the 21st century', in Peter H. Gleick (ed.), *Water in Crisis: a Guide to the World's Fresh Water Resources* (Oxford: Oxford University Press), pp. 105–113.

Grover, V.I. (2006), 'Water governance and water economics', in V.I. Grover (ed.), *Water: Global Common and Global Problems* (Enfield: Science Publishers), pp. 221–234.

Kaya, Ibrahim (2003), *Equitable Utilization: the Law of Non-navigatinal Uses of International Watercourses* (Aldershot: Ashgate Publishing Limited).

Legge, J. (translator) (1968), *The Sacred Books of China: the Texts of Confucianism* (Delhi: Motilal Banarsidass).

Lozán, J.L. et al. (2007), 'The water problem of our Earth: from climate and the water cycle to the human right for water', in José L. Lozán et al. (eds), *Global Change: Enough Water for All?* (Hamburg: Wissen Schaftliche Auswertungen), pp. 13–18.

McNeill, J.R. (2000), *Something New under the Sun: an Environmental History of the Twentieth Century World* (London: W.W. Norton and Company).

Murray, Geoffrey and I.G. Cook (2004), *The Greening of China* (Beijing: China Intercontinental Press).

Roughton, Geoffrey E. (2007), 'The ancient and the modern: environmental law and governance in Islam', *Columbia Journal of Environmental Law* 32, 99–139.

Saeijs, H.L.F. and M.J. van Berkel (1997), 'The global water crisis: the major issue of the twenty-first century: a growing and explosive problem', in Edward H.P. Brans et al. (eds), *The Scarity of Water: Emerging Legal and Policy Responses* (London: Kluwer Law International), pp. 3–20.

Shiklomanov, Igor A. (1993), 'Global fresh water resources', in Peter H. Gleick (ed.), *Water in Crisis: a Guide to the World's Fresh Water Resources* (Oxford: Oxford University Press), pp. 13–24.

Siebert, Stefan and Petra Döll (2007), 'Irrigation water use – a global perspective', in José L. Lozán et al. (eds), *Global Change: Enough Water for All?* (Hamburg: Wissen Schaftliche Auswertungen), pp. 104–107.

Storer, David (1995), 'The potential role of privatization in the management of water resources in the

Middle East', in J.A. Allan and Chibli Mallat (eds), *Water in the Middle East: Legal, Political and Commercial Implications* (London: Tauris Academic Studies), pp. 261–269.

Tarlock, A. Dan (2005), 'Water transfers: a means to achieve sustainable water use', in Edith Brown Weiss et al. (eds), *Fresh Water and International Economic Law* (Oxford: Oxford University Press), pp. 35–59.

Weeramantry, C.G. (1998), 'Separate opinion in the case concerning the Gabcikovo–Nagymaros Project', *International Legal Materials* 37, 204–217.

2. The conceptual framework of sustainability

INTRODUCTION

The notion of sustainability, or at least the notion of the careful and prudent use of water, has long been recognised. Only recently has it become institutionalised and formalised. There has been much debate about what it means, what is its function, what part it plays within the legal system and how it can be applied and achieved. In any event it is and will remain for the foreseeable future the pivot around which water resources are governed. While sustainability sits at the apex of discourse about water resources governance, it is the most general and least precise expression used in this context. It is sustainable development that has emerged as the most commonly used expression. There is even now little consensus about the meaning of this expression and, more importantly, how it can be put into effect.

Judge Weeramantry of the International Court of Justice, it has already been noted, has suggested that the key to good governance of water resources is sustainable development. The idea of sustainable development is quite simple and straightforward. It is to integrate the social, economic and ecological aspects of the use and development of natural resources for present and future generations in making decisions about these resources and in undertaking operations in relation to them. It is difficult to apply in specific sets of circumstances. It is more difficult to construct a system of governance for the sustainable use and development of water resources. It is perhaps impossible to create a set of legal arrangements for doing so effectively.

There has been over recent years, as we shall see, a diversity of sophisticated but by no means consistent analyses of sustainable development from a range of different perspectives. It is common ground nevertheless that sustainable development became the basis for the discourse about managing natural resources after the publication in 1987 of the report of the World Commission on Environment and Development chaired by Gro Harlem Brundtland. Sustainable development, however, did not emerge from nowhere. It may be seen to some extent as the synthesis of sets of values that had been competing with each over the previous 40 years. The focus of the international community had been first on economic development and then on environmental protection as a counterpoint to economic development. To this there has now been added social justice – at least social justice for current generations in developing countries and social justice for future generations worldwide.

Until the middle of the 1960s the priority was economic development. The focus – particularly in the developed countries – then switched to protecting the environment to some extent for its own sake. The beginning of the 21st century witnessed the increasing relevance of social justice and human rights. Environmental protection was not an innovation of the 1960s. An ethic of conservation had evolved during the second half of the 19th century itself as a counterpoint to industrialisation. The context out of which sustainable development has emerged, therefore, was an underlying but unfocused ethic of conservation seeking to align economic development with environmental protection, and then with social justice. The meanings of these words are essentially simple. But they have become overlaid with values emanating from different sources.

THE THEORETICAL FOUNDATIONS OF SUSTAINABLE DEVELOPMENT

(a) The Language of Sustainable Development

A useful point of commencement is to consider the literal meanings of the relevant expressions.[1] Let us begin with development. First of all, the development of land. Literally this means using or transforming the use of land as an asset or resource to achieve the objectives of the person or entity with the power to control the land – traditionally its owner. Now go one step further. Economic development or development of an economy is the use or transformation of the use of the resources of the community in accordance with the priorities set by the person or entity with the power to control these resources. The resources developed in this way are elements of the environment whether the environment is the geosphere, the hydrosphere, the atmosphere or the biosphere. The environment in this sense is the place where these resources are used or their use transformed to facilitate economic development, and the place where they are deposited once they have been developed in this way and have subsequently become useless waste. Environment in this sense is a neutral expression.

Protection of the environment adds a value to a term which is otherwise neutral. Protection of the environment is the function of protecting the environment from the risk of degradation to an extent incompatible with its intrinsic and human- attributed values. Now take this literal translation to its conclusion. At some point economic development – the development of the resources of the environment – will degrade the environment to an extent from which it cannot be protected. Societies have coped with this more or less successfully over many years. But the pressures on the environment over the last 100 years or so, and more particularly over the last 50 years, have prompted a range of responses from these societies. The first was conservation and the most recent has been sustainable development. What do these words mean?

The literal meaning of the noun conservation is given as preservation. This is not particularly helpful. It is more helpful to consider the meaning of the verbs conserve and preserve. To conserve is to keep from harm, decay or loss especially with a view to later use. To preserve is similarly to keep safe but to preserve also means to maintain a state of things or to retain a quality or a condition. These expressions therefore have two connotations. The first is keeping safe from harm, decay or loss. This is akin to protecting from risk. The second is contemplating future use or maintaining a state of things or retaining their quality or condition. It is this connotation of future use and maintenance or retention for the future that is particularly important. Significantly it is an element of both conservation and preservation.

If this linguistic approach is applied to the resources of the environment, then these resources must be not only kept safe from harm, decay or loss but also maintained in that state for the future. The reason for this commonality of meaning is their shared etymological derivation from the Latin verb *servare*. Conservation thus has two aspects. The first is protection and the second is use. Accordingly economically and ecologically valuable resources are protected and maintained for present and future use. In addition ecologically valuable resources are maintained not only for their value for humans but also as elements of ecosystems.

Then there is sustainable development. Development, as we have seen, is the use or transformation of the resources of the environment. 'Sustainable' means 'able to keep going continuously'. In relation to the resources of the environment, therefore, it means that these resources are able to be kept continuously in their current state. If this is linked to development, it means that the use or transformation of the resources of the environment is able to be continued – that is, for the future. This applies, no doubt, to the resources of the environment of whatever value and for whatever use.[2]

Sustainable development thus incorporates a combined notion of the development of resources and of their conservation in the literal senses already discussed. Economic or resource development had already by 1987 been aligned with protection of the environment. It remained unclear whether protection of the environment was anthropocentric or ecocentric or both. By 2002, however, the relevance and importance of social justice had been formally recognised.[3] Thus there emerged the three elements of sustainability:

- economic – the perspective of resource development;
- ecological – the perspective of environmental protection;
- social – the perspective of conservation for present and future human needs.

(b) Sustainable Development as a Cultural Value

So much for the literal meaning of words. Just as important are the nature, function and status of sustainable development as part of the governance system for natural resources. It may be a value – a moral value perhaps – which implicitly or even explicitly underlies the system. It may be the overriding goal set for strategists and policy makers. It may be one of a number of factors in making decisions about particular developments. It may ultimately be a rule of law and enforceable as such. Let us begin by considering sustainable development at its most generalised level as a broad idea or concept.

The idea of sustainability has developed historically in a rather fragmented way since the late 1960s.[4] While aspects of sustainability appear among the principles stated in the Stockholm Declaration of 1972,[5] it was probably not until some eight years later, on the publication of the World Conservation Strategy in 1980, that an attempt was made to formalise the meaning of sustainable development. It was therein stated to be 'the integration of conservation and development to ensure that modifications to the planet do indeed secure the survival and well-being of all people'.[6] This approach is entirely consistent with the literal meanings of conservation and development already discussed. The strategy went on to explain how development may be sustainable. Thus, 'for development to be sustainable it must take account of social and ecological factors, as well as economic ones; of the living and non-living resource base; and of the long term as well as the short term advantages and disadvantages of alternative action'.

This explanation incorporates the notion not only of conserving resources for the future but also of incorporating social, ecological and economic factors in making decisions and undertaking activities in relation to resources. Subsequent international instruments, including the Rio Declaration of 1992 and the Johannesburg Declaration of 2002, have in rather subtle ways modified and developed the manner in which these expressions have been used. In any event, whatever the subtleties of meaning, sustainable development undoubtedly incorporates an assessment of economic, ecological and social perspectives. The integration of these three perspectives brings together not only three different disciplines and their methodologies for assessing behaviour and decisions but also three different sets of underlying values. This presents major challenges for the governance of water resources and for its supporting legal structures.

In this sense sustainability represents one of a set of competing values to which society has committed itself. This clearly includes the international community as well as local communties. These competing values have been afforded different priorities at different times in the evolution of the society in question. As such a value, 'sustainability is best understood as a higher order social goal akin to other goals widely supported in a given society, such as democracy, equity, religious conformity, rule of law or justice'.[7] Islamic communities have traditionally acknowledged the centrality of water in their culture. The most succinct description of sustainablity yet coined has been ascribed to the principles underpinning Islam: in this case 'to cultivate their world as if they would live forever'.[8] Sustainability in this sense is implicit rather than explicit in the arrangements of a society for its governance.

The recognition of sustainable development, although itself a contested concept, may be seen alternatively as a way of making sustainability more explicit in the governance of resources. For example 'the international community's strong commitment to the concept of sustainable development has had and will continue to have a considerable evolutionary impact on existing international environmental law and on the development of new law'.[9] These developments are occurring not only at the international level but also – even more significantly for recognition as well as compliance – at regional and national levels.

It is the debate about the values underpinning sustainable development, it has been suggested, which has become critical. In other words 'sustainable development is a discourse, not a concept, and still less a scientific concept'.[10] For this purpose a discourse is composed of 'shared concepts, categories, and ideas that enable actors to understand situations'.[11] In this sense, sustainability is not so much a value in itself but rather a collection of ideas and concepts out of which a value or values will emerge as a result of informed discussions. In similar fashion it has been described as a 'paradigm'.[12] As a paradigm it is 'elucidated through practice'.[13] What happens becomes more important than what is said. Custom in preference to law: informality in preference to formality.

Despite disagreements about its specific meaning, the international community has taken sustainability to be 'the overarching framework for improving quality of life throughout the world'.[14] More specifically, it is 'the idea that the current generation wants future generations to be able to benefit from [natural] resources in much the same way that we have'.[15] Consistently with these approaches, a critique of sustainability from an Islamic perspective has suggested the concept of sustainable development should be seen 'as an evolving philosophy' which has 'severe policy implications'.[16] These implications include discourse about the foundations of the neo-classical economic paradigm[17] and about the move towards intergenerational distributive justice.[18] Perhaps in this way the concept of sustainable development acknowledges social justice and ecological justice as counterpoints to economic development.

The perception of sustainability as a discourse, a paradigm, a philosophy, a framework or a high order social goal acknowledges its importance as an indicator of the cultural values of the community. It is this all-embracing nature of sustainability that is its hallmark and its challenge. Thus it has been described as:

> A political, socio-economic, even potentially moral, objective which may in turn have certain normative consequences, but its origins lie not so much in jurisprudence as in the argument that economic, environmental and social considerations must be integrated if environmental protection and development are to be mutually supportive.[19]

Consistently with this, the idea of sustainability has been reduced to 'essentially a contested discourse rather than a set of reified policy concepts and management procedures',[20] or to a set of aspirations:

> Sustainable development has evolved into a multifaceted concept which seeks to incorporate the aspirations of developed and developing countries; and integrate economic, social and environmental interests at the local, national, regional and international levels.[21]

It has been suggested that it is the principle of sustainability that makes sustainable development 'operable' as a legal norm. 'In other words, the concept of sustainable development can only perform its normative functions in so far as it incorporates the idea of ecological sustainability.'[22] Accordingly, this has two consequences:

- Sustainable development is given meaning and direction
- Existing treaties, laws and legal principles need to be interpreted in the light of the principle of sustainability.[23]

The first is a matter of substance and the second is one of methodology. In addition, it has been suggested, sustainable development directs, at least for developed countries, 'only one political goal: any use of natural resources has to be sustainable'.[24] Thus sustainability is the principle and sustainable development is the goal. In neither case is it a rule of law in the traditional – positive – sense. But, as we shall see, it is increasingly recognised to have normative value. Finally it has been concluded that 'sustainability represents the foundational concept of the emerging "sustainability law" based on ecological justice, human rights and institutions'.[25] This approach constitutes the theoretical basis for our subsequent analysis of the relevant instruments as sets of legal and paralegal rules.

(c) Sustainable Development as a Moral Value

The Johannesburg Declaration of 2002 recognised at paragraph 12 'the deep fault line that divides human society between the rich and the poor and the ever-increasing gap between the developed and the developing worlds' and it went on to assert at paragraph 15 'a collective responsibility to advance and strengthen the interdependent and mutually reinforcing pillars of sustainable development'. To some extent at least this presents sustainable development as an idea – if not more – based on moral values. Indeed it has been suggested that 'a good argument can be made that sustainability is an ethical concept rooted in ecocentrism'.[26] This is particularly important in the context of the assumption of collective responsibility. Each and every person and entity is under a moral obligation to pursue sustainable development. This moves sustainability closer to a set of legal arrangements.

While responsibility is perhaps a moral rather than a juridical notion, the two are linked. For example, a duty to protect the natural environment has been analysed from a deontological perspective. Such an obligation is 'an obligation to our cultural tradition, to the values which we have cherished'.[27] It is not, in other words, an obligation to the state, to any particular institution or to any particular person. There is therefore no corresponding right which inheres in or is vested in any other institution or person. In this sense protection of the environment is morally justified, and this carries with it the moral responsibility of everyone to do what they can to discharge this obligation even in the absence of any personal or institutional benefit.[28] Thus there emerges what has been described as a regime of shared responsibilities in relation to the environment. According to this analysis, 'it is perfectly proper to do what one can – within limits – to force others to shoulder their share of a responsibility that is jointly shared'.[29]

The quintessentially moral foundations of sustainability have been recognised by Justice Charles Gonthier writing extrajudicially. His Honour has delineated some of the differences between law and morality. Thus:

> The real protection of our moral and physical environment must come from the responsibility people are prepared to assume in exercising their freedom. We cannot simply assume that legal rights will solve everything. We must put greater emphasis on the duties and responsibilities individuals have to each other and their community.[30]

A critical question is whether and to what extent such duties and responsibilities can and should be formalised within the legal system. Similarly, but from an Islamic cultural perspective, it has been stated categorically that 'living sustainably also involves accepting a duty to seek harmony with other people and with nature'.[31] The basis for this is sustainability as a moral value.[32] Thus:

> Broadly speaking, sustainability means a condition in which economic, social and natural systems survive and thrive together indefinitely. By virtue of being human, we live with a sense of our bonds to others. These moral obligations are the ultimate grounding for the call to sustainability.[33]

Significantly a call for collective responsibility to achieve the goal of environmental protection had been included as early as 1972 in paragraph 7 of the preamble to the Stockholm Declaration. This 'will demand the acceptance of responsibility by citizens and communities and by enterprises and institutuions at every level, all sharing equitably in common efforts'. Again, the question is when and how the moral commitment to sustainability should be reflected in the institutional and juridical arrangements put in place by society.

THE OPERATIONAL FOUNDATIONS OF SUSTAINABLE DEVELOPMENT

(a) Sustainable Development as a Function

Whether sustainable development is a discourse, a paradigm, an evolving philosophy, an overarching framework, or a fundamental moral value invoking collective responsibility, it is more than a mere platitude. It performs a function or functions in the governance of resources despite its indeterminate and variable substance. One of the fundamental issues is whether sustainable development is a matter of procedure, of substance or of both.[34] This is a point of practical significance. If it is a matter of procedure, it is a statement about how decisions are made or functions performed. If it is a matter of substance, it is either the criterion which directs the merits of a decision or the undertaking of an activity or one of the criteria which guide the making of a decision or the undertaking of an activity. As a matter of procedure it has been described variously as a multidimensional process[35] or as an ongoing process of change.[36] More specifically, 'sustainable development can most constructively be viewed as a process, which is participatory, incremental and interactive'.[37] Process is a significant aspect of governance. But so is substance. While process is usually a justiciable issue, matters of substance are traditionally – at least for the most part in common law jurisdictions – beyond the scope of judicial review.

Whether sustainable development is process or substance, it performs a function within international as well as national legal arrangements. It has been described as 'an umbrella concept'.[38] The details of the arrangements protected by the umbrella are not at all clearly defined. More specifically, it has been suggested that sustainable development is:

> A legal term that refers to processes, principles and objectives as well to a large body of international agreements on environment, economics and civil and political rights.[39]

There was in addition a perceived need for all of these elements to be integrated within each other. Sustainable development had by the early 1990s 'become a major policy goal of the international community'.[40] It has consequently evolved as 'a bridging principle that combines the developmental aspects of international economic law with the environmental considerations of development. In doing so, this principle seeks to reconcile the developmental aspirations of states with the need for environmental protection while carrying out their development projects'.[41] It has become, therefore, the link between on the one hand how decisions are made and how activities are undertaken and on the other hand what decisions are made and what activities are undertaken. Thus process and substance are merged.

In the context of customary law and sustainable development, their relationship has been described as interdependent in this way. Sustainable development has become the 'meta-goal', while customary precautionary principles are 'instruments to provide good solutions to that end'.[42] The detail of the arrangements comprising the 'umbrella' has been revealed. 'The attainment of sustainable development requires the balancing up of various factors, principles and rules of international law with a view to achieving the reconciliation of the legitimate interest of states in pursuing their economic development and the protection of the environment.'[43] Sustainable development thus has a function to perform, not only nationally but also internationally, by drawing together the several elements of the legal system. Some of these are process and some of these are substance. Sustainable development may perform as an objective, a principle or a rule. Accordingly, the law itself is beginning to describe, and in appropriate circumstances to prescribe, the function performed by sustainable development.

Although much depends on how the arrangements are structured, a closer relationship is perhaps emerging between process and substance. Thus sustainable development may perform a number of functions in the process of making decisions or undertaking activities. For example, sustainable development may be:

- the overriding objective in assessing what high level strategy is the most appropriate policy option;
- the objective to be achieved in making specific decisions and undertaking specific activities;
- the objective to be given priority in making such decisions and undertaking such activities;
- one among a number of objectives so to be achieved;
- the principle that must guide the making of decisions and the undertaking of activities;
- a principle that must be considered in making decisions and undertaking activities;
- a factor to be considered in making decisions and undertaking activities.

The formal function to be performed by sustainable development depends upon the language of the instrument which states the function – whether the instrument has legal status or not. For example, principles are often formulated without reference to persons, institutions or states. Rights normally are linked to persons, institution or states. And so are duties and obligations. Sustainable development is sometimes described only as such. Increasingly it is supported, complemented, explained or expanded by reference to statements or propositions described variously as, for example, factors, considerations, guidelines, practices or even principles. These give it lateral substance. One or two examples describing the international arrangements will demonstrate the point.

The Rio Declaration of 1992 – the formal catalyst for developing the concept of sustainable development – was the basis for a description of the four substantive and the two procedural elements of sustainable development.[44] The four elements of substance:

- the sustainable utilisation of natural resources;
- the integration of environmental protection and economic development;
- the right to development;
- the pursuit of equity in the allocation of resources both within the present generation and between present and future generations.

And the two elements of procedure:

- public participation in decision making;
- environmental impact assessment.

Four core elements have been similarly identified but in different language:

- the needs of present and future generations must be taken into account (intergenerational equity);
- the needs of the world's poor must receive priority and abject poverty eliminated (intragenerational equity);
- the environment needs to be preserved at least to a significant degree;
- economic, social and environmental policies must be integrated.[45]

Perhaps the most comprehensive and authoritative statement is the New Delhi Declaration of 2002, put together by the International Law Association. It contains not only a statement of seven principles but also an extended description and explanation of them. The seven principles are:

- the duty of states to ensure sustainable use of natural resources;
- the principle of equity and the eradication of poverty;
- the principle of common but differentiated responsibilities;
- the principle of the precautionary approach to human health, natural resources and ecosystems;
- the principle of public participation and access to information and justice;
- the principle of good governance;
- the principle of integration and interrelationship, in particular in relation to human rights and social, economic and environmental objectives.

The language used in the New Delhi Declaration is indicative of some of the functions performed by sustainable development. The last principle, for example, refers to the relationship between human rights and social, economic and environmental objectives. While much depends upon the function performed by objectives in general, the relationship between an objective and a right creates not only linguistic but also juridical difficulties. The principle of common but differentiated responsibilities introduces a different linguistic phenomenon – responsibilities. The explanation accompanying this statement of principle declares that all states are under a duty to cooperate in the achievement of global sustainable development and the protection of the environment. While on the face of it this separates protection of the environment from sustainable development – of which it is otherwise a part – it perhaps does so to emphasise the significance of environmental protection. Particularly so since the principle assumes in this explanation the form of a duty – albeit a duty to cooperate. The first principle itself assumes the form of a duty. The accompanying explanation indicates that states are under a duty to manage natural resources in a rational, sustainable and safe way. While the New Delhi Declaration is cast within a framework of principles, the language is to some extent that of rights, responsibilities and duties. In any event these are the declared principles of international law relating to sustainable development. Their provenance, however, indicates that this statement of these rights, responsibilities and duties is not a formal part of international law.

(b) Sustainable Development as a Legal Norm

The functions capable of being performed by sustainable development are as diverse as its conceptual foundations. It is for those using and managing natural resources in the first instance to determine what sustainable development is and what function it performs in their use and management systems. But that is the end rather than the beginning of the matter. Sustainable development now performs a function, or perhaps a range of functions, within the legal system at international, regional and national levels. Its precise function within the legal system is a matter of controversy. It is important therefore to analyse the range of juridical approaches to it.

Judge Weeramantry has positively affirmed that sustainable development is a principle with normative value.[46] But what is a norm? The dictionary tells us relevantly that it is a standard or customary behaviour. In other words how people should behave or how they do behave. The word itself derives from the Latin noun *norma*, which means a carpenter's square. Its etymological origins thus give it a connotation of certainty, precision and specificity. But sustainable development is as much about making decisions as about standards of behaviour. Norms of decision-making are intrinsically different from norms of behaviour. Perhaps sustainable development is a rule of law the content of which responds differently to the several contexts in which it appears. In the context of the legal system this effectively means that the standard of decision-making or actual decision-making or behaviour must be able to be identified sufficiently precisely to be able to be recognised and enforced. This is the challenge for much of environmental law, and in particular for sustainable development.

Whether sustainable development is a legal rule has attracted a variety of responses. On the one hand, it has been suggested, the trend is against sustainable development 'having any normative status within international law'.[47] On the other hand, sustainable development is arguably 'vested with some degree of normative rather than merely exhortatory status in international law'.[48] Consistently, sustainable development has been regarded as a helpful 'principled approach which is both substantive and procedural'.[49] Similarly, the view has been expressed that 'the sustainability discourse has coalesced into a set of fairly general normative principles, rather than a precise formula for environmental policy'.[50] A principle, however, guides rather than determines decision-making and behaviour. In this sense a principle is, on the face of it, insufficiently precise in normative value to constitute an obligation which is an enforceable rule of law.

Let us take the analysis one step further. Is it futile to search for a structure or form that ensures the enforceability of sustainable development as a rule of law? We need to bear in mind that sustainable development may be seen as a matter of substance, as a matter of procedure or as a matter of procedure verging on a matter of substance. Such a search has produced this response first of all:

> Normative uncertainty, coupled with the absence of justiciable standards for review, strongly suggest that there is as yet no international legal obligation that development must be sustainable, and that decisions on what constitutes sustainability rest primarily with individual governments.[51]

In other words sustainable development is unlikely to raise a justiciable issue. In this event whether a particular development is sustainable is an issue which calls for the factual prediction of an outcome by the person or the institution with the competence to control the development. However, there is a second point to consider:

> A more plausible argument is that although international law may not require development to be sustainable, it does require development decisions to be the outcome of a process which promotes sustainable development.[52]

In this case there is a close link between a justiciable process and an outcome which is by itself not justiciable. But is the relationship between the process and the outcome – the methodology of decision-making – justiciable?

In any of these contexts sustainable development has a part to play within the legal arrangements for the governance of natural resources. Even if the substantive decision on the merits – the outcome – is not justiciable, the incorporation of sustainable development in the deliberative processes of decision-making – principally by government in one form or other – may or may not raise a justiciable issue depending upon how the arrangements describing the process are structured. Indeed, as it has been clearly pointed out, one of the conceptual deficiencies in relation to sustainable development is that, if it were to be considered as an obligation, 'doubts would arise whether it represents a duty of achievement or a duty of performance'.[53] In other words, if it were an obligation, is it a duty to achieve sustainable development or is it a duty to perform functions in relation to the achievement of sustainable development? The first is a substantive issue and the second is a process issue, but closely linked to the substantive issue. A fine distinction, no doubt, but one which reflects not only the nature of sustainable development but also the way in which the obligation is structured as a matter of language, and hence of law.

The normative dynamism of sustainable development may be analysed further. Norms assume different forms and structures within the legal system simply because they perform different functions. Norms emerge as rights and duties of various kinds. Some norms impact on decision-making and others on behaviour. Just as the relationship between process and outcome is critical for the enforceability of sustainable development, so is the relationship between decision-making and behavioural norms. Aspects of these relationships have been scrutinised according to three distinct analytical approaches.

There has been considerable discussion over the years about the differences between principles and rules or, in our terminology, between paralegal rules and legal rules. 'Principles and rules', it has been suggested, 'are eqully valid legal norms, but with different functions.'[54] In other words, the normative arrangements constituted by legal rules and paralegal rules are different in nature, in substance and in status. The question is what is the normative status of sustainable development. The response has been this:

> The concept of sustainable development is located somewhere in the spectrum of (non-legal) policies and (legal) norms. We can perhaps exclude two extremes. On the one hand, the concept is more than a mere political ideal as it has been referred to repeatedly in hard law and, in much detail, in soft law. On the other hand, the concept is less than a rule as it lacks direct legal consequence. The proximity to policies seems closer than to a rule, but how close? This depends on the degree of political and legal guidance the concept is capable of providing.[55]

Let us take this theoretical analysis further by considering in more detail the nature and function of the normative arrangements described in the two other approaches.

The second of these three approaches raises initially two interesting issues:

- whether environmental law is supported by a *Grundnorm* as a hypothesis of juristic thinking consistent with the Kelsen theory[56]
- whether environmental law discloses a set of jural correlatives in accordance with the Hohfeld theory.[57]

If this is extrapolated to sustainable development, the first question is whether sustainable development is a *Grundnorm* of the legal system. The second question is where does sustainable development fit within the set of jural correlatives described first as claim and duty, and second as competence and subjection. Or, more simplistically, does sustainable development create a justiciable issue? What normative value does sustainable development have? Where is the enforceable duty located?

This analysis later adopted a juridical approach to obligations – particularly international obligations – defined by different types of norms. These included norms of conduct[58] and norms of competence.[59] These are similar to but not necessarily the same as behavioural norms and decision-making norms. This analysis then went on to consider how such norms are structured so as to create enforceable obligations. It was prefaced by the creation of a threefold classification for the creation of rules:

- stipulations of objectives
- rules on permissibility of various conduct (ie norms of conduct)
- rules on implementation (eg concerning reporting, monitoring, supervision, enforcement and sanctions).[60]

This language is interesting. A reference is made to 'stipulations' of objectives. The issue, then, is whether and when the stipulation of an objective becomes an obligation that assumes the form of a rule. Finally, and most importantly, there are a number of ways according to which obligations may be defined so as to give effect to these several norms. The three ways particularly relevant for present purposes are these:

- obligations defined by balancing norms
- obligations defined by fixed norms
- obligations defined by goal-oriented norms.[61]

Although the context of this analysis was environmental law generally and not sustainable development, it is interesting to investigate the extent to which sustainable development falls within these classifications.

What does each entail? A balancing norm is described as 'a particular kind of regulatory technique, a legal approach, for defining obligations, whereby the balance takes place 'within' the norm'.[62] This involves balancing different interests against each other in light of the relevant considerations. If sustainable development were a balancing norm, it would require economic, ecological and social interests to be balanced against each other in reaching the solution. A fixed norm is one which 'itself denotes the solution of the normative problem in an "all-or-nothing fashion"'.[63] Specifically this eschews a balancing approach. In this event sustainable development is the standard – fixed by the legal system – which determines the solution. A goal-oriented norm is one which stipulates 'the goal to be achieved, in more or less precise terms, without defining the means to reach the goal'.[64] Such norms are not without legal effect. Goal-oriented norms 'reveal a framework structure and need to be supplemented by other norms'.[65] Thus sustainable development as a goal or an objective is not by itself sufficient. It needs to be supported, in other words, by other elements – supporting norms – of the overall structure of which it is a part.

The significance of this approach is this. Although the types of norms thereby identified may well overlap, there has emerged a theoretical foundation for environmental law in general. This is very much a reflection of its nature. It is an accommodation of public rights and public duties together with private rights and private duties within a complex matrix of instruments, some of which are substantive, some procedural, some reactive and some proactive. Each plays a part in the overall system and each has some normative value. The normative value will vary depending upon how the norm is structured. It may be – albeit unlikely – that sustainable development has emerged as a *Grundnorm* of the system. If so, it would be a fixed norm but one of relatively variable substance. Sustainable development is more likely to be identified as a balancing norm or a goal-oriented norm. If it is such a norm, it may not constitute an enforceable obligation by itself. In any event it is linked to a range of other norms which themselves may be, and indeed are likely to be, enforceable.

Although the context of the third of these three analytical approaches is rather different, much of the analysis is consistent with the first approach. The point of this third approach is that sustainable development is not a binding norm of international law within the traditional meaning of this expression.[66] But this does not necessarily mean that sustainable development is not a norm. Thus:

> There is a sense in which the concept of sustainable development exemplifies another species of normativity which is of great potential value in the handling of concepts of international environmental law.[67]

If sustainable development does not have the 'fundamentally norm-creating character'[68] required to be an enforceable rule of law, what is it? The answer is clear:

> It is a convenient, if imprecise, label for a general policy goal which may be adopted by states unilaterally, bilaterally, or multilaterally. It is clearly entitled to a place in the Pantheon of concepts that are not to be questioned in polite company, along with democracy, human rights, and the sovereign equality of States.[69]

But it is more than that:

> Sustainable development can properly claim a normative status as an element of the process of judicial reasoning. It is meta-principle, acting upon other legal rules and principles – a legal concept exercising a kind of interstitial normativity, pushing and pulling the boundaries of true primary norms when they threaten to overlap or conflict with each other.[70]

The concept of interstitial normativity – the relationship between legal and paralegal rules – is critical. So much so that this expression has been adopted in this context and will appear from time to time in this analysis.

There is no doubt a range of norms of this kind. They link substantive issues with process issues. They link norms of competence with norms of behaviour. They constitute an approach to the methodology of decision-making in particular cases. They provide what is called 'coherence and consistency'[71] in making decisions in particular cases. They do not, it is emphasised, 'regulate the conduct of legal persons directly, and they are not addressed to those persons'.[72] Clearly so. This is not their purpose. They are designed to set the parameters of decision-making and of behaviour and to ensure – perhaps even in an enforceable sense – that these decisions and this behaviour do not stray beyond these parameters. While these norms are directed at no one in particular, they in reality apply to everyone in general including strategists, planners, decision-makers and those undertaking operational activities.

There is a very clear example of this:

> The concept of sustainable development precludes the possibility of a tribunal deciding a case by upholding a completely unfettered right of a property-owner to utilize the property in a particular way without any regard for the serious and irreversible environmental harm caused by that particular use.[73]

If sustainable development has normative value in this way, it becomes – it is suggested – a part not only of judicial reasoning in the context of a dispute but also a part of the methodology of decision-making on the part of whoever is responsible for making the decision, whether as an administrator or an operator in either the public or the private sector. Sustainability has thus evolved and continues to evolve as an element, and potentially an enforceable element, of the juridical structure for managing natural resources.

THE SUSTAINABLE DEVELOPMENT OF WATER RESOURCES

So far the discussion has centred upon the sustainable development of natural resources at large. Water is a natural resource. It is part of the environment. So any reference to natural resources or the environment necessarily includes water and water resources unless they are specially excluded. Is the foregoing general discussion about sustainability and sustainable development able to be translated to water resources? There is in principle no reason why not. Indeed, 'the concept of sustainable development plays a pervasive role in strategies to avoid and alleviate water scarcity'.[74] The way in which sustainable development applies to water resources depends not only upon how the law describes the function of sustainable development but also upon the characteristics of water as a resource and the circumstances of any particular case.

The Stockholm Declaration of 1972 and the Rio Declaration of 1992 were concerned with the environment and with natural resources generally. Neither declaration dealt specifically with water, although the Stockholm Declaration specifically included water as one of the natural resources of the earth. Similarly neither the Convention on Climate Change of 1992 nor the Convention on Biological Diversity of 1992 – which emerged from the 1992 Rio Conference – apply in specific terms to water or water resources. Nevertheless, the climate system which is protected by the Convention on Climate Change includes within the definition in Article 1(3) the hydrosphere. Similarly the Convention on Biological Diversity includes in Article 2 aquatic ecosystems within the meaning of biological diversity.

Immediately prior to the Rio Conference in 1992 the international community had directed its attention to water resources at a conference convened in Ireland earlier that year. The international community thus subscribed to the Dublin Statement on Water and Sustainable Development of 1992. It noted in its preface that 'concerted action is needed to reverse the present trends of overconsumption, pollution, and rising threats from drought and floods'. Four guiding principles were the basis for future action. These were:

- fresh water is a finite and vulnerable resource, essential to sustain life, development and the environment;
- water development and management should be based on a participatory approach involving users, planners and policy-makers at all levels;
- women play a central part in the provision, management and safeguarding of water;
- water has an economic value in all its competing uses and should be recognised as an economic good.

The first and fourth are essentially substantive and the second and third essentially procedural. The procedural principles are important because they include within them functions to be performed by all persons and institutions involved in and affected by the management of water resources. To this extent it is an element of procedural integration. The two substantive principles incorporate the three elements of sustainable development: social, economic and environmental. A number of benefits were expected to accrue as a result of the application of these principles. These were:

- alleviation of poverty and disease;
- protection against natural disasters;
- water conservation and reuse;
- sustainable urban development;
- agricultural production and rural water supply;
- protecting aquatic ecosystems;
- resolving water conflicts.

It was contemplated that the preparation and implementation of integrated management plans would be critical to the achievement of sustainable development based upon these principles. The importance, in other words, of the principle of integration.

The plan of action arising out of the 1992 Rio Conference – Agenda 21 – deals specifically with the protection of the quality and supply of freshwater resources. Not surprisingly, the concept of integrated water resources management is critical. Thus, according to paragraph 18.8 of Agenda 21:

> Integrated water resources management is based on the perception of water as an integral part of the ecosystem, a natural resource and a social and economic good, whose quantity and quality determine the nature of its utilization. To this end, water resources have to be protected, taking into account the functioning of aquatic ecosystems and the perenniality of the resource, in order to satisfy and reconcile needs for water in human activities. In developing and using water resources priority has to be given to the satisfaction of basic needs and the safeguarding of ecosystems.[75]

Water and water resources are, therefore, approached internationally – and the same is true of regional and national arrangements – either as a resource in its own terms or as one among a number of natural resources. This is an issue of some significance. Much depends upon whether the set of legal arrangements applies specifically to one resource or one set of circumstances or – as would be expected in relation to sustainable development – to one resource as part of a much wider set of relationships either of an economic, social or ecological character.

Sustainability and sustainable development – whatever their meaning and whatever their function – apply to water and water resources as much as they apply to any other natural resource. It has already been suggested that the idea of sustainable development is quite simple. Namely to integrate the social, economic and ecological aspects of the use and development of natural resources for present and future generations in making decisions about these resources and undertaking operations in relation to them. Its application in particular sets of circumstances is no doubt difficult. The function that it performs within the legal system is not always clear. But this does not detract from the essential simplicity of the idea.

We have looked at sustainable development from a number of different perspectives. There is no doubt about its relevance for water and water resources. It may be useful to conclude from a different perspective by looking at what the sustainable use and development of water resources might involve from a practical perspective. A positive response is this:

> Sustainability must seek to balance the water available at any particular point in time and space with the demand for water for various 'uses', and the need for enough water to safeguard human health and the aquatic ecosystem. Underpinning this, the water available must be of sufficient quality to satisfy the different users of water including again safeguardng human and other life.[76]

A combination of anthropocentric and ecocentric approaches in relation to both quantity and quality of water. But how is sustainability to be assessed? By reference, it has been suggested in the context of regulation by licensing, 'to the resource availability and the effect that overuse on non-sustainable abstraction may have on the waters'.[77] The definition of sustainable development incorporated in the Brundtland Report has been translated specifically to water resources in these words:

> Sustainable development of water resources encompasses those patterns of water resources utilization that will enhance social and economic benefits for the present and future generations, without impairing the eco-hydrological processes. This implies that any use of water resources which either leads to or has the potentiality to lead to, the degradation of the eco-hydrological system is ipso facto unsustainable and hence unacceptable.[78]

Clearly this encapsulates the three elements of sustainable development. Its significance lies in the specific reference to eco-hydrological systems as distinct from ecosystems generally. And if substance and process are linked:

> Development cannot be sustainable if it exacerbates conditions of water scarcity, and development is virtually guaranteed to be unsustainable unless it is based upon basin-wide, inegrated planning and management for freshwater resources.[79]

Sustainable water resources management has been described as a practice. As such it:

> Avoids irreversible (and quasi-irreversible) damages to the resource water and all resources linked to it (eg soil and ecosystems). It conserves in the long run the ability of the resource to extend its services including ecological services.[80]

This is a rather narrow approach with a distinct focus on water as a resource. Nevertheless it recognises, at least indirectly, the relationship between water as a resource and the other values associated with the management of water resources.

Another similarly practical approach is cast more in the form of a number of objectives. These objectives, however, tend to reflect the immediate rather than the longer term future goals of water resources management. The four objectives are these:

- to provide clean water and safe sanitation for all people;
- to secure sufficient water for industry and agriculture;
- to provide effective water management, particularly measures for water conservation, water efficiency and water safety;
- to improve international cooperation on water, particularly to allocate sufficient financial and organisational resources for a global water strategy.[81]

Water is a natural resource. It is used individually and locally. It is sourced from a catchment or an aquifer. The catchment or aquifer is managed in accordance with the values of the community. Increasingly these values are directed at the sustainable use and development of these water resources. For this purpose the generic meaning and function of sustainable development apply to water resources as to any other resource. Water discloses features and characteristics of its own. These must be taken into account in analysing a system of water resources governance. We have considered the notion of sustainable development and what it may mean in a number of different contexts. The introduction of a management as well as a legal perspective invites an analysis not so much of management as of governance in relation to natural resources generally and then more specifically – once again – to water resources.

CONCLUSION

While the notions of sustainability and sustainable development are in themselves simple ideas, their translation into effective institutional and legal arrangements is as difficult as their application in particular sets of circumstances. This is so not only for the use and development of water resources but also for the use and development of any natural resources. The international community has accepted sustainable development as a goal. Its function within the legal system remains something of a mystery. It is certainly emerging as an element of the context within which water resources are used and developed. In this sense it informs and guides decision-making and activities. Whether it does so as a process or in relation to substance remains equivocal. But process and substance are merging. The one impacts upon the other – at least indirectly, if not directly – and any decision or activity is informed by both. In this way sustainable development may be a norm of behaviour or conduct; a norm of performance; a norm of methodology; or a norm of outcome or result. Depending upon how it is structured and upon the function that it is designed to perform, sustainable development may emerge with the status of interstitial normativity or as an enforceable rule of law. While its normative status is likely to remain somewhat ambiguous, sustainable development undoubtedly in one form or another has a function to perform within the legal arrangements for water resources governance.

NOTES

1. The source is the *Oxford English Dictionary*.
2. In 1983 when a national conservation strategy was being developed for Australia development was described as 'the modification of the biosphere and the application of human, financial, living and non-living resources to satisfy human needs and improve the quality of human life' and conservation as 'the management of human use of the biosphere so that it may yield the greatest sustainable benefit to present generations while maintaining its potential to meet the needs and aspirations of future generations'. See Fisher (1987, p. 214).
3. The Johannesburg Declaration 2002 included at para. 2 a commitment 'to build a humane, equitable and caring global society cognizant of the need for human dignity for all'.
4. Weeramantry (1998, pp. 206 and 207), Rieu-Clarke (2005, pp. 31–46), Cordonier Segger and Khalfan (2006, pp. 15–43), Magraw and Hawke (2007, pp. 614– 618).
5. E.g. Principle 2 (intragenerational and intergenerational equity), principle 3 (renewable resource conservation) and principle 5 (non-renewable resource conservation).
6. See Pinto (1995, pp. 72 and 73).
7. Dovers and Connor (2006, p. 31).
8. Allan (1995, p. 344).
9. Boyle and Freestone (1999, p. 6).

10. Dryzek (2007, p. 56).
11. Dryzek (2007, p. 46).
12. Magraw and Hawke (2007, p. 614).
13. Orebech and Bosselman (2005, p. 14).
14. Magraw and Hawke (2007, p. 614).
15. Orebech and Bosselman (2005, p. 12).
16. Iqbal (2005, p. 1).
17. Iqbal (2005, p. 7).
18. Yousri (2005, p. 30).
19. French (2005, p. 35).
20. Richardson and Wood (2005, p. 13).
21. Rieu-Clarke (2005, p. 46).
22. Bosselmann (2008, p. 41).
23. Bosselmann (2008, p. 41).
24. Bosselmann (2008, p. 41).
25. Bosselmann (2008, p. 41).
26. Bosselmann (2006, p. 130).
27. Sagoff (1974, p. 265).
28. Sagoff (1974, p. 265).
29. Goodin (1995, p. 314).
30. Gonthier (2005, p. 36).
31. Iqbal (2005, p. 9).
32. Iqbal (2005, p. 10).
33. Iqbal (2005, p. 10).
34. Boyle and Freestone (1999, p. 8).
35. Mutembwa (2000, p. 262).
36. Ginther and de Waart (1995, p. 9).
37. Adande et al. (2000, p. 291).
38. Fitzmaurice and Elias (2004, p. 159).
39. Sands (1994, p. 379).
40. Subedi (2005, p. 14).
41. Subedi (2005, p. 14).
42. Ørebech and Bosselman (2005, p. 21).
43. Fuentes (1998, p. 126).
44. Boyle and Freestone (1999, pp. 8 and 9).
45. Magraw and Hawke (2007, p. 619).
46. Weeramantry (1998, p. 204).
47. Rieu-Clarke (2005, p. 56).
48. Schrijver (2005, p. 551).
49. Cordonier Segger (2005, p. 592).
50. Richardson and Wood (2006, p. 14).
51. Boyle and Freestone (1999, p. 16).
52. Boyle and Freestone (1999, p. 17).
53. Lang (1995, p. 93).
54. Bosselmann (2008, p. 54).
55. Bosselmann (2008, p. 54).
56. Ebbesson (1996, pp. 49 and 50).
57. Ebbesson (1996, pp. 66–68).
58. Norms of conduct tell what a subject must do, may do, may refrain from doing or must refrain from doing: Ebbesson (1996, p. 79).
59. Norms of competence concern the legal capability – the power – to take action so as to change a legal position: Ebbesson (1996, p. 79).
60. Ebbesson (1996, pp. 84 and 85).
61. Ebbesson (1996, p. 85).
62. Ebbesson (1996, p. 87).
63. Ebbesson (1996, p. 90).
64. Ebbesson (1996, p. 90).
65. Ebbesson (1996, p. 91).
66. Lowe (1999, p. 21).
67. Lowe (1999, p. 21).
68. Lowe (1999, p. 31).

69. Lowe (1999, pp. 30 and 31).
70. Lowe (1999, p. 31).
71. Lowe (1999, p. 33).
72. Lowe (1999, p. 33).
73. Lowe (1999, p. 34).
74. McCaffrey (1997, p. 54).
75. Johnson (1993, p. 334).
76. European Environment Agency (1999b, p. 10).
77. Chave (2001, p. 104).
78. Aswathanarayana (2001, p. 388).
79. McCaffrey (1997, p. 54).
80. Kinzelbach et al. (2007, p. 291).
81. Simonis (2007, p. 329).

GENERAL REFERENCES

Adande, Nana et al. (2000), 'Policy foundations for sustainable development', in Jim Whitman (ed.), *The Sustainability Challenge for Southern Africa* (Basingstoke: Macmillan Press), pp. 291–295.

Allan, J.A. (1995) 'Striking the "right price" for water? Achieving harmony between basic human need, available resources and commercial viability', in J.A. Allan and Chibli Mallat (eds), *Water in the Middle East: Legal, Political and Commercial Implications* (London: Tauris Academic Studies), pp. 325–346.

Aswathanarayana, U. (2001), *Water Resources Management and the Environment* (Lisse: A.A. Balkema Publishers).

Bosselmann, Klaus (2006), 'Ecological justice and law', in Benjamin Richardson and Stepan Wood (eds), *Environmental Law for Sustainability: a Reader* (Oxford: Hart Publishing), pp. 129–163.

Bosselmann, Klaus (2008), *The Principle of Sustainability: Ethics, Law and Governance* (Aldershot: Ashgate Publishing).

Boyle, Alan and David Freestone (1999a), 'Introduction', in Alan Boyle and David Freestone (eds), *International Law and Sustainable Development* (Oxford: Oxford University Press), pp. 1–18.

Chave, Peter A. (2001), *The EU Water Framework Directive* (London: IWA Publishing).

Cordonier Segger, M.C. (2005), 'Governing and reconciling economic, social and environmental regimes', in M.C. Cordonier Segger and C.G. Weeramantry (eds), *Sustainable Justice: Reconciling Economic, Social and Environmental Law* (Leiden: Martinus Nijhoff Publishers), pp. 561–592.

Cordonier Segger, M.C. and Ashfaq Khalfan (2006), *Sustainable Development Law: Principles, Practices and Prospects* (Oxford: Oxford University Press).

De Villiers, M. (2001), *Water Wars: is the World's Water Running Out*? (London: Phoenix Press).

Dovers, Stephen and Robin Connor (2006), 'Institutional and policy change for sustainability', in Benjamin Richardson and Stepan Wood (eds), *Environmental Law for Sustainability: a Reader* (Oxford: Hart Publishing), pp. 21–60.

Dryzek, John S. (2007), 'Paradigms and discourses', in Daniel Bodansky et al. (eds), *The Oxford Handbook of International Environmental Law* (Oxford: Oxford University Press), pp. 44–62.

Ebbesson, Jonas (1996), *Compatibility of International and National Environmental Law* (London: Kluwer Law International).

European Environment Agency (1999b), *Sustainable Water Use in Europe: Part 2: Demand Management* (Luxembourg: European Communities Official Publications).

Fisher, D.E. (1987), *Natural Resources Law in Australia: a Macro-Legal System in Operation* (Sydney: Law Book Company).

Fitzmaurice, Malgosia and Olufemi Elias (2004), *Watercourse Co-operation in Northern Europe: a Model for the Future* (The Hague: TMC Asser Press).

French, Duncan (2005), *International Law and Policy of Sustainable Development* (Manchester: Manchester University Press).

Fuentes, Ximena (1998) 'Sustainable development in the equitable utilization of watercourses', *British Yearbook of International Law* 69, 119–200.

Ginther, Konrad and Paul de Waart (1995), 'Sustainable development as a matter of good governance: an introductory view', in Konrad Ginther et al. (eds), *Sustainable Development and Good Governance* (Dordrecht: Martinus Nijhoff Publishers), pp. 1–14.

Gonthier, Charles D. (2005), 'Fraternity: a global value underlying sustainable development', in M.C. Cordonier Segger and C.G. Weeramantry (eds), *Sustainable Justice: Reconciling Economic, Social and Environmental Law* (Leiden: Martinnus Nijhoff Publishers), 33–46.

Goodin, Robert (1995), *Utilitarianism as a Public Philosophy* (Cambridge: Cambridge University Press).

International Union for the Conservation of Nature (1980), *World Conservation Strategy* (Gland: International Union for the Conservation of Nature).

Iqbal, Munawar (2005), 'Introduction', in Munawar Iqbal (ed.), *Islamic Perspectives on Sustainable Development* (Basingstoke: Palgrave Macmillan), pp. 1–21.

Johnson, Stanley P. (1993), *The Earth Summit: the United Nations Conference on Environment and Development (UNCED)* (London: Graham & Trotman).

Kinzelbach, Wolfgang et al. (2007), 'Sustainable water resources management in arid and semi-arid regions', in José L. Lozán et al. (eds), *Global Change: Enough Water for All?* (Hamburg: Wissenschaftliche Auswertungen), pp. 291–295.

Lang, Winfried (1995), 'How to manage sustainable development', in Konrad Ginther et al. (eds), *Sustainable Development and Good Governance* (Dordrecht: Martinus Nijhoff Publishers), pp. 90–104.

Lowe, Vaughan (1999), 'Sustainable development and unsustainable arguments', in Alan Boyle and David Freestone (eds), *International Law and Sustainable Development* (Oxford: Oxford University Press), pp. 19–37.

Magraw, D.B. and L.D. Hawke (2007), 'Sustainable development', in Daniel Bodansky et al. (eds), The *Oxford Handbook of International Environmental Law* (Oxford: Oxford University Press), pp. 613–638.

McCaffrey, Stephen (1997a), 'Water scarcity: institutional and legal responses', in Edward H.P. Brans et al. (eds), *The Scarcity of Water: Emerging Legal and Policy Responses* (London: Kluwer Law International), pp. 43–58.

Mutembwa, Ammon (2000), 'Toward a sustainable water management strategy for Southern Africa', in Jim Whitman (ed.), *The Sustainability Challenge for Southern Africa* (Basingstoke: Macmillan Press), pp. 239–264.

Ørebech, Peter and Fred Bosselman (2005), 'The linkage between sustainable development and customary law', in Peter Ørebech et al. (eds), *The Role of Customary Law in Sustainable Development* (Cambridge: Cambridge University Press).

Pinto, M.C.W. (1995), 'Reflections on the term sustainable development and is institutional implications', in Konrad Ginther et al. (eds), *Sustainable Development and Good Governance* (Dordrecht: Martinus Nijhoff Publishers), pp. 72–79.

Richardson, Benjamin and Stepan Wood (eds), *Environmental Law for Sustainability: a Reader* (Oxford: Hart Publishing), pp. 1–18.

Rieu-Clarke, Alistair (2005), *International Law and Sustainable Development: Lessons from the Law of International Watercourses* (London: IWA Publishing).

Sagoff, Mark (1974), 'On preserving the natural environment', *Yale Law Journal* 84, 205–267.

Schrijver, Nico (2005), 'The New Delhi Declaration of principles of international law relating to sustainable development', in M.C. Cordonier Segger and C.G. Weeramantry (eds), *Sustainable Justice: Reconciling Economic, Social and Environmental Law* (Leiden: Martinus Nijhoff Publishers), pp. 549–552.

Sands, Philippe (1994), 'International Law in the Field of Sustainable Development', *British Yearbook of International Law* 65, 303–81.

Simonis, Udo E. (2007), 'Water related conflicts – and the need for a global water strategy', in José L. Lozán et al. (eds), *Global Change: Enough Water for All?* (Hamburg: Wissen Schaftliche Auswertungen), pp. 329–333.

Subedi, Surya P. (2005), 'Regulation of shared water resources in international law: the challenge of balancing competing demands', in Surya P. Subedi (ed.), *International Watercourses Law for the 21st Century* (Aldershot: Ashgate Publishing Ltd), pp. 7–18.

Weeramantry, C.G. (1998), 'Separate opinion in the case concerning the Gabcikovo–Nagymaros Project', *International Legal Materials* 37, 204–217.
Yousri, Abd al-Rahman (2005), 'Sustainable development: an evaluation of conventional and Islamic perspectives', in Munawar Iqbal (ed.), *Islamic Perspectives on Sustainable Development* (Basingstoke: Palgrave Macmillan), pp. 22–57.

LEGAL INSTRUMENTS

Agenda 21 (1992): Action Plan arising from the United Nations Conference on Environment and Development 1992, in S.P. Johnson (1993), *The Earth Summit: the United Nations Conference on Environment and Development (UNCED)* (London: Graham Trotman), pp. 125–503.
Dublin Statement on Water and Sustainable Development (1992), in Salman M. A. Salman and Daniel D. Bradlow (2006), *Regulatory Frameworks for Water Resources Management: A Comparative Study* (Washington DC: World Bank), pp. 169–174.
International Union for the Conservation of Nature (1980), *World Conservation Strategy* (Gland: International Union for the Conservation of Nature).
Johannesburg Declaration (2002): Declaration of the World Summit on Sustainable Development in Johannesburg, 2002, UNDoc.A/CONF.199/L.6/Rev.3.
New Delhi Declaration (2002): Declaration of Principles of International Law relating to Sustainable Development (2002), in Alistair Rieu-Clarke (2005), *International Law and Sustainable Development* (London: IWA Publishing), pp. 181–187.
Rio Declaration (1992): Declaration of the United Nations Conference on Environment and Development (1992), *International Legal Materials* 31, 874–880.
Stockholm Declaration (1972): Declaration of the United Nations Conference on the Human Environment (1972), *International Legal Materials* 11, 1416–1421.
United Nations Convention on Biological Diversity (1992), *International Legal Materials* 31, 818–841.
United Nations Framework Convention on Climate Change (1992), *International Legal Materials* 31, 849–873.

3. Governance for sustainable development

THE THEORETICAL FOUNDATIONS OF GOVERNANCE

(a) The Challenge of Governance for Sustainability

The way water resources are managed is a matter of interest and concern for the international, national and local communities. If the challenges facing the management of water resources are to be successfully confronted, then the responses need to be local, national, international and global. The key is often said to be integration. But what is integration?

It has been suggested that there are at least three aspects to integration:

- the systematic consideration of the various dimensions of water – surface and groundwater, quantity and quality;
- water as a system and as a component of other systems interacting with them;
- the relationships between the use and development of water on the one hand and social and economic development on the other hand.[1]

But it is more than this. The integration of objectives – social, economic and environmental – to be effective requires institutional as well as normative, strategic and operational integration. At these three levels the questions are:

- what ought to be done – normative;
- what can be done – strategic;
- what will be done – operational.[2]

It is the function of the law to establish arrangements according to which integration can be achieved in all of these ways. Not an easy task. This is particularly challenging if a further level of integration is included: namely the harmonisation of international, national and local responses.

It was commented in 2004 that 'current governance arrangements remain a crazy quilt of overlapping activities'.[3] It was nevertheless suggested that institutional fragmentation was 'not necessarily incoherent'.[4] Consistently with this, 'governance for sustainable development faces a number of obstacles';[5] for example, the procedures associated with the relevant institutions. 'The international architecture for sustainable development is highly fragmented' simply because of the different functions performed by a variety of institutions in accordance with their 'different organizational missions and goals'.[6] In other words – the formal arrangements for achieving sustainable development must follow the function of sustainable development, which is to bring together in an integrated fashion the relevant social, economic and environmental values in making decisions and in undertaking activities. Consequently, even if the function of sustainable development is clear enough, the formal arrangements for doing so are not.

Governance for sustainability, it has been suggested,[7] invites three responses: innovation; reconciliation; and creativity and adaptability. The first involves all groups in society becoming involved in 'new interorganizational, cross-scale arrangements'. The second involves aligning multi-level governance structures with the principles of sustainability. The third involves the creation of the relevant 'scientific, regulatory and economic infrastructures'. Significantly there is a role for the law to be creative and adaptive – as a regulatory infrastructure.

If this approach is taken further, it leads to five 'principal elements' of a governance structure.[8] Each of the five elements focuses upon their essential multiplicity. There is the multi-level character of policy implementation; the multi-actor character of implementation; the multi-faceted objectives of policy implementation; the multi-instrumental character of policy strategies; and a complex multi-resource base for implementation. In slightly more simple terminology, a structure of governance for sustainability involves all participants in the public and private sectors, acknowledges the diverse range of ways and means of achieving sustainable development, and the enormity of the intellectual, technical and financial resources needed to do so.

The relevant legal arrangements are a small part – but an important part – of such a response. Rights and duties in the form of legal and paralegal rules are such instruments. But at a more general level, 'transparency, accountability and other characteristics of the rule of law in good governance provide conditions that enable stakeholders to act more effectively to protect their interests'.[9] Accordingly rule making and norm development are two 'of the more salient aspects of governance'.[10] It is the notion of 'coherence' that underpins this approach to governance for sustainability.[11] And it is coherence both vertically and horizontally across all the relevant systems. In the context of the legal system it is ultimately the notion of harmonisation of the relevant international, national and local legal arrangements.

(b) The Function of Governance

If sustainable development creates a new paradigm for global, national and local communities at all levels and therefore invokes a process of change, how should this paradigm be institutionalised and what processes of change are necessary to do so? Even if sustainable development is no more than a statement of overarching policy, it is at the very least a statement of public policy at the highest level. In other words, it is ultimately in the overall public interest that development is sustainable. But sustainable development – as a statement of public policy – has implications across the whole of society, whether those involved are in the public sector or the private sector or are interested and concerned individuals within the community. In other words sustainable development is the responsibility of every person and every entity within the community whatever is their function and from whatever perspective they perform their functions.

The creation of arrangements constituting governance for sustainable development has been described as the challenge of adapting form to function.[12] The form of governance appropriate to sustainable development must in this sense reflect the function of sustainable development. If this is to be successful, 'it must build on a premise of rational democratic governance'.[13] These are difficult concepts. First, the system must be based upon reason.

Second, it is democratic in the sense that it reflects 'the power and authority of the community membership'.[14] Third, it is a system of governance. But what is governance?

The function of governance has characteristics similar to the functions of management, administration and government. These words are often used institutionally as well as functionally. For example, the government is an institution but government is a function. We are presently concerned however with the function and not the institution. Institutions must be designed appropriately to fulfil functions, rather than the other way round. Management is often used to describe a function of the private sector; administration a function of the bureaucratic sector; and government a function of the political sector. Governance is altogether different: at least as it is used in this context for present purposes. It seeks to incorporate each and every one of these functions.

Governance in one sense involves steering the community in the right direction – in this case the direction described as sustainable development. In other words:

> Linguistically, 'governance' has traditionally connoted 'the act or process of governing', with the latter primarily associated with governmental steering by regulation and sanctions.[15]

But governance now comprehends not only regulation but also instruments of other kinds. For example, decisions are made and activities undertaken not in consequence of bureaucratic direction but as a result of economic and commercial incentives and disincentives. Thus:

> Governance has currently come to indicate the totality of 'mechanisms' and 'instruments' available for influencing social change in preordained directions.[16]

These mechanisms and instruments include strategic decision-making by the political sector, traditional regulation by the bureaucratic sector, operational decisions made by the private sector, and involvement on the part of concerned as well as interested members of the community.

The involvement of all sectors of the community in managing its affairs is by no means a novel idea. It is the use of the word governance to describe it that is relatively new. Thus:

> The main purpose of this relatively new term seems to be the wish to distinguish between governance systems (regimes) which rely for their realization either on governments or on governance systems (regimes) which are able to function without governments.[17]

The evolving notion of governance thus distinguishes between systems of governance which rely for their effectiveness on governments as institutions and systems of governance which function without governments as institutions; in other words, the distinction between government as an institution and government as a function. In practice there is likely to be a range of functions performed by governments as institutions and a range of functions performed by institutions other than governments. It is the combination of these functions that constitutes governance in the sense under discussion.

A number of empirical studies or observations have lent support to such a notion of governance. The point of commencement is the idea that governance is steering the community in a specific direction. Thus:

> In general 'governance' refers to the fact that steering no longer is the privilege of governmental

> agencies, but is *de facto* (and in many cases also *de jure*) the common responsibility of a variety of agencies, representing governmental bodies, market agencies and civil society organisations.[18]

Thus governance has become – certainly in the environmental context – a multi-dimensional concept and a multi-level activity. It is these multi-levels of involvement on the part of those performing functions that lead inexorably to the notion of their 'common responsibility'. In other words, in slightly more juridical language, the performance of functions is not simply the exercise of individual or institutional rights and powers, but also the acceptance of responsibility for the consequences of the exercise of these rights and powers in common with all other individuals and institutions within the system.

(c) Governance as a Juridical Concept

So much for the concept of governance. What does it mean from a juridical perspective? As an inchoate element of the legal system, governance has been described as:

> The exercise of economic, political and administrative authority to manage a country's affairs at all levels and on all issues. It comprises the mechanisms, processes and institutions through which citizens and groups articulate their interests, exercise their legal rights, meet their obligations and mediate their differences.[19]

Clearly so: but what about the quality of such a system of governance. The notion of 'good governance' has been emerging in practice, but not as a matter of law.[20] The quality of governance may be analysed from two perspectives: procedural and substantive: in other words, how institutions perform and to what ends institutions perform. Examples of the former include the principles of public participation, access to information and access to the justice system. In relation specifically to sustainable development these include the precautionary principle and environmental impact assessment. Values associated with procedure have also been taken to include transparency, responsiveness, consensus orientation, and accountability.[21] If the rule of law is added – as it sometimes is – it may include judicial supervision of decision-making to ensure that objective and rational decisions are made according to the formalities and methodologies of the legal system. This takes governance closer to what is decided rather than how it is decided.

Arguably a system of good governance encapsulates all of these values and approaches. A model has been described which brings together several facets of governance.[22] This model is based upon a juridical approach, an economic-managerial approach and a political and civic approach. The juridical approach deals with the fundamental elements of the rule of law. This approach ensures not only legal security but also accountability and confidence in the system. The economic-managerial approach originates in the culture of the private sector, but it can be – and has been recently – transposed to the public sector. The political and civic approach, however, by comparison 'highlights values in the structure and culture of democracy itself'.[23] The economic-managerial approach is one of flexibility on the one hand and autonomy on the other hand. The political and civic approach provides responsiveness on the one hand and political primacy on the other hand.

The values underlying each of these three approaches are quite different, a point acknowledged when the three approaches were synthesised:

> While the economy needs flexibility (in capital, labour and knowledge), the juridical dimension requires legality, certainty and predictability. While the market needs autonomous decision-making (exclusion) and free access of suppliers, the political dimension urges collective decision-making (inclusion) and authoritative value-allocation (political primacy). While the political dimension requires responsiveness to all those social problems which are legitimately recognised, the juridical approach requires involvement under conditions of accountability, trust and truth, security and maintenance.[24]

If such an approach is applied to sustainable development, what picture emerges? Much depends upon the nature of sustainable development and its function within the overall system. One of the problems is that the achievement of long term objectives, of which sustainable development is one, is often problematical for political reasons.[25] Decisions in democratically arranged circumstances are often constrained by immediate budget and election issues. Similarly, short-term considerations often influence decisions 'in less democratic countries'.[26] Since sustainable development involves the making of decisions for the future and over long time spans, what formal arrangements can be put in place which satisfy on the one hand the values of democratic politics and on the other hand the need for long-term decision making? In other words:

> Are there political approaches that enable societies to adopt – and implement – strategies that authoritatively bind future decision-making within the perceived boundaries of ecologically sustainable development, but still retain the legitimacy of democratic politics?[27]

The response has been what is called management by objectives and results. In this case sustainable development is clearly the objective and, one hopes, the actual result of the decision-making process. Any such set of arrangements accordingly needs to satisfy five conditions, of which two are substantive in nature and three procedural in nature.[28] The three procedural elements are considerable discretion, participation at all levels and feedback to enable accountability. Most important from the juridical point of view are the two substantive conditions: namely the establishment of explicit goals with legal backing and an institutionalised distribution of responsibility for goal implementation. In other words, the creation of sustainable development as a goal-oriented norm and perhaps even as a fixed norm.

A set of arrangements that might achieve this purpose has been suggested. These include a constitutional mandate giving special status to sustainable development, together with a carefully constructed set of integrated arrangements involving the legislative, the executive and the judicial processes.[29] This takes us back to our point of commencement: the need to adapt form to function so as to ensure that sustainable development functions effectively within a formal and institutionalised system structured specifically to ensure that sustainable development performs its intended function.

(d) Governance for Sustainable Water Resources Development

Do water resources fit within this paradigm of governance for sustainable development? This definition of water governance has been provided:

> The range of political, organizational and administrative processes through which communities articulate their interests, their input is absorbed, decisions are made and implemented, and decision makers are held accountable in the development and management of water resources and delivery of water services.[30]

While this relates particularly to national arrangements, it has relevance – suitably modified – in the international context. The principles underpinning water resources governance are in the first instance the same as those underpinning governance for sustainable development generally. These include, for example, accountability, effectiveness, efficiency, transparency, participation, and the rule of law. In the context of water supply management, five principles have been isolated:

- protection of public health and safety;
- accountability for stewardship and performance;
- transparency;
- participation;
- equity, efficiency and effectiveness.[31]

These have been stated to be coherent and ranked in order of priority. Coherence again emerges as a fundamental principle. This involves 'a high degree of internal consistency between the different principles'.[32]

These principles support a set of generic governance models. These have been described as a planning model, a commercial model and a community model.[33] The first focuses upon the management of water resources by the public sector. The second focuses upon the rights and duties of those using and developing water resources – particularly in the private corporate sector. And the third focuses upon the rights and duties of the users of water resources within particular communities. In practice there is likely to be a combination of two or perhaps even all of these models. In each case there is likely to be a combination of public rights and duties together with private rights and duties across each of the models. It is important therefore to identify the function of each of the participants in each of the models, to ensure that the formal design of the model follows the function to be performed by it and to ensure consistency and coherence within and across each of these forms and functions. The legal arrangements giving effect to such a coherent structure will themselves assume the form of paralegal rules as well as legal rules. This will effectively create a system of textual and contextual integrity and cohesion. To ensure harmony between national and international legal arrangements makes a complex task even more complicated. Are there any such models?

MODELS OF GOVERNANCE FOR SUSTAINABLE DEVELOPMENT

(a) Introduction

Sustainability, and more particularly sustainable development, have been considered from a number of different perspectives. Not surprisingly – certainly from a practical perspective – an issue is the meaning of the words themselves. This is particularly so in the context of the evolution of the concept of sustainable development. It emerged linguistically from development of natural resources, tempered by conservation, linked to environmental protection and maturing as sustainable development. In this sense it evolved out of the ethic of conservation but it may be underpinned now by what may be described as an ethic of ecological justice.

Some would assert that there is a general consensus on some of the core ideas underlying sustainable development.[34] Thus, 'these ideas form part of the concept of ecological justice'.[35] This requires justice to be linked to the 'non-human world'.[36] This requires recognition of ecological integrity. Thus there emerges the idea of ecological justice. It has three elements:

- intragenerational justice – concern for the poor on the part of existing generations;
- intergenerational justice – concern for future generations;
- interspecies justice – concern for the non-human natural world.[37]

These ideas are implicit in the concept of sustainability, and therefore in sustainable development in its institutionalised and formalised manifestation. When the ethic of ecological justice is translated into a definition of sustainable development, it emerges as:

> The procedural and substantive requirement to accommodate, reconcile or integrate economic growth, human rights and environmental protection, for participatory, equitable improvement in our collective quality of life that can last over the long term.[38]

Significantly this is couched as a requirement. It is a requirement to adopt a certain methodology in reaching solutions and the methodology is directed towards the goal of improvement in the quality of life. This definition discloses a number of characteristics and these reflect those of a fixed norm, a balancing norm and a goal-oriented norm. Are there any examples of such an approach? For present purposes, three examples suffice: one national, one regional and one international.

(b) New Zealand

The Resource Management Act of 1991 of New Zealand in sections 5–8 states its purpose, provides a definition of sustainable management, indicates the objectives to be achieved and imposes a number of duties upon all persons performing functions under the Act. The purpose is to promote the sustainable management of natural and physical resources. Sustainable management means managing the use, development and protection of natural and physical resources. The objectives to be achieved in doing so reflect the economic, social and ecological foundations of sustainable development and, hence, in an institutionalised fashion the ethic of ecological justice. The duties include a duty to recognise and provide for a range of matters of national importance and a duty to have particular regard to a range of other matters.

Critical is section 5 which states the purpose and defines sustainable management. The Environment Court of New Zealand has said this about section 5:

> The proper application of section 5 involves an overall broad judgement of whether or not a proposal promotes the sustainable management of natural and physical resources. Such a judgement allows for a comparison of conflicting considerations and the scale or degree of them, and their relative significance in the final outcome.[39]

The other elements of the legislation that constitute its framework were stated to 'inform and assist the purpose of the Act'. More particularly, 'the subsequent sections must not be allowed to obscure the sustainable management purpose of the Act'.[40] The purpose stated in section 5 is thus a type of *Grundnorm*, and the supporting provisions in the framework part of the legislation may be seen as goal-oriented norms and balancing norms. While section 5 is not structured as an obligation – and hence not immediately enforceable – it provides strategists, planners, administrators and those undertaking operational activities with a clear statutory methodology for performing their functions. A failure to comply with this methodology is likely to raise a justiciable issue. Whether or not the achievement of the purpose of the Act is a justiciable issue is another question. In this sense the New Zealand legislation contains a duty of performance but perhaps not a duty of achievement.

(c) The European Union

Almost since its inception, the European Community has displayed a concern for the environment. The formal arrangements currently in force place considerable emphasis upon the environment. The structural arrangements proposed by the Constitution for the European Union will place even greater emphasis upon the environment. Article 3(1)(l) of the Treaty establishing the European Community currently requires a community policy in the sphere of the environment. This policy is required by Article 174(1) to contribute to the pursuit of a number of stated objectives. These include the protection of the environment, the protection of human health and the prudent and rational utilisation of natural resources. The aim of the policy according to Article 174(2) is a high level of protection and it is required to be based upon four widely recognised principles:

- the precautionary principle;
- the principle of prevention;
- the principle of rectification of damage at source;
- the polluter pays principle.

In addition the Council is required by Article 175(2)(b) to adopt measures in particular affecting the quantitative management of water resources or measures affecting, directly or indirectly, the availability of water resources. Significantly the structure of these arrangements is fourfold:

- a policy;
- a set of objectives;
- a set of principles;
- a requirement for measures in specific circumstances.

It is against this structural framework that the directives of the European Community have been formulated, interpreted and applied. The 'ecological imperative' in some of these directives, it has been suggested, 'can be overridden only in the most exceptional circumstances'.[41]

On the commencement of the Treaty establishing a Constitution for Europe, the arrangements currently in place will become part of the wider structural framework created by the constitution: in particular the objectives prescribed for the European Union and the set of fundamental rights declared for it. Interestingly the constitution states in Article I–2 the values on which the European Union is founded. While these do not include any reference to the environment or sustainable development, they specifically include respect for human dignity, freedom, democracy, equality, the rule of law, and respect for human rights. However one of the objectives of the Union stated in Article I–3(3) is to work for the sustainable development of Europe based on balanced economic growth and price stability, a highly competitive social market economy, aiming at full employment and social progress, and a high level of protection and improvement of the quality of the environment. While the focus of this objective appears to be economic development and environmental protection, some of the other elements of sustainable development are incorporated less directly and more interstitially.

More direct is the declaration in Article II–97 of environmental protection as a fundamental right of the Union. Thus:

> A high level of environmental protection and the improvement of the quality of the environment must be integrated into the policies of the Union and ensured in accordance with the principle of sustainable development.

Just as positive is the requirement in Article III–119 linked to the functioning of the Union – in particular the relationship between the achievement of environmental protection and the achievement of the other objectives of the Union. Thus:

> Environmental protection requirements must be integrated into the definition and implementation of the policies and activities referred to in this Part, in particular with a view to promoting sustainable development.

Significantly, therefore, while sustainability or sustainable development is not treated as a value of the European Union, it is an element of the Union's objectives; it guides the integration of environmental protection into the policies of the Union; and its promotion is the stated outcome of incorporating environmental protection with the other policies and activities of the Union. How these constitutional arrangements will be interpreted and applied is presently a matter of speculation. What is clear, nevertheless, is that sustainable development is likely to play a formal part in these arrangements, and hence at least influence the development and implementation of European Union legislation.

(d) The Draft International Covenant

One of the most comprehensive sets of potential arrangements for the achievement of sustainable development at the level of international law is the Draft International Covenant on Environment and Development put together on the initiative of the Commission on Environmental Law of the International Union for the Conservation of Nature. The approach adopted by the Draft Covenant is both substantive and procedural. The substantive measures are structured as an objective, a set of fundamental principles and then a series of general and then more specific obligations. The procedural matters deal with implementation, enforcement and compliance.

The objective in Article 1 is critical. It is to achieve environmental conservation as an indispensable component of sustainable development through establishing integrated rights and obligations. Although environmental protection and environmental conservation are together a part of sustainable development – at least according to its historical evolution – they are specified separately in this objective. Perhaps this is to emphasise the importance of environmental conservation even if it is an essential element of sustainable development. In this sense the objective is perhaps in the nature of a *Grundnorm*. And as such a norm, it is supported and explained by the integrated set of rights and obligations created by the Draft Covenant. The integration of these rights and obligations in a juridical rather than a management sense is fundamental to the form and structure of the Draft Covenant.

Then there are the fundamental principles stated in Articles 2–10. It is their function to support and explain the objective, the rights and the duties. An example is the way the right to development is stated. Although it describes itself as a right, it functions in Article 8 as one of the fundamental principles which guide the achievement of the objective. The right exists by itself. It is thus of general relevance and application and has effect *erga omnes*. But Article 8 goes further. It states:

> The implementation of the right to development entails the obligation to meet the developmental and environmental needs of humanity in a sustainable and equitable manner.

This not only juxtaposes a right and an obligation. It also juxtaposes developmental and environmental needs. It also provides substance to the obligation by referring to the requirement to do so in a sustainable and equitable manner. The juxtaposition, once again, of sustainable and equitable emphasises the necessary relationship between the economic, the social and the ecological values associated with sustainable development. Significantly neither the objective nor the principles are linked to persons, institutions or states. This is entirely consistent with the function performed by objectives and principles in any set of juridically significant structural arrangements. In other words, the objective and these principles perform 'interstitial normativity' by impacting upon other normative elements of the overall arrangements such as duties.

Let us turn, then, to the obligations. These are directed not only at states – potential parties to a Covenant – but also persons in general. This emphasises the notion that it is humanity at large that is responsible for the achievement of sustainable development. However, as a Covenant would be an international instrument, most of the obligations are directed at states parties to it. Perhaps the most important obligation that illuminates the intrinsic nature of sustainable development is the obligation in Article 13 to integrate policies related to environment and development. The first element of the obligation requires parties to pursue policies aimed at eradicating poverty, changing consumption production patterns, and protecting and managing the nature resource base as overarching objectives of and essential requirements for sustainable development. This is clearly aimed at the strategic and planning functions of states. Equally clearly it integrates the social, economic and ecological perspectives of sustainable development.

The second element of the obligation requires parties to ensure that environmental conservation is treated as an integral part of the planning and implementation of activities at all stages and at all levels, giving full and equal consideration to environmental, economic, social and cultural factors. This element of the obligation is directed not only at planning but also at implemention by the undertaking of activities. It is operational rather than strategic in character. This obligation is likely to be discharged through the creation and implementation of national legal arrangements.

The Draft Covenant is no more than a draft. It is nevertheless an influential model which is having an impact upon the development of legal doctrine in consequence of its logical and coherent structure. Essentially it is:

- a matrix of an objective, fundamental principles and obligations;
- a collocation of norms of behaviour and norms of competence;
- an arrangement of goal-oriented norms, balancing norms and fixed norms;
- a set of correlative rights and duties and powers and liabilities.

Significantly there is one objective – in essence sustainable development. It is a goal-oriented norm and perhaps even a *Grundnorm* in the sense of the Kelsen theory. It is achieved by a set of balancing norms – the fundamental principles – in conjunction with the set of obligations or duties in the form of fixed norms. An extravagant hypothesis perhaps – but one towards which the three examples considered here are moving. A model, that is, for the future.

CONCLUSION

Although water is a renewable source, it is a finite resource. Its availability at any point in time and the location of its availability are very unpredictable. Water resources are used for a variety of purposes, some of which are critical to the fundamental health and welfare of communities. Water resources are being used at an increasing rate to the point of overexploitation in some instances, and the quality of the water resource in many cases is heavily degraded. If water resources are used and developed sustainably, the expectation is that a number of these difficulties will be solved.

Whether this is so or not, if water resources are to be used and developed sustainably, much depends upon the arrangements society puts in place for the governance of these resources. These may be local, national, regional or global. The concepts of sustainable development and governance demand assessment and analysis before a formal set of arrangements can be designed for sustainable water resources governance. Sustainable development and governance are both multi-dimensional and multi-faceted concepts. It has been suggested that – given their nature and function – formal arrangements must follow where the function of sustainable development leads them rather than fit sustainable development within existing arrangements. This applies to all levels of involvement from international to local.

The sustainable use and development of water resources – like sustainable development generally – have emerged as a common concern and a common interest of human kind. They are also seen by many as the common responsibility of human kind. While it would no doubt be premature to suggest that at any level there is a right that water resources must be developed sustainably, there is what may be described as a growing cluster of diverse norms around which common but differentiated responsibility for sustainable water resources governance is beginning to crystallise. These norms assume different forms and perform different functions. Some are enforceable while others are not. These norms together with the host of institutional arrangements linked to them are an integral part of the overall system of governance for the sustainable use and development of water resources.

The challenge for the legal system at all levels is:

- to recognise sustainable development of water resources for what it is – the outcome of a process;
- to recognise and respond to the particular characteristics of water as a resource;

- to recognise that sustainable development of water resources is achieved through an integrated matrix of norms;
- to recognise that some of these norms relate to substance and some to process;
- to recognise that norms of substance and norms of process perform different functions;
- to recognise that norms of substance and of process are essential elements of the overall system;
- to create a set of structurally integrated norms that are part of a system of governance for the sustainable use and development of water resources;
- to ensure that these norms are structured so as to be enforceable through the judicial system.

NOTES

1. Mitchell (1990, p. 1).
2. Mitchell (1990, p. 2).
3. Haas et al. (2004, p. 279).
4. Haas et al. (2004, p. 279).
5. Bradnee Chambers and Green (2005, p. 2).
6. Bradnee Chambers and Green (2005, p. 2).
7. Martin et al. (2007, p. 95).
8. Martin et al. (2007, p. 101).
9. Bruns (2007, p. 37).
10. Haas et al. (2004, p. 268).
11. Bradnee Chambers and Green (2005, pp. 5–7) and Grover (2006, p. 222).
12. Lafferty (2004a, p. 3).
13. Lafferty (2004a, p. 3).
14. Lafferty (2004a, p. 4).
15. Lafferty (2004a, p. 5).
16. Lafferty (2004a, p. 5).
17. Lang (1995, p. 91).
18. Leroy and Arts (2006, p. 12).
19. Cordonier Segger and Khalfan (2006, p. 166).
20. De Waart (1997, pp. 116–118); Cordonier Segger and Khalfan (2006, p. 170).
21. Cordonier Segger and Khalfan (2006, pp. 170 and 171).
22. Arts and Goverde (2006, pp. 76–79).
23. Arts and Goverde (2006, p. 78).
24. Arts and Goverde (2006, p. 79).
25. Lundqvist (2004 pp. 95 and 96).
26. Lundqvist (2004, p. 96).
27. Lundqvist (2004, p. 96).
28. Lundqvist (2004, p. 101).
29. Lafferty (2004b, pp. 207 and 208).
30. Grover (2006, p. 221).
31. Grover (2006, p. 222).
32. Grover (2006, p. 222).
33. Grover (2006, p. 224).
34. Bosselmann (2006, p. 150).
35. Bosselmann (2006, p. 150).
36. Bosselmann (2006, p. 150).
37. Bosselmann (2006, pp. 150–152).
38. Cordonier Segger (2005, p. 590).
39. *Genesis Power case* (2005, p. 554).
40. *Genesis Power case* (2005, p. 554).
41. Fisher (2008, p. 111).

GENERAL REFERENCES

Arts, Bas and Henri Goverde (2006), 'The governance capacity of (new) policy arrangements: a reflective approach', in Bas Arts and Pieter Leroy (eds), *Institutional Dynamics in Environmental Governance* (Dordrecht: Springer), pp. 69–92.

Bosselmann, Klaus (2006), 'Ecological justice and law', in Benjamin Richardson and Stepan Wood (eds), *Environmental Law for Sustainability: a Reader* (Oxford: Hart Publishing), pp. 129–163.

Bradnee Chambers, W. and J.F. Green (eds) (2005b), *Reforming International Environmental Governance: From Institutional Limits to Innovative Reforms* (Tokyo: United Nations University Press).

Bruns, Bryan (2007), 'Community priorities for water rights: some conjectures on assumptions, principles and programmes', in B. van Koppen et al. (eds) (2007), *Community-based Water Law and Water Resource Management Reform in Developing Countries* (Wallingford: CAB International), pp. 28–45.

Cordonier Segger, M.C. (2005), 'Governing and reconciling economic, social and environmental regimes', in M.C. Cordonier Segger and C.G. Weeramantry (eds), *Sustainable Justice: Reconciling Economic, Social and Environmental Law* (Leiden: Martinus Nijhoff Publishers), pp. 561–592.

Cordonier Segger, M.C. and Ashfaq Khalfan (2006), *Sustainable Development Law: Principles, Practices and Prospects* (Oxford: Oxford University Press).

De Waart, P.J.I.M. (1997), 'Securing access to safe drinking water through trade and international migration', in Edward H.P. Brans et al. (eds), *The Scarcity of Water: Emerging Legal and Policy Responses* (London: Kluwer Law International Ltd), pp. 100–118.

De Villiers, M. (2001), *Water Wars: is the World's Water Running Out*? (London: Phoenix Press).

Fisher, D.E. (2008), 'Legal and paralegal rules for biodiversity conservation: a sequence of conceptual, linguistic and legal challenges', in Michael Jeffery et al. (eds), *Biodiversity Conservation: Law + Livelihoods* (Cambridge: Cambridge University Press), pp. 94–131.

Grover, V.I. (2006), 'Water governance and water economics', in V.I. Grover (ed.) (2006), *Water: Global Common and Global Problems* (Enfield: Science Publishers), pp. 221–234.

Haas, Peter M. et al. (2004), 'Conclusion: institutional design and institutional reform for sustainable development', in Norichika Kanie and P.M. Haas (eds) (2004), *Emerging Forces in Environmental Governance* (Tokyo: United Nations University Press), pp. 263–281.

International Union for the Conservation of Nature, (2004), *Draft International Covenant on Environment and Development* (3rd edn, Gland: International Union for the Conservation of Nature).

Lafferty, William M. (2004a), 'Introduction: form and function in governance for sustainable development', in William M. Lafferty (ed.), *Governance for Sustainable Development* (Cheltenham: Edward Elgar), pp. 1–27.

Lafferty, William M. (2004b), 'From environmental protection to sustainable development: the challenge of decoupling through sectoral integration', in William M. Lafferty (ed.), *Institutional Dynamics in Environmental Governance* (Cheltenham: Edward Elgar), pp. 191–219.

Lang, Winfried (1995), 'How to manage sustainable development', in Konrad Ginther et al. (eds), *Sustainable Development and Good Governance* (Dordrecht: Martinus Nijhoff Publishers), pp. 90–104.

Leroy, Pieter and Bas Arts (2006), 'Institutional dynamics in environmental governance', in Bas Arts and Pieter Leroy (eds), *Institutional Dynamics in Environmental Governance* (Dordrecht: Springer), pp. 1–20.

Lundqvist, Lennart J. (2004), 'Management by objectives and results: a comparison of Dutch, Swedish and EU strategies for realising sustainable development', in William M. Lafferty (ed), *Governance for Sustainable Development* (Cheltenham: Edward Elgar), pp. 95–127.

Martin, Andrew et al. (2007), 'Towards a model of governance for sustainability', in Suzanne Benn and Dexter Dunphy (eds), *Corporate Governance and Sustainability: Challenges for Theory and Practice* (Abingdon: Routledge), pp. 94–121.

Mitchell, Bruce (ed.) (1990), *Integrated Water Management: International Experience and Perspectives* (London: Belhaven Press).

Sands, Philippe and Paolo Galizzi (eds) (2006), *Documents in European Community Environmental Law* (2nd edn, Cambridge: Cambridge University Press).

LEGAL INSTRUMENTS

International Union for the Conservation of Nature (2004), *Draft International Covenant on Environment and Development* (3rd edn, Gland: International Union for the Conservation of Nature).

Resource Management Act 1991 of New Zealand.

Treaty establishing a Constitution for Europe, in Philippe Sands and Paolo Galizzi (eds) (2006), *Documents in European Community Environmental Law* (2nd edn, Cambridge: Cambridge University Press), pp. 51, 52, 58, 61 and 62.

Treaty establishing the European Community, in Philippe Sands and Paolo Galizzi (eds) (2006), *Documents in European Community Environmental Law* (2nd edn, Cambridge: Cambridge University Press), pp. 6, 7, 22 and 23.

CASES

Genesis Power case: *Genesis Power Ltd v Franklin District Council* [2005] New Zealand Resource Management Appeals, 541–590.

PART II

The formal structure of water resources governance

The relationships between humans and their institutions on the one hand and between humans, their institutions and the resources of their surrounding environment on the other hand depend upon the values of the society in question, the norms reflecting these values, the practices and customs of the society in question, and the legal and paralegal rules in place to render these values and norms effective and enforceable. When these arrangements are transformed into enforceable legal rules, they assume the form of rights and obligations of various kinds. A system of water resources governance thus involves a matrix of values, norms and rules. In this Part we consider the norms and the rules which constitute the formal structure of water resources governance. Formal in this sense describes the rules which identify the holders of rights in relation to water rather than the rules which state the criteria for the exercise of these rights.

The legal system comprises norms of behaviour, conduct or decision-making which prescribe what individuals and institutions may, must or must not do and norms of competence which describe the individuals and institutions able to make decisions and undertake activities and the scope of these decisions and activities. There emerges in consequence an arrangement of protectable rights and enforceable obligations of various kinds. In this Part we focus upon norms of competence which assume the form of the elusive notions of sovereignty and property.

The formal structures created by the legal system are intended to achieve the outcomes desired by society and its leaders. Water available for human use comes from a diversity of sources – as diverse as the purposes for which water is used. The water cycle is complex and the relationships between its elements difficult to identify in particular sets of circumstances. If the formalities of the law are to be effective, then they need to respond quite specifically to the challenge of the water cycle.

In this Part we consider:

- the challenges of the water cycle;
- how the fundamental doctrines of the legal system – particularly sovereignty and property – relate to these challenges;
- the range of functions performed by nation states in exercise of their rights of sovereignty;
- the nature and sources of rights in general in relation to water;
- the range, scope and function of entitlements to specific water resources.

Chapter 4 looks at the concept of sovereignty and the concept of property in some detail before discussing the concept of property in water. In particular it looks at water as a resource and water as a commodity in the context of water as common property and water as private property. Chapter 5 considers the range of functions performed by states and the nature of the rights held by states in relation to water, including rights of sovereignty, rights of property and rights to control water resources. Chapter 6 reviews the range of rights of individuals in relation to water, including human rights, constitutional rights and customary rights. These rights may or may not be enforceable in a traditional sense. Accordingly it is Chapter 7 which considers in some detail water entitlements in the sense that these constitute entitlements to use water from specific sources and for particular purposes. These rights are intrinsically protectable by the legal system and the corresponding duties enforceable. This part thus considers in detail the fundamental concepts and doctrines that support international and national legal arrangements in relation to water.

4. The doctrinal foundations of water resources law

THE CHALLENGES OF THE WATER CYCLE

(a) Introduction

Since there is the assumption that form follows function, a system of rights and obligations in relation to water must reflect the water cycle, the hydrological characteristics of water, the sources and destinations of water, its uses and abuses, and its relationship with humans and their institutions. In this context there is a distinction between water as a resource on the one hand and water as a fungible commodity on the other hand. The former refers to water in its original location and condition and the latter to water in a location and condition in which it can be used. Water in its original location may be water which lies or gathers naturally on the surface of the land; water which flows across the surface of the land either in an indiscriminate fashion or in a naturally occurring watercourse; or water which lies, gathers or flows underground. In consequence of the volatility and fluidity of water and of the phenomena of precipitation and evaporation, it is not possible to identify any particular particle of water compared with any other over either a short or long time span.

Add to this the complexity of human interference with water. It may be stored in a reservoir formed after the construction of a dam; it may be collected in an artificially constructed depression on the surface of land; it may be collected in an artificially constructed tank or other receptacle; it may be extracted from a flowing watercourse and either used immediately, conveyed by pipe or channelled to some other point of use; or it may be harnessed to produce energy and then returned to or stored in an artificial receptacle. Water is thus transformed from water as a resource to water in a form in which it is used for human purposes. It becomes an asset, a product or a good useful for human purposes. Once it is no longer required for such useful purposes, it may then be returned to or discharged back into the environment, but not necessarily the same environment out of which it originally emerged or was taken. Any system of governance must reflect each and every one of these circumstances.

The different functions performed by water in these ways lead almost inevitably to a variety of formal arrangements for the governance of water. The most important question to be addressed in this chapter is who has the power to control water as a resource and to control and use water as a fungible commodity. This question is critical at each and every stage of the natural water cycle and of the artificially engineered water cycle. The response may be rules of behaviour: how members of the community must behave towards each other in relation to water; or rules of competence and validity: not only by whom but also how decisions must be made that regulate and control the behaviour and conduct of members of the community. These are subsequently transformed into rights and obligations which emerge from legal doctrines associated with sovereignty and property. Whatever their specific meaning and function, these concepts quintessentially connote the idea of power and control vested exclusively in one legally identifiable entity. In relation to sovereignty the entity is normally a nation state exercising rights of sovereignty within its territorial boundaries as they have evolved over the last 300–400 years. In relation to property it is the owner of the resource acknowledged in accordance with the rules operating within the nation state.

(b) Water as a Common-pool Resource

Water may be seen either as a resource in its natural state or as a resource transformed into a form in which it can be used for social and economic purposes. This distinction is no more than a recognition of the range of functions performed by water at different parts of the water cycle. Water in its natural state is often described as a common-pool resource. Common-pool resources are essentially natural resources – including water – the benefits of which are shared among a number of persons or institutions. In economic jargon, there is a capital element and an income or benefit element. Thus, 'most natural resource systems used by multiple individuals can be classified as common-pool resources. Common-pool resources generate finite quantities of resource units, and one person's use subtracts from the quantity of resource units available to others'.[1] For example, groundwater basins and irrigation systems are common-pool resources and water is the resource unit derived from them. Common-pool resources have similarly been described as 'natural or human-made facilities or stocks that generate flows of useable resource units over time'.[2] Again, an irrigation system is given as an example of a renewable resource system and water is the resource flow unit from such a system.

The relationship between water as a common-pool resource and the resource units derived from it has been described in more dramatic fashion in these words:

> While water in a stream channel, a lake or sea, or an underground aquifer appears as an undifferentiated mass resource – the quintessential 'common-pool' – water extracted from the surface or underground body is an alienable quantity. It can be stored, measured, transported, even bought and sold as a commodity, and consumed. A quantity of water captured for consumption by one user is unavailable to another user, at least until it is returned to the body of water from which it was taken.[3]

A similar approach may be adopted in relation to natural resources the benefits of which are shared among nation states. Thus:

> Freshwater resources shared by states are common pool resources. This is because the use of the resource by a state subtracts from the use by other states. The pollution of a river, for instance, by an upstream state affects the use of water downstream. Extensive use of water by upstream states means that downstream states would have to either undertake the costs of purifying the water, for drinking or agricultural use, or forfeit use or together.[4]

Whether water resources are shared among nation states or among persons or institutions within nation states, the challenges for their governance are much the same. It seems to be recognised generally that some system of governance – incorporating rules of one kind or another – is desirable and probably necessary for their use and development.

THE FUNDAMENTAL DOCTRINES OF THE LEGAL SYSTEM

(a) Introduction

If water resources are regarded as common-pool or shared resources – even perhaps as an element of the common heritage of mankind – each and every nation state and person within the global community has an interest in how they are used and developed. But the mere recognition of water resources in this way does not constitute a legal right or an entitlement to do so. Arrangements are therefore put in place – formally or informally – to recognise, protect and manage the interest of the community in these shared resources. These arrangements are currently based upon the concepts of sovereignty and property. Each is associated with a power to control. In a sense each is a norm of competence. But neither is an uncontrolled power to control. The exercise of this norm of competence, in other words, is itself subject to a set of normative arrangements.

Sovereignty and property may be translated back into their equivalents in the Latin language: *imperium* and *dominium*. The former is the exercise of political power: to make decisions for the public benefit. The latter is the exercise of individual power: to make decisions for the benefit of the individual. The significance of this is that the power is vested exclusively in one clearly identifiable institution: the *imperator* on the one hand and the *dominus* on the other hand. In contemporary parlance these are the institutions of the public sector and those of the private sector. The former includes the legislative, the executive and the judicial branches of government and the latter includes the corporate as well as all those in the non-government sector.

Although rights are associated with sovereignty and property, neither, it is suggested, is a right *per se*. Each is a concept or a doctrinal statement which describes in its own way the relationships between and among nation states, between and among nation states and their citizens, and between and among the citizens themselves. The subject matter of these relationships includes water resources. These relationships certainly involve rights, but importantly they are also the media through which duties are imposed and enforced. The law, in other words, performs a restrictive as well as a facilitative function. And increasingly so in relation to water resources. Let us consider sovereignty first.

(b) The Concept of Sovereignty

Sovereignty is the basis upon which nation states organise the way in which their citizens and the inhabitants of their territory conduct their affairs. It is simultaneously the basis upon which their relationship with other nation states is conducted. Sovereignty in this sense emerged in Europe about the middle of the 17th century. Critical was the notion of enclosure. Nation states began to be recognised in accordance with artificially created and recognised territorial boundaries. A set of rights exclusive to the nation state thus became available to it within its territorial boundaries. Each and every other nation state enjoyed the benefit of precisely the same set of exclusive rights within its own territorial boundaries. The concept of sovereignty thus carried with it an obligation on the part of each nation state to recognise reciprocally the rights of sovereignty of all other nation states.

It can be analysed externally, therefore, as well as internally. Thus:

> Internally it means that the government of a state is considered the ultimate authority within its borders and jurisdiction.[5]

And:

> External sovereignty means that a state is not subject to the legal power of another state or of any other higher authority, and stands in principle on an equal footing with other states.[6]

One of the functions of international law has been to protect the sovereignty of nation states. Increasingly it has become a function of international law to indicate how rights of sovereignty should be exercised as a consequence of the implied duty of nation states to cooperate with each other because of their status as equals within the international community. Equality, in other words, carries with it responsibilities.

One of the rights associated with sovereignty is the right of a nation state to control its natural resources and its environment. This has been a critical factor in the rise and subsequent demise of political and economic colonialism on the part of European states. The inclusion of a colony as a territorial extension of the nation state was designed in some circumstances to give to the nation state the benefit of the natural resources of its colony. The second half of the 20th century saw the gradual demise of European colonialism and the reversion of sovereignty to the former colonies. Another of the rights associated with sovereignty is the right of a nation state to enter into an agreement with one or more other nation states. This has been particularly important in relation to shared water resources and there have been, as we shall see, many instances of such agreements.

Nation states are, of course, subject to their responsibilities not only under such agreements but also under the rules of customary international law and of international law generally. These rules are for the most part cast in fairly general terms, and it is the responsibility of the nation state to ensure compliance with these rules through the relevant sets of legal arrangements within the national system. Even so, it has been suggested that the international community is increasingly embracing a number of universal norms and values. A number of these relate to the use and development of natural resources and the protection of the environment. While the emergence of these international norms restricts the way in which the right of a nation state to control its natural resources is exercised, the concept of sovereignty remains intact. It has been described as 'an organisational principle for the realization of certain universal norms and values'[7] including those underpinning human rights and sustainable development. Significantly:

> Sovereignty functions, therefore, not merely as the basis of rights but also as the source of responsibility, accountability and liability and as the basis of international cooperation. International law not only delineates the borders of state territory and national jurisdiction but also increasingly defines the substance of the responsibilities of states, internationally as well as nationally.[8]

International law thus no longer merely facilitates what nation states may do within their territorial boundaries. It also increasingly restricts what they may do within their territorial boundaries. This is particularly important in relation to water resources generally and shared water resources in particular.

(c) The Concept of Property

Property may in like fashion be described as 'the organisational principle' around which rights are recognised or granted and accordingly it becomes also 'the source of responsibility, accountability and liability'. Property means different things in different contexts. While it is often loosely used to describe the thing owned by the owner, more accurately in legal terms it describes the relationship between the owner of the object or the asset and the object or the asset itself. The owners – the holders of the rights associated with ownership – are generally acknowledged to have available to them a number of rights:

- to use;
- not to use;
- to permit someone else to use;
- to alienate.

These rights are exclusive to their holder. Despite their exclusivity in this sense, their exercise may well impinge upon the exercise of their rights by others, for example, where there is a community of interest in the case of a shared resource – as in the case of water resources – or where the use of the resource is so closely related physically to the use of another resource. In this sense, the mutuality of interest requires that the exercise of the right must not interfere or interfere unreasonably with the exercise by another of a similar right. It has been argued, for example, that 'the definition of property as an enforceable claim of a person to some use or benefit of something' is sufficient to include all types of rights associated with property,[9] in this case the enforceable claim of one person that another person must not interfere with the use or benefit accruing to the first person of the object or asset in question. If this is applied to shared water resources, one person must not use or benefit from the use of these resources so as to deprive another interested person of their use or the benefit of their use.

It does not matter for the purposes of this analysis who or what institution is the owner. In other words, as it is often expressed, who is the holder of the title. It may indeed be the state itself; an agency of the state; a corporation; a group of persons acting together formally as an association or informally as a group of individuals; or individuals. The uses to which the object or the asset may be put may vary depending upon the functions able to be performed lawfully by the owner. The expression *dominus* identifies the holder of the title. The expression *dominium* describes the relationship between the holder of rights of ownership and the subject matter of these rights. The significance of use rather than title emerges if we add the relevant adjective to the noun: in other words *dominium publicum* and *dominium privatum*. The use to which the object or the asset may be put may be for the benefit of the public on the one hand or the benefit of an individual on the other hand. This demonstrates the legality of the use of the object or the asset rather than the legality of the owner – in the sense of title – of the object or the asset. This is demonstrated by the expressions *res publica* and *res privata*. In other words an object or asset owned by the state on the one hand or by an individual on the other hand. In both cases it is ownership in the sense of title. It is ownership in this sense that expresses exclusivity of control. Again, to use the vernacular, the object or asset is in my ownership, your ownership or their ownership. But the analysis changes dramatically in the context of use.

The use by an owner of that person's object or asset may impact upon – perhaps either beneficially or detrimentally – the use of another object or asset by its owner. It depends upon the nature of the object or the asset. Clearly the use of land or the use of water resources, and perhaps more generally the use of natural resources, may impact upon the rights and interests associated with the *dominium publicum* or the *dominium privatum* in relation to *res publica* or *res privata* in the ownership of others. These issues have been addressed by legal arrangements going back hundreds of years; long before the emergence of the concept of sovereignty of nation states as it has emerged since the middle of the 17th century. Indeed the political leader of the relevant community – whether a tribal elder, a small group or even a larger group of elected representatives, a self-appointed leader, or a hereditary monarch – would have available rights associated with *dominium publicum* or *dominium privatum* about the use of the *res publicae*. This enabled control of land and also water resources.

The relationships involved are neatly demonstrated again by the relevant Latin maxim, namely, *sic utere tuo ut non alienum laedas*. Take the example of land. Essentially this imposes a duty upon the owner of land not to use it so that land in someone else's ownership is harmed. This can be and indeed has been applied in relation to water resources. It is doctrinally based upon a recognition that there is a community of interest about the way land and water resources are used. Clearly it depends on the circumstances, the relative location of the resources and their mutual impacts of use. This is clear from the description of the relevant legal arrangements as the law of neighbourhood or the law of *voisinage*.

It is an interesting question whether a doctrine such as this applies to *res publica* as well as *res privata*. In other words, are the uses of land and water for public purposes subject to the same rule as the uses of land and water for the private purposes of individual users? In principle, in this context public uses and private uses are no different. Significantly, therefore, this illustrates the critical distinction between the exclusivity of interest conferred by the subject matter of ownership and the commonality of interest acknowledged by the use of the subject matter of the interest. It is this community of interest which justifies the existence of duties and potential liabilities imposed upon the way the rights to use the resource associated with its ownership are exercised. But how far does this community of interest extend? Immediately adjacent landowners, those with access to the same source of water, all riparians of a particular watercourse, all users of water within a watershed or water basin, everyone within the community, everyone within the nation state or perhaps even the global community. The tendency over recent years has been to extend the community of interest but not without limits.

(d) The Nature and Function of Rights in Relation to Water

While the concepts of sovereignty and property are not in themselves rights, they are certainly the source of rights. But the word right is an equivocal term. It may comprise a statement or a proposition about:

- a value or values;
- a freedom;
- a liberty;
- a privilege;
- a power;

- an entitlement;
- a claim.

If it is a value, for example, it may be spiritual or secular; philosophical or cultural; moral or utilitarian. Values sourced in these ways are sometimes included in legal instruments. Mostly they are not. If they are so included, they tend to be preambular rather than operational in nature. In any event they are influential in the sense that they represent the aspirations of the society in question.

Freedoms, liberties and privileges are similarly statements of influential generality but perhaps more directly relevant in a legal context. A freedom is the capacity to do something which would otherwise be restricted or a capacity to do something free from interference by someone else. A liberty probably connotes a total absence of restrictions. A privilege is the capacity of one person to do something which others are generally prevented from doing. While these rights are essentially personal in nature, they affect what others within the community may or may not do. Unless these freedoms, liberties or privileges are limited to a particular person or a particular group of persons, they are less likely to be relevant in the context of water resources management because of the essential community of interest in water resources that we have identified. It is this community of interest that in effect restricts the exercise of rights associated with not only property but also sovereignty. In this context, therefore, rights tend to assume the form of powers, entitlements and claims. What are these?

A power is the capacity to engage in a particular activity or to make a particular decision in this case in relation to water: for example, a power to use water or a power to control its use. The exercise of such a power, however, is almost necessarily going to be restricted because of the potential impacts of its exercise. An entitlement is more specific. It tends to represent an interest with which certain powers are associated: for example, a share in a water resource coupled with a power to use it in accordance with specific conditions. Then there is a claim. This almost necessarily involves another person in the sense that another person has interfered with or is likely to interfere with the rights and interests of the claimant. The relationship thus becomes three dimensional rather than two dimensional. The interest of the claimant; the subject matter of the interest; and the interest of the second party. This relationship is thus more dynamic and less static, and at the same time more confrontational and less harmonious.

It is however the relationship between these powers, entitlements and claims on the one hand and on the other hand the more general statements of value within which they are placed which represents in practice the concept of property which underpins much of the legal arrangements for the management of water resources. Each of these types of rights performs a different function within the overall system and is accordingly differently structured. While the concept of property remains critical, it has been and continues to be modified and adapted to fit the circumstances facing any particular community or society at any particular time.

These rights and duties derive from a number of sources. They may be based upon custom and practice. They may be based upon rules of law recognised and enforced by the judicial system because of their intrinsic authority or their incorporation in a code. Alternatively rights and duties may be granted and imposed not in general, but in specific circumstances through the exercise of powers granted to either the executive or judicial branch of government in accordance with a set of constitutional or statutory arrangements. In other words, rights and duties in relation to water resources may simply exist as a matter of law or they may be granted in accordance with the law. The former arrangement is likely to be more rigid, while the latter more flexible. But much depends upon how the relevant rules are structured.

Let us consider rules relating to entitlements. A right of ownership may be protected by a duty correlatively imposed upon everyone not to interfere or affect in any way that right of ownership. In this sense the title to the object or asset is protected. But what about use? A right of use may be protected by the imposition of a duty upon everyone else not to interfere with or affect that right of use in any way whatsoever. But such a duty would undermine the community of interest which otherwise exists in natural resources – certainly water resources. So the entitlement rule is modified to take account of this. It is modified by introducing the criterion, for example, of whether the use is reasonable or unreasonable or whether the impact of the use is reasonable or unreasonable.

The right of ownership has been described as a rigid entitlement and the right of reasonable use as a flexible entitlement. Thus:

> Together, this bundle of rigid and flexible entitlements constitutes the 'property right'.[10]

And consequently:

> The holder of the property right is secure, in that no one else may legitimately enforce a co-extensive claim to the property, yet is at the same time constrained not to use the property in an unreasonable manner and forced to endure reasonable interferences with his/her use and enjoyment of the property.[11]

In other words, a clear distinction once again between title and use as distinct but nevertheless linked elements in the concept of property.

This has produced according to one analysis a set of three paradigmatic models of water rights which combine rigid and flexible entitlement rules. These are the functional assignment model, the volumetric assignment model and the private exchange model. There are, as we shall see, examples of each of these models based either upon a set of rules of law or a set of allocations granted executively. In more detail:

> The functional assignment model tempers a stable core-entitlement to the productive use of water with flexible entitlements to water for domestic use. The volumetric assignment model rigidly protects entitlement to a circumscribed volume of water, but imposes a regime of sharing in respect of surplus water. In contrast, the private exchange model pursues adaptability through allowing the right to use water to gravitate to the most efficient use, and to remain there only for as long as it remains the most efficient use.[12]

The usefulness of any model depends upon the circumstances of the relevant hydrological cycle and the wider circumstances in which it operates.

Rights in relation to water may arise in consequence of an administrative grant or allocation. The rules in this case are determined not by the legal system but by an administrative agency acting in accordance with its source of competence. This may be either a constitution, a code, the doctrine of the public domain – *dominium publicum* – or its statutory mandate. In any of these cases the allocation may be made in accordance with an open discretion unfettered by any rules of law, a discretion exercised in accordance with a set of clearly stated criteria, or a methodology of decision-making constrained by a set of procedural and substantive rules of law. Thus, just as rules of law may be intrinsically more or less rigid or flexible, so administrative allocation systems may be more or less rigid or flexible. In both cases it depends ultimately upon how the rules are structured, and this in turn depends upon the language used to formulate their structure and describe their content. In other words, the approach that form follows function and not the other way around.

THE CONCEPT OF PROPERTY IN WATER

(a) The Ambiguity of the Concept

If a system of governance is designed to achieve the sustainable use and development of common-pool resources such as water resources, then what form should the relevant institutional structures take? Or, more specifically, what set of rights and of obligations complementary to these rights should be included within such a system? The answer traditionally has been some form of property regime.

But property is an ambiguous term the meaning of which varies in accordance with the disciplinary context in which it is used – for example economics, policy analysis or law.[13] This is particularly important in relation to water because the answer to the question who owns – and hence controls – water depends very much upon the particular part of the water cycle to which it is addressed. For example, who owns water lying on the surface of land; who owns water flowing through a natural watercourse; who owns water percolating underground; and perhaps even who owns water falling to the surface of the earth as precipitation?

Although notions of property and ownership are not particularly helpful in relation to water, they are and always have been used in this context. In the English language property is the expression often used to describe the thing owned. It also describes, relevantly for present purposes, the regime of rights and obligations which relate to the thing owned. In the Latin language – and in Roman law – the relevant type of ownership is connoted by the adjective which is attached to the noun and which describes the thing owned. The approach of Roman law is significant not only for its own sake but also because it has influenced the development not only of the civil law but also of the common law of England.

The Latin language and Roman law are a useful point of commencement for any analysis of the notions of property and ownership.[14] Traditionally there have been four relevant types of property regimes. With the precision of the Latin language these have been described using the singular rather than the plural as *res publica, res communis, res propria*, and *res nullius*. In other words, the thing belongs respectively to the state, to the community, to a particular person and to nobody. The significance of this classification is this. *Res publica* and *res propria* belong to a clearly identifiable entity: in one case the state and in the other case a particular person. The identity of the owner is unequivocal. In the case of *res nullius* there is, of course, no owner.

More complicated is *res communis*. The thing in question belongs to the community. But what is the community? Is it the residents of a village or of a town? Is it the members of an identifiable group such as a group of irrigators, a group of farmers, a group of industrialists, the residents of a nation state or perhaps even the global community? For practical purposes, however, in the context of water resources it is likely to be the members of the group who have access in a physical sense to the water resources in question. Although the expression *res communis* may be the more precise way of describing it, it is often referred to as a regime of common property. But the critical question – let it be emphasised – is who controls the thing. The adjective or adjectival phrase answers the question:

- *res publica* – the state – public property;
- *res communis* – the community – common property;
- *res propria* – a particular person – private property;
- *res nullius* – no person – ownerless.

Does this doctrinal approach sit comfortably with the perception of property as an economic instrument?

(b) Property as an Economic Instrument

If water is taken to be a common-pool resource, the point of commencement is the meaning and function of property from the perspectives of economics, ecology and culture.[15] In this context, property has been defined 'as the institution that, by allocating things to people, gives people liberty over things'.[16] This connotes three elements: an allocator, a receiver and a thing. In economic terms it is this 'liberty' that is important for it removes 'the obstacles to allocating resources by bargaining and voluntary exchange';[17] in other words, a regime of competitive markets. But if property is in this sense an economic institution, what form does it take as an institution of the law?

Presumably it must either be assumed to exist by the legal system or established by the legal system. Either way it involves an institution of the state performing legislative, executive or judicial functions. Thus, the right may be created by the legislature; it may be granted by the executive in exercise of powers conferred by the legislature; or it may be recognised as an inherent element of the legal system by the judiciary. From the legal perspective it has been suggested that:

> The state not only has to decide whom to entitle, but it must also simultaneously make a series of equally difficult second order decisions. These decisions go to the manner in which entitlements are protected and whether an individual is allowed to sell or trade the entitlement.[18]

And if the right is in the form of property:

> An entitlement is protected by a property rule to the extent that someone who wishes to remove the entitlement from its holder must buy it from him in a voluntary transaction in which the value of the entitlement is agreed upon by the seller.[19]

It is this form of entitlement 'which gives rise to the least amount of state intervention'.[20]

Some form of institutional arrangements seems to be essential. If this approach is adopted in relation to common-pool resources such as water, then the scenario has been described in these words:

> Thus managing common-pool resources comes down to limiting access (who will be excluded and how) and affecting subtractability (how much the taking by one person will impinge on the taking by another). Managing common-pool resources involves various ways of limiting access (deciding who will be excluded and how) and affecting harvesting efforts (how much one actor can take).[21]

This involves some form of property regime as the doctrinal foundation of a governance system. It has been stated unequivocally that 'the basic functions of resource management – coordinating users, enforcing rules, and adapting to changing environmental conditions – cannot be fostered without a specified system of property rights'.[22] But the institution of property – and the rights and obligations associated with it – performs different functions and assumes different forms within any legal system. The functions and forms need to be described precisely.

(c) Water as Common Property

Despite the nuances of the English and Latin languages, the function of a property regime depends upon the adjective attached to the noun property or the noun *res*. In policy and economic terms there are three such regimes: private property, state property and common property. In the case of the first two, an identifiable person or institution controls the thing in question. Common property – like a common-pool resource – is a more complex concept. It has been suggested that a private property regime – and perhaps also a state property regime – may not by itself be sufficient to ensure sustainability. Thus:

> Even when resources are divisible and can be privately owned, ensuring environmental sustainability to protect the well-being of future generations cannot be accomplished by individuals acting out of self-interest alone. Even if environmental services could be entirely privately owned, we would need intergenerational commons institutions to ensure sustainability.[23]

The justification for such an approach is the common responsibility of all human kind for the achievement of sustainable development. So common responsibility in the normative sense is aligned with common property, while private – or individual – responsibility is aligned with private property. Ultimately the challenge is a system of governance which incorporates both common rights and private rights with common obligations and private obligations. But the immediate question is what is common property.

Common property has been described in a number of different ways. For example, it means that 'the resource is held by an identifiable community of users who can exclude others and regulate use'.[24] Significantly this includes the characteristics of identification, exclusion and regulation. The inclusion of regulated rather than unfettered use is important. But it probably implies a system of self-regulation by the community rather than regulation by an independent set of institutional arrangements. A more juridical approach emerges from this description:

> The term 'common property' as employed here refers to a distribution of property rights in resources in which a number of owners are co-equal in their rights to use the resource. This means that their rights are not lost through non-use. It does not mean that the co-equal owners are necessarily equal with respect to the quantities (or other specification) of the resource each uses over a period of time.[25]

If this is applied to common-pool resources, it means that the common-pool resource – the water as the resource in question – is owned by a community of identifiable owners. Each of these owns a share of the common-pool resource. This share carries with it the right to use the resource in accordance with the arrangements for the management of the common-pool resource. In the context of this description, we return again to the fundamental distinction between water as a resource in its original location and water in a form in which it can be used for a specific purpose. The water resource is in this sense common property and the water resource share is private property. The relationship between the two is governed by a set of norms, customs, practices and rules. Perhaps it even amounts to a form of social compact among the members of the group.

(d) Water as Private Property

Let us now switch the focus from water as a common-pool resource to water in the form in which it can be used for economic, commercial, social or even environmental purposes. In this context water has often been described as a good. The noun good in this sense presumably means the satisfaction of some need, the realisation of a desirable outcome or the production of a benefit or a usufruct. Once this has been achieved – for example when water is immediately available for use by turning a tap – water is increasingly regarded as a commodity, and as such is able to be bought and sold.

The economic, commercial, social and environmental benefits flowing from the use of water resources are quite different. Social benefits accrue to members of a community as individuals and as human beings. Economic benefits are those accruing to the community at large as a result of the way the assets of the community – in this case its water resources – are used and developed. Economic and social benefits need to be distinguished from commercial benefits. Commercial benefits may be described as those benefits which accrue to individuals or institutions that are engaged in business activities and that use water during their business or commercial activities. This benefit is neither social, in the sense that it relates to an individual as a human being, nor economic, in the sense that it relates to the use of the assets of the community. It is nevertheless a benefit accruing to an individual or an institution operating in a commercial manner for the advantage of that individual or that institution. There is in other words no direct benefit to the community or the public. Economic benefits, on the other hand, accrue to the community at large, and only incidentally thereby to individual members of the community.

Environmental benefits, quite differently, are outcomes directly advantageous to the environment as an institution in its own right. But the environment does not exist as an institution within the legal system. Environmental benefits are produced and managed by humankind in accordance with the traditional institutions of the legal system. Although environmental outcomes are regarded as public benefits, they are intended to benefit the environment as much as the public interest in protecting the environment. So the use and development of water resources for economic, commercial, social and environmental purposes involve the existence or creation of sets of rights and obligations which are designed to achieve simultaneously and in an integrated fashion a range of benefits or outcomes that reflect different normative values.

The idea of water as a good leads to the notion of water as a commodity. This goes beyond the idea of water as beneficial to individual members of society, the economy or the environment. It includes the notion of tradeability. In other words the benefit which belongs to one person may well be able to be transferred to another person or institution. Water in its natural state probably cannot be regarded as a commodity. But water in some other form may well be seen as a commodity. Flowing water, it has been stated, 'cannot be treated as a separate commodity because it is in a 'constant state of diffusion' or of movement, and a precise unit thus cannot be allocated to a single individual'.[26] Flowing water, accordingly, cannot be *res propria*. Similarly in relation to underground water:

> In the case of pools of underground water, property rights can be defined as an area of land above them, but it will not be clear whether the water extracted really comes from the area below this surface because of the fluid nature of the resource. Pumping from another area of land may in fact be extracting the water underneath another owner's patch of ground.[27]

The nature of water as a resource thus suggests that it will likely be managed more effectively as common property than as private property.[28] But what about water which has been either naturally or artificially transformed in ways that make it immediately or perhaps even in the longer term socially, economically or environmentally beneficial?

At some point in the hydrological cycle water may become a commodity. If so, it is likely to be the subject of ownership or property rights. But at what point of the hydrological cycle does water assume this characteristic? Consider some examples:

- water collected and stored in an artificial reservoir created by the construction of a dam that has obstructed or reduced the flow of water in a watercourse;
- water extracted from water flowing in a watercourse and collected and stored for immediate or future use in an artificial channel or pipe;
- water supplied to a consumer through a reticulation system that enables the consumer to use the water at the discretion of the consumer but for a prescribed purpose;
- water returned to the supplier after use by the consumer through a reticulation system enabling reuse of the water;
- water extracted from the sea, transformed by a desalination process and supplied to a consumer through a reticulation system.

In each of these examples the water has been transformed so that it performs a function which is socially, economically, environmentally or possibly even commercially beneficial. Water transformed in these ways may well be *res propria* and taken to be in the ownership of the person or institution in lawful possession of it. For example, water in the lawful possession of the operators of a desalination process is clearly under their control. While the water is in their possession and under their control in this way, it may be regarded as *res propria* and in this sense owned by them as a commercial commodity. Ultimately it is a question of control. While issues of ownership may arise in certain circumstances, it is suggested that the most helpful question to ask from the legal point of view is not who owns the water but who controls the water at the various stages of its evolution through the hydrological cycle, whether natural or artificial.

CONCLUSION

The answer to this question who controls water requires an analysis of the origins, sources and characteristics of rights in relation to water and of how such rights are acquired. In this context, sovereignty, property and ownership are not in themselves rights. Rather they are descriptive of the arrangements of rights – and, as we shall see, their correlative duties –

which exist in relation to water. However sovereignty, property and ownership may well be sources of rights in the sense that the coalescence of a number of rights available conjunctively may constitute what is described by these concepts.

While sovereignty and property operate as concepts in different contexts, they disclose some common characteristics. Neither is absolute. Each is restricted in its own way. Exclusive rights derived from sovereignty became effective when nation states were confined within their territorial boundaries and exclusive rights derived from property became effective when the holders of these rights were confined within their cadastral boundaries. The exercise of rights of sovereignty and of property has been tempered to the extent that conflicting values such as human rights, environmental protection and sustainable development have been incorporated within the relevant legal arrangements.

Fundamentally, rights are either original and inherent on the one hand or contingent and derivative on the other hand. In other words, rights exist either because of their own rational authority or because of the rational authority of their source. The relationship between original and contingent rights or between inherent and derivative rights is important. Do they represent a hierarchy of values or are they of equal relevance and significance in any particular set of circumstances? For example, if there is an inherent or original human right to water, does it take precedence over any contingent or derivative right to take and use water for agricultural, industrial or other commercially linked purposes?

The doctrinal foundation of the rights of nation states is the notion of sovereignty; that of the rights of individual and institutional members of a nation state is some form of property regime. Are these rights original and inherent or contingent and derivative, or perhaps a mixture of both? A system of water resources governance is based to some extent upon an arrangement of rights such as these. Let us now consider in relation to water:

- the sources and nature of the rights of nation states as members of the international community;
- the sources and nature of the rights of nation states as institutions for water resources governance within their territory;
- the sources and nature of the rights of individual and institutional citizens and members of nation states;
- how the rights and entitlements of such persons are acquired and used.

NOTES

1. Ostrom (2001, p. 18).
2. Ostrom and Schlager (1996, p. 129).
3. Schlager and Blomquist (2001, pp. 134 and 135).
4. Louka (2006, p. 88).
5. Schrijver (1999, p. 70).
6. Schrijver (1999, p. 71).
7. Schrijver (1999, p. 97).
8. Schrijver (1999, p. 98).
9. Macpherson (1978, p. 4).
10. du Bois (1995, p. 113).
11. du Bois (1995, p. 113).
12. du Bois (1995, pp. 113 and 114).
13. Fisher (2004b, p. 201–204).

14. See e.g. Borkowski and du Plessis (2005, pp. 153 and 154).
15. See generally Yandle (1983, p. 8), Hanna et al. (1996, pp. 2–5), Costanza and Folke (1996, pp. 26–28), Berkes (1996, pp. 89, 94–101), McCay (1996, pp. 112–117), Ostrom and Schlager (1996, pp. 129–141), McKean (1996, pp. 225–227) and Ostrom (2001, pp. 18–22 and 28–31).
16. Cooter and Ulen (1988, p. 117).
17. Cooter and Ulen (1988, p. 117).
18. Calabresi and Melamed (1972, p. 1092).
19. Calabresi and Melamed (1972, p. 1092).
20. Calabresi and Melamed (1972, p. 1092).
21. Burger et al. (2001, p. 3).
22. Hanna et al. (1996, p. 4).
23. Burger et al. (2001, p. 9).
24. Berkes (1996, p. 89).
25. Ciriacy-Wantrup and Bishop (1975, pp. 714 and 715).
26. Luterbacher and Wiegandt (2005, p. 14).
27. Luterbacher and Wiegandt (2005, p. 14).
28. McKean (1996, pp. 227–232).

GENERAL REFERENCES

Berkes, Fikret (1996), 'Social systems, ecological systems and property rights', in Susan Hanna et al. (eds), *Rights to Nature* (Washington: Island Press), pp. 87–107.

Borkowski, Andrew and Paul du Plessis (2005), *Textbook on Roman Law* (3rd edn, Oxford: Oxford University Press).

Burger, Joanna et al. (2001), 'Common-pool resources and commons institutions', in Joanna Burger et al. (eds), *Protecting the Commons* (Washington, DC: Island Press), pp. 1–15.

Calabresi, Guido and A.D. Melamed (1972), 'Property rules, liability rules and inalienability: one view of the cathedral', *Harvard Law Review* 85, 1089–1128.

Ciriacy-Wantrup, S.V. and R.C. Bishop (1975), 'Common property as a concept in natural resources policy', *Natural Resources Journal* 15, 713–729.

Cooter, R.D. and T.S. Ulen (1988), *Law and Economics* (New York: Scott Foresman and Company).

Costanza, Robert and Carl Folke (1996), 'The structure and function of ecological systems in relation to property-rights regimes', in Susan Hanna et al. (eds), *Rights to Nature* (Washington: Island Press), pp. 13–34.

Du Bois, François (1995), 'Water law in the economy of nature', in J.A. Allan and Chibli Mallat (eds), *Water in the Middle East; Legal, Political and Commercial Implications* (London: Tauris Academic Studies), pp. 111–123.

Fisher, D.E. (2004b), 'Rights of property in water: confusion or clarity', *Environmental Planning Law Journal* 21, 200–226.

Hanna, Susan et al. (1996), 'Property rights and the natural environment', in Susan Hanna et al. (eds), *Rights to Nature* (Washington, DC: Island Press), pp. 1–10.

Louka, Elli (2006), *International Environmental Law* (Cambridge: Cambridge University Press).

Luterbacher, Urs and Ellen Wiegandt (2005), 'Cooperation or confrontation: sustainable water use in an international context', in Edith Brown Weiss et al. (eds), *Fresh Water and International Economic Law* (Oxford: Oxford University Press), pp. 11–34.

Macpherson, C.B. (1978), 'Meaning of property', in C.B. Macpherson (ed.), *Property: Mainstream and Critical Positions* (Oxford: Basil Balckwell), pp. 1–13.

McCay, Bonnie J. (1996), 'Common and private concerns', in Susan Hanna et al. (eds), *Rights to Nature* (Washington, DC: Island Press), pp. 111–126.

McKean, Margaret A. (1996), 'Common-property regimes as a solution to problems of scale and linkage', in Susan Hanna et al. (eds), *Rights to Nature* (Washington, DC: Island Press), pp. 223–243.

Ostrom, Elinor (2001), 'Reformulating the commons', in Joanna Burger et al. (eds), *Protecting the Commons*, Washington, DC: Island Press, pp. 17–41.

Ostrom, Elinor and Edella Schlager (1996), 'The formation of property rights', in Susan Hanna et al. (eds), *Rights to Nature* (Washington, DC: Island Press), pp. 127–156.

Schlager, Edella and William Blomquist (2001), 'Water resources: the southwestern United States', in Joanna Burger et al. (eds), *Protecting the Commons* (Washington, DC: Island Press), pp. 133–159.

Schrijver, Nico (1999), 'The changing nature of state sovereignty', *British Yearbook of International Law* 70, 65–98.

Yandle, Bruce (1983), 'Resource economics: a property rights perspective', *Journal of Energy Law and Policy* 5, 1–19.

5. The range of functions performed by states

RIGHTS OF SOVEREIGNTY

(a) Introduction

Sovereignty – like property – is an equivocal concept, but the nature of the rights associated with it are fundamental to the way water resources are governed, not only nationally but also internationally. On the face of it sovereignty describes the range of exclusive rights available to a nation state within its territorial boundaries. It is generally accepted that this approach emerged – principally in Europe – about the middle of the 17th century. In this way the territory of a state, and hence the natural resources under its control, were delineated by artificial boundaries, and this has clearly had significant impacts upon the way water as a common-pool resource has been governed. Rivers, for example, may flow through the territories of several nation states and underground aquifers may similarly underlie the territorial surface of several nation states.

The right to control natural resources has been a critical factor in the rise and subsequent demise of political and economic colonialism. Indeed in 1974 the General Assembly of the United Nations confirmed the importance of sovereignty by resolving that 'every State has and shall freely exercise full permanent sovereignty, including possession, use and disposal, over all its wealth, natural resources and economic activities'. While this response may have been inspired by the emerging tensions between developed, undeveloped and developing nation states, it was tempered by international arrangements for the protection of the environment and subsequently by the emergence of sustainable development.

These two doctrinal elements of international law – environmental protection and sustainable development – represent in effect restrictions upon the exercise of full permanent sovereignty by nation states over their natural resources. An equally important restriction upon sovereignty may be emerging in relation to the governance of water resources – namely the recognition of a human right to water. In essence, sovereignty, as a bundle of rights inhering in a nation state, is being increasingly limited by the emergence of a number of potentially conflicting values:

- groups of human rights;
- rights to cultural, social and economic development;
- protection of the environment;
- sustainable development.[1]

How does this impact upon water as a common-pool resource and upon water as an economic, social and environmental good?

(b) Rights of Sovereignty in Relation to Water Resources

Since most nation states are not islands, many share water resources. But water is not only a resource but also a good and a commodity which moves between or among states or is transported between or among states. How do these phenomena impact upon the rights of sovereignty of a nation state? It has been suggested there are four doctrines that may underlie the international legal arrangements for the governance of water resources:[2]

- absolute territorial sovereignty;
- absolute territorial integrity;
- limited territorial sovereignty;
- community of interests.

The difference between the two 'absolute' or extreme approaches has been described in these words:

> While the doctrine of absolute territorial sovereignty insists upon the complete freedom of action of the upstream state, that of absolute and territorial integrity maintains the opposite: that the upstream state may do nothing that might affect the natural flow of the water into the downstream state.[3]

Neither of these two absolute approaches has received overwhelming support. Thus:

> The doctrine of 'limited territorial sovereignty' is probably the prevailing theory of international watercourse rights and obligations today. According to this theory, the sovereignty of a state over its territory is said to be 'limited' by the obligation not to use that territory in such a way as to cause significant harm to other states.[4]

This approach has been incorporated, for example, in Article 11(1) of the Draft International Covenant on Environment and Development. Thus:

> States have, in accordance with the Charter of the United Nations and the principles of international law, the sovereign right to utilize their resources to meet their environmental and sustainable developmental needs, and the duty to ensure that activities within their jurisdiction or control do not cause potential or actual harm to the environment of other States or of areas beyond the limits of national jurisdiction.

To this is linked a right as well as a duty to protect the environment under their jurisdiction from significant harm caused by activities outside their national jurisdiction. If there is such harm there is an entitlement to a remedy. Significantly this is a right to protect a state's internal environment from external activities. Then there is the doctrine of a community of interests in water resources. This is akin to the notion of water resources as common-pool resources. The idea has had a long history – particularly before the territorial enclosure of nation states within boundaries during the 17th century.[5] The doctrine is based rationally upon all of the characteristics of water as a resource. Thus:

> The notion that there is a community of interest of riparian states in the waters of an international watercourse is a venerable one. It derives from the vital nature of water to all life, the unity of a watercourse, and the importance of watercourses as means of transportation, communication, and socio-economic development.[6]

In other words it recognises the nature of the hydrological cycle, the characteristics of water resources, a human right to water and the relevance of water as a social, economic and commercial good and perhaps even as an environmental good.

(c) The Exercise of Rights of Sovereignty

States within the international community are entitled to exercise their rights of sovereignty over their natural resources in whatever way they decide, but subject to their responsibilities and obligations under international law. It has been suggested that the role of a state may be functional or discretionary,[7] functional in the sense that states should ensure satisfactory outcomes for their citizens in compliance with international law, or discretionary in the sense that states can set their own largely uncontrolled outcomes. This raises the question whether the norms of international law impose effective restrictions or not. One of the most important of these restrictions is the emerging principle of sustainable development. In any event, the arrangements within nation states for exercising their rights of sovereignty vary enormously. Much depends upon the constitutional and institutional arrangements according to which their legislative, executive and judicial functions are severally performed.

The fundamental question to be addressed is how the inherent right of a nation state to control the use and development of its water resources is distributed within its jurisdiction. In this context the rights of nation states tend to assume the form of powers exercisable in various ways and for different purposes. The emerging responsibility of nation states is to manage their water resources sustainably. This responsibility becomes in the context of a national legal system a set of self-imposed arrangements for achieving this outcome. It can be achieved in a number of ways:

- by creating fundamental or constitutional norms relating to sustainable development;
- by creating or recognising a set of enforceable rules designed to achieve sustainable development;
- by creating a system for granting rights in relation to water with the objective of achieving a prescribed outcome;
- by creating a system according to which the state itself – acting directly, through its agencies or through licensed institutions – either comprehensively or selectively manages the water resources of the state so as to ensure the achievement of the intended outcomes.

These are in synoptic form a regime of *Grundnorms*; a regime of appropriation by enforceable rules; a regime of allocation by administrative grants; and a regime of direct or indirect public management. In practice, a set of legal arrangements is likely to comprise a mixture of one or more of these approaches. Let us consider them separately.

A REGIME OF *GRUNDNORMS*

Rights of ownership and control of natural resources vested in the state may be the fundamental *Grundnorm* of the legal arrangements according to which these natural resources are managed. Alternatively these arrangements may be formally acknowledged in a constitution or created by the legislature without specific constitutional or other fundamental protection. Or these *Grundnorms* may be acknowledged by the legal system – in practical terms the courts – to be the assumption upon which the system is based. Consider some examples.

According to Roman law perennial rivers – *flumina* – were *res publica*, and thus under the control of the state.[8] Water on or under *ager privatus* – land owned privately – was under the control of the landowner.[9] More ambiguous doctrinally,[10] flowing water – *aqua profluens* – was *res communis* and thus accessible by no one in particular but by the relevant community at large.[11] The French doctrine of the public domain – including its application to public water resources – was based upon the exercise of *la puissance publique* over the resources in question.[12] While the public domain is not strictly a form of ownership, it has been described as *une propriété administrative* akin somewhat to ownership but essentially a form of public control.[13] In Japan the legislature has provided that watercourses belong to the public domain.[14] In Islamic law all members of the Moslem community were considered to have a right of access to water resources.[15] However under the Ottoman Civil Code 'all waters were declared as vested in the state, the crown or incorporated in the public domain, the state thus taking the place of the Moslem community'.[16]

Constitutional arrangements vary enormously. Some are structural, while others are normative. In the United Kingdom, for example, the fundamental principle has traditionally been the sovereignty of Parliament – but subject now to the legal regime of the European Union. In the federations of Australia and Canada the constitutional arrangements distribute structurally the power to legislate between the Commonwealth and the states in Australia and between the Dominion and the provinces in Canada. In some instances the constitutional arrangements include fundamental norms in the form of human rights sometimes supported by the imposition of obligations upon the state to protect and give effect to these rights. A human right, for example, may relevantly take the form of a right to life, and hence to water. This may be expanded by a duty imposed on the state to realise this right. Or it may be a statement of aspiration or principle in the form of what has been described as a paralegal rule. For example, Article II(15) of the Constitution of the Philippines requires the state to protect and promote the right to health of the people.[17] Article 21 of the Constitution of the Union of India confers on every person the fundamental right to life; Article 48–(A) declares that one of the directive principles of state policy is to endeavour to protect and improve the environment; and Article 51–(A) places a fundamental duty on every citizen to protect and improve lakes and rivers.[18]

Somewhat differently, the Constitution of the People's Republic of China by Article 9 uses ownership as a technique of control. Thus:

> All mineral resources, waters, forests, mountains, grasslands, unreclaimed land, beaches and other natural resources are owned by the state, that is, by the whole people, with the exception of the forests, mountains, grasslands, unreclaimed land and beaches that are owned by collectives in accordance with the law. The state ensures the rational use of natural resources.

Thus all waters are owned by the state and none by collectives. In effect, the Chinese approach is not unlike the doctrine of the public domain. While the criterion for the exercise by the state of its right is the rational use of natural resources, it is so structured as to be effectively unenforceable. These examples do no more than confirm the fundamental right of a nation state to manage its water resources as it sees fit, but consistently with its perceptions of its international legal responsibilities. Their effectiveness depends upon how the courts interpret and apply these constitutional *Grundnorms* and the capacity of the courts to protect and enforce them.

A REGIME OF APPROPRIATION BY ENFORCEABLE RULES

While custom may be a source of law in some circumstances, it also provides the doctrinal basis upon which more formal and enforceable sets of rules are formulated. While a legislature can no doubt acknowledge custom, it is more likely to be a function of the judicial system to recognise or create rules of law which are based upon custom and practice. Indeed, custom performed a function in the early development of Roman law before codification.[19] Similarly custom was and remains an important source of Islamic law notwithstanding the formulation of rules in accordance with Shari'ah law.[20]

In England, the formal body of common law grew out of custom. This was however a slow process in relation to water rights and obligations. It has, for example, been suggested that:

> Before 1600 the bulk of water disputes would have fallen to be decided in local courts outside the common law. Local decision-making would have been at the discretion of court and jury, with scant reference to legal rules, though common-law entitlements as developed in the royal courts may have served as a base-line for claims. What is striking is a steep increase in common-law water litigations from around 1600, prompted by factors such as the increasing density of industry on watercourses, a general rise of business of the common-law courts, and also the burgeoning common-law hostility to traditional monopolies. The common-law increasingly resorted to natural rights to promote effective solutions and at the same time to enhance jurisdiction.[21]

A rules based system was thus created in response to the judicial settlement of disputes about the use of water. In Roman law similar ad hoc and fragmented responses were eventually codified so as to create a principled basis upon which to make decisions about the application of rules and the settlement of disputes. Not so in English law. The rules based system in England remained in a state of doctrinal flux until the second half of the 19th century when it too was overtaken gradually: first by ad hoc and subsequently by more comprehensive intervention by the legislature.

A rules based system has been and probably always will be a significant part of water resources governance. The rules may be recognised or created by the common law or the civil law or they may be created by the legislature. In either case these rules are the basis for trading in water rights – not a new phenomenon but one increasingly preferred in some jurisdictions. But a rules based system created by the courts is essentially haphazard rather than planned in nature. This is not surprising because the courts are not equipped to perform planning functions. The courts nevertheless reflect the values of their time and are no doubt influenced by them in providing reasons for their decisions – at least by implication if not explicitly. For these reasons it is unlikely but not impossible for the judiciary to respond positively and creatively to the principle of sustainable development. This has become a function of formal planning for water resources use and development by the state.

A REGIME OF ALLOCATION BY ADMINISTRATIVE GRANTS

The holders of rights in relation to water may, if so minded, put in place – no doubt informally – plans for the exercise of their rights and the use of the water. In this event these plans and the consequential actions are likely to be dominated by what are perceived to be the best interests of the holders of these rights. Unless, of course, there are legal restrictions upon how these rights may be exercised. For present purposes, however, the point is the nature of the power of the nation state in exercise of its rights of sovereignty to create a system for the grant of rights in relation to water to members of the community; in effect an allocation system based upon planning for the best use of the water resources in the public interest.

If the state has rights of ownership – as in the case of the People's Republic of China – it may use these rights as the basis for creating such a system. In this event the arrangements are essentially executive in character. Alternatively, as in the federated states in Australia, the legislature may create such a system that enables the executive to grant rights in relation to water in accordance with the scheme created by the legislature. Much therefore depends upon how the scheme is structured. It may give, for example, an open and unfettered discretion to the executive to do what the executive wishes. It may restrict the executive procedurally in terms of how it makes decisions. Or it may restrict not only how the executive makes decisions but what decisions the executive may make.

In any event the granting of rights is likely to be preceded by and based upon a process for planning how water resources should be used in the public interest and how rights in relation to water should be distributed accordingly. Increasingly the outcome to be achieved by such a planning process and by such an allocation process is sustainable development. How sustainable development in this sense is incorporated within the system is a matter for further consideration. For present purposes, three examples will demonstrate the different characteristic of allocation systems: the Ottoman Civil Code, the Water Law of the People's Republic of China, and the statutory arrangements in Australia.

The Ottoman Civil Code vested all waters in the state. The arrangements contained in the code contained three other elements:

- government (as agent for the state) was given control of the use of water (other than for drinking or animal watering purposes) not regulated in accordance with earlier legislation or Shari'ah law;
- water commissions were set up to survey and recognise established water rights;
- land registers were compiled in order to keep a written record of duly recognised land and water rights.[22]

These are the characteristics of an incipient regulatory system for allocating water resources based upon the acquisition of relevant information and the publication of water rights. On the face of it, the exercise of these powers by the agencies of the state was for the most part uncontrolled.

The right of ownership of all waters conferred upon the state in China by Article 9 of the Constitution of the People's Republic of China is explained and developed by the more detailed provisions of the Water Law of 2002. Article 3 of the Water Law provides:

> Water resources are owned by the State. The State Council, on behalf on the State, exercises the right

> of ownership of water resources. The water of ponds belonging to rural economic collectives and the water of reservoirs built and managed by such collectives shall be used by the collectives respectively.

Article 7 provides:

> For water resources, the State applies, in accordance with law, the system of licensing for water-taking and the system of compensation for use of water, except for water of the ponds and reservoirs belonging to rural economic collectives that is used by such collectives and their members. The administrative department for water resources under the State Council is responsible for making arrangements for implementing the system of licensing for water-taking and the system of compensation for use of water throughout the country.

While the state owns water resources and collectives own water in their ponds and reservoirs, individual rights in relation to the taking of water arise as a consequence of a licensing system. How the licensing system operates is the responsibility of the State Council. However – and most significantly – Article 1 states why the law has been enacted:

> This law is enacted for the purposes of rationally developing, utilising, conserving and protecting water resources, preventing and controlling water disasters, bringing about sustainable utilization of water resources, and meeting the need of national, economic and social development.

A planning function, that leads, one hopes, to the sustainable utilisation of water resources. It is the responsibility of the State Council – an executive body – to achieve this. But this is a responsibility unlikely to be capable of judicial enforcement.

The distinction between the exercise of ownership rights and statutory rights in relation to water resources is demonstrated by the constitutional arrangements in the colonies and subsequently the states that comprise the Commonwealth of Australia. Under the common law of England all land was held by the Crown until it was alienated. On the establishment of a colony all land in the colony was owned by the Crown and managed on its behalf by the Governor. This enabled the Governor in this capacity to manage the land and any water resources associated with the land as the Governor saw fit. The right to use and to control land and its associated water resources was thus an incident of the right of ownership vested in the Crown.[23] This changed substantially upon the enactment of a constitution for a colony, which subsequently became the constitution of the state upon federation. According to these constitutional arrangements the management and control of the waste lands of the Crown – that is the unalienated lands of the Crown – was vested in the legislature. Effectively this deprived the Crown of its right of management and control of land as owner of the land and transferred it to the legislature. The consequence was that the lands of the Crown could only be managed in accordance with legislation enacted by the legislature.

For most of the 19th century the rules of the common law of England applied to the management of water resources in the colonies and then the states in Australia. But by the beginning of the 20th century each of the states had enacted legislation for the use and management of water resources. The nature of this legislation has changed over the years. And so has its impact. Originally it was directed mostly at the unrestricted use and development of water resources, but now at their sustainable development.[24] Significantly the current legislation is structured not in terms of ownership but in terms of use and control. Take the legislation of Queensland as an example.

According to section 19 of the Water Act of 2000 of Queensland all rights to the use, flow and control of all water are vested in the state. This is complemented by a prohibition on taking or supplying water or interfering with water without authorisation. This is the foundation of a system of allocation based upon a comprehensive set of arrangements for planning the use and development of water resources in the public interest, the grant of rights in relation to water, and the regulation of the way in which water is supplied to the community. Perhaps even more importantly, legislation of this kind – certainly in jurisdictions like Australia – restricts the way in which the rights of the state are able to be exercised. These restrictions are directed at the achievement of the sustainable use and development of water resources.

A REGIME OF DIRECT OR INDIRECT PUBLIC MANAGEMENT

While a state and its agencies may be empowered to grant rights in relation to water to members of the community, the state either itself or acting through its agencies may manage its water resources directly or permit another institution to do so on its behalf. A power of this kind is not a power of direct regulation as in the case of a system of allocation, or a power of indirect regulation as in the case of a system of appropriation. Rather it is a system of direct operational involvement in each and every activity that comprises the management of water resources from upstream operations – storing, diverting or abstracting water – through downstream activities to the point of supply or distribution to the consumer or other beneficial user. The exercise of such a power involves designing and constructing a legal regime that includes all of the functions of water resources governance:

- collecting relevant information;
- analysing such information;
- identifying the best uses of the resource in the public interest;
- planning for the use and development of the resource accordingly;
- implementing the plans for doing so by building and operating the relevant infrastructure;
- delivering the water to consumers and other users;
- reviewing, maintaining and operating the whole system.

The desirability of such a comprehensive approach is dependent, no doubt, upon a range of political, economic, and cultural considerations. In practice different institutions may variously perform one or more of these functions. For example, the planning function may be separated from the operational function or the upstream activities separated from the downstream activities. Central government may perform the planning function while an operational agency of central government may construct and operate the upstream infrastructure. Local government may perform the function of receiving water from the upstream operator and distributing it to consumers and users within its area. Similar functions may be carried out by statutory bodies acting on behalf of central government or perhaps independently of central government.

An alternative is for all or some of the operational functions to be delegated to a private sector operator who is thus licensed to operate and construct the infrastructure and provide a service by way of supply or delivery of the water to consumers and other users. Such a licensed operator is expected, or perhaps even required, to operate as an effective and efficient member of the private sector acting in a commercial manner and producing a profit on the use of its assets. The manner in which these licensed operators provide their services by exercising their rights in relation to water may itself be regulated by the law, and this is a matter for further consideration.

CONCLUSION

The way in which a nation state exercises its rights of sovereignty and discharges its international responsibilities is essentially a matter for itself. These ways include:

- state ownership of water resources;
- a rules based system of appropriation;
- a planning based system of allocation;
- a system of public enterprise;
- a system of enterprise delegated to the private sector.

But if there is to be a regime of sustainable water resources governance – particularly if it incorporates legal sanctions – then this represents a very significant restriction upon how a nation state performs these functions. Undoubtedly international legal arrangements and legal arrangements within nation states are moving – in some respects perhaps imperceptibly – in this direction. To identify the extent to which this may be happening requires not only an examination of the nature and sources of rights in relation to water but also an assessment of how the exercise of these rights is constrained by the expectation – if not the rule – that water resources should be used and developed sustainably. We need to consider how rights in relation to water are recognised, appropriated or granted before we consider the criteria according to which they are exercised.

NOTES

1. See generally Hildering (2004, pp. 143–167) and Rieu-Clarke (2005, pp. 60–90).
2. McCaffrey (2007, p. 112).
3. McCaffrey (2007, p. 126).
4. McCaffrey (2007, p. 135).
5. McCaffrey (2007, pp. 147 and 148) and Hildering (2004, p. 67).
6. McCaffrey (2007, p. 161).
7. Hey (1995, pp. 128–130).
8. Caponera (1992, pp. 31, 32, 35 and 41) and Borkowski and du Plessis (2005, p. 154).
9. Caponera (1992, pp. 31 and 32).
10. Perruso (2002, pp. 71–75).
11. Borkowski and du Plessis (2005, p. 154).
12. Economic Commission for Asia and the Far East (1968, p. 177).
13. Economic Commission for Asia and the Far East (1968, p. 177).
14. Caponera (1992, p. 106).
15. Economic Commission for Asia and the Far East (1968, p. 211).
16. Caponera (1992, p. 106).
17. See Fisher (2003, p. 78).

18. See Fisher (2003, p. 83).
19. Caponera (1992, p. 30).
20. Caponera (1973, p. 39) and Maktari (1971, p. 29).
21. Getzler (2006, pp. 119 and 120)
22. Caponera (1992, p. 71).
23. See Fisher (1987, pp. 62 and 63).
24. See generally Fisher (2004c, 2005 and 2006a).

GENERAL REFERENCES

Borkowski, Andrew and Paul du Plessis (2005), *Textbook on Roman Law* (3rd edn, Oxford: Oxford University Press).

Caponera, D.A. (1973), *Water Laws in Moslem Countries* (Rome: Food and Agriculture Organisation of the United Nations).

Caponera, D.A. (1992), *Principles of Water Law and Administration* (Rotterdam: A.A Balkema Publishers).

Economic Commission for Asia and the Far East (1968), *Water Legislation in Asia and the Far East*, Water Resources Series No. 35 (New York: United Nations).

Fisher, D.E. (1987), *Natural Resources Law in Australia* (Sydney: Law Book Company).

Fisher, D.E. (2004c), 'Land, water and irrigation: hydrological and legal relationships in Australia (Part 1)', *Water Law* 15, 229–237.

Fisher, D.E. (2005), 'Land, water and irrigation: hydrological and legal relationships in Australia (Part 2)', *Water Law* 16, 14–22.

Fisher, D.E. (2006a), 'Water law and policy in Australia – an overview', *Environmental Law Reporter (USA)* 36, 10264–10276.

Getzler, Joshua (2006), *A History of Water Rights at Common Law* (Oxford: Oxford University Press).

Hey, Ellen (1995), 'Sustainable use of shared water resources: the need for a paradigmatic shift in international watercourses law', in Gerald H. Blake et al. (eds), *The Peaceful Management of Transboundary Resources* (London: Graham and Trotman), pp. 127–152.

Hildering, Antoinette (2004), *International Law, Sustainable Development and Water Management* (Delft: Eburon Academic Publishers).

Maktari, A.M.A. (1971), *Water Rights and Irrigation Practices in Lahj* (Cambridge: Cambridge University Press).

McCaffrey, Stephen C. (2007), *The Law of International Watercourses: Non-Navigational Uses* (2nd edn, Oxford: Oxford University Press).

Perruso, Richard (2002), 'The development of the doctrine or *res communis* in medieval and early modern Europe', *Tijdschrift voor Rechtsgeschiedenis* 70, 69–93.

Rieu-Clarke, Alistair (2005), *International Law and Sustainable Development: Lessons from the Law of International Watercourses* (London: IWA Publishing).

LEGAL INSTRUMENTS

International Union for the Conservation of Nature (2004), *Draft International Covenant on Environment and Development* (3rd edn, Gland: International Union for the Conservation of Nature).

United Nations General Assembly Resolution (1974): Resolution No. 3281 (xxix), *International Legal Materials* 14, pp. 251–261.

Constitution of the People's Republic of China: *Constitution of the People's Republic of China* (Beijing: Foreign Language Press, 2004), pp. 8–85.

Water Law of the People's Republic of China (2002): *Water Law of the People's Republic of China* (Beijing: Law Press China, 2004), pp. 103–165.

Water Act 2000 of Queensland, Australia.

6. Individual water rights

INTRODUCTION

So much for the rights of sovereignty of a nation state and how they are exercised. Let us turn now to the nature of the rights of individuals in relation to water and the sources of these rights. For this purpose individuals are those who have an interest either personally or institutionally in water resources and in the water itself. Water rights are different from water entitlements. The former exist or are recognised generally. The latter are available to identifiable individuals in specific sets of circumstances. Rights in relation to water may be original and inherent on the one hand or contingent and derivative on the other hand. A right in relation to water is original or inherent if it is a natural right or a human right. It exists because of its own rational strength. It is a contingent or derivative right if it is created by a legal instrument such as a constitution or an enactment of a legislature or if it is acquired or granted in accordance with a set of formally recognised or created norms, rules and procedures. Rights in relation to water do not always fit comfortably within this classification. There are in practice arrangements in which rights of one kind overlap with those of another kind. It is nevertheless helpful to begin our analysis in this way. Let us commence by considering natural and human rights in relation to water.

NATURAL AND HUMAN RIGHTS IN RELATION TO WATER

A natural right – a right arising by virtue of *ius naturale* – is a right which is common to all members of humankind rather than a right enjoyed by members of a particular community or even of a state. The idea was derived from Aristotle who described a natural right as one which had the same force among all men.[1] In accordance with Roman law flowing water – *aqua profluens* – was regarded by *ius naturale* as common to all: that is, *res communis*. The reason for this was its 'primitive state of innocence'.[2] It was thus the nature of flowing water that justified its acknowledgement as the subject matter of a natural right.

The common law of England adopted a similar approach. Thus, 'the right to running water has always been properly described as a natural right, just like the right to the air we breathe; they are the gifts of nature and no one has a right to appropriate them'.[3] While the right to running water was a natural right, it was not a right of ownership. The right was required to be exercised 'in a reasonable manner, with as little injury to his neighbour's rights as may be'.[4] In any event, it was a natural right because the subject matter of the right was a gift of nature.

A human right is subtly different. A human right is a right which inheres in a human being because the person so entitled is a human being. While a human right has substance – for example a right to life – it is not the subject matter of the right – as in the case of a natural right – that gives it its quality. It is the unique condition of the holder of a human right that gives the right its quality as a human right. Thus, while it is substance that gives a natural right its characteristics, it is the characteristics of a human right as a human right that give it its substance.

There has been much debate over whether water is a right or a need. The World Health Organisation has reported the Secretary-General of the United Nations to have said that 'access to safe water is a fundamental human need and, therefore, a basic human right'.[5] In this sense it is the need for water – a matter of substance – that justifies its description as a human right. Although a right to water is not included in terms in any of the international instruments dealing with human rights, a human right to water is increasingly inferred from these instruments. This is in a sense a return to the traditional values of earlier civilisations. Thus:

> Long before human rights were codified, people in many of the world's cultures regarded water as common good, to which every person was entitled. Withholding water from someone was tantamount to refusing them the right to life. This is mirrored in the world's religions, not least of all in the world religions of Judaism, Christianity and Islam, which arose in the arid, desert regions of the Middle East. In these religions as well as numerous religions in Africa, for instance, water the life-giving element is obviously held in very high regard. It is a sacred gift from God or the Gods and must be handled with care.

And therefore:

> The drive to anchor the right to water as a human right ties in with these cultural and religious ideas and derives from them a substantial part of its cogency for people from the most diverse societies.[6]

But is a need for water a right to water? Arguably a right to water may be inferred from Articles 11 and 12 of the International Covenant on Economic, Social and Cultural Rights of 1966. These Articles express respectively a right to an adequate standard of living (including adequate food) and a right to the enjoyment of the highest attainable standard of physical and mental health. The United Nations Committee on Economic, Social and Cultural Rights has formally commented on the concept of a right to water, and this comment probably represents the most authoritive statement on this issue. In particular paragraph 10 states this:

> The right to water contains both freedoms and entitlements. The freedoms include the right to maintain access to existing water supplies necessary for the right to water, and the right to be free from interference, such as the right to be free from arbitrary disconnections or contamination of water supply. By contrast, the entitlements include the right to a system of water supply and management that provides equality of opportunity for people to enjoy the right to water.

The use of the word entitlement is critical. The creation of an entitlement as distinct from the declaration of a right connotes the existence of a rule of law which not only protects the entitlement but also provides a remedy if it is infringed. In this sense, paragraph 10 does not describe a right to be supplied with water, for it contains no duty to supply water. It contemplates, however, access to water supplies and the existence of systems of water supply and management. Consistently, paragraph 12 contemplates that water will be available for personal and domestic uses; that water will be of an appropriate quality for personal or domestic uses; and that water will be accessible to everyone.

The realisation of such a human right to water is a major challenge. This is because it requires a number of substantial obligations to be placed upon the institutions responsible for these arrangements. The report of the Fourth World Water Forum held in 2006 'stressed the importance of the International Covenant on Economic and Social Rights as a mechanism requiring governments to implement the right to sufficient and clean water'.[7] A human right to water in this sense is an extension of a right to life and a right to health. These rights are themselves linked to an actual or inchoate right to an environment of an appropriate standard.

ENVIRONMENTAL RIGHTS IN RELATION TO WATER

An environmental right is a complex idea. It incorporates human as well as ecological values. This is certainly so in relation to water. Water is an element of the natural environment, and as such embraces its own intrinsic values. Water, of course, simultaneously provides humans with a range of goods and services. These include water as an essential requirement for life and health. However, if the intrinsic values of water are not recognised, then the benefits of water for humans may be more difficult to realise or perhaps even disappear totally. The achievement of this balance depends upon the recognition not only of a human but also of an environmental right to water. Since neither the environment nor water is afforded any status as such within the legal system, an environmental right to water can exist only if it is incorporated constructively within the legal system.

One attempt to do so at the international level has been the Draft Declaration of Human Rights and the Environment. The Draft Declaration commences by stating in Article 1 that human rights, an ecologically sound environment, sustainable development and peace are interdependent and indivisible. It then formulates an extensive list of human rights relating to the environment. These are linked to a duty imposed on all persons to protect and preserve the environment and a duty imposed on all states to respect and ensure the environmental right declared by the Draft Declaration. The most significant right in relation to the environment – stated in Article 2 – is the right of all persons to a secure, healthy and ecologically sound environment. This is supported by the right – stated in Article 5 – of all persons to freedom from pollution, environmental degradation and activities which – among others – threaten life, health or sustainable development within, across or outside national boundaries. The juxtaposition of life, health and sustainable development is significant. Equally significant is the extended reference to activities and the consequences of activities within, across or outside national boundaries.

More particularly, Articles 6 and 8 provide that all persons have the right to protection and preservation of – among others – water and the right to safe and healthy food and water adequate to their well-being. The Draft Declaration thus recognises the human right for water to be protected and preserved and the human right to safe and healthy water. While the environment and the values of the environment are clearly one of the beneficiaries of these arrangements, they are structured as human rights in relation to the environment rather than as environmental rights in terms of the environment. This is consistent with the notion that the legal system does not afford status to the environment or to water as such. While the Draft Declaration is no more than that, it is consistent with the gradual evolution of a human right to water in the form of a right to safe and healthy water adequate to the well-being of all persons.

CONSTITUTIONAL RIGHTS IN RELATION TO WATER

Environmental rights and water rights may similarly be linked in a national legal system through the provisions of a constitution which contains a Bill of Rights. There are at least three possible aspects:

- a right to life or to health;
- a right to an environment of a specific quality;
- a right to water.

The fundamental declaration of these rights is in some cases supported by the creation of a duty imposed on the state to realise and protect these rights and a duty imposed on the citizens to recognise and respect these rights and to protect the environment. The existence of these rights and their inchoate capacity to be protected and enforced move them closer to entitlements in the strict sense.

It is not uncommon for a right to life to be given constitutional recognition and protection. The approach in India and the Philippines is positive. Article 21 of the Constitution of the Union of India confers on every person the fundamental right to life. Protection of the environment is one the directive principles of state policy and one of the fundamental duties imposed upon every citizen of India. The Constitution of the Philippines places a duty upon the state to protect and promote the right to health of the people, and in relation to the environment it imposes a duty on the state to protect and advance the right of the people to a balanced and healthful ecology in accord with the rhythm and harmony of nature. The approach in Japan is more cautious. Article 13 of the Constitution of Japan states that all of the people shall be respected as individuals and their right to life shall, to the extent that it does not interfere with the public welfare, be the supreme consideration in legislation and in other governmental affairs. None of these examples refers specifically to water. However, in both doctrinal and practical terms, a right to life may be interpreted to include a right to water while a right in relation to the environment – whether or not coupled with a duty in relation to the environment – may be interpreted to include a right to water of an acceptable quality rather than simply a right to water itself. Much depends upon how the right is structured, how it is linked to any correlative duties and how the judiciary interprets and applies these constitutional arrangements.

A particularly interesting model is South Africa. According to Section 24 of the Constitution of the Republic of South Africa:

> Everyone has the right –
> to an environment that is not harmful to their health or wellbeing; and to have the environment protected, for the benefit of present and future generations, through reasonable legislative and other measures that –
> (i) prevent pollution and ecological degradation;
> (ii) promote conservation; and
> (iii) secure ecologically sustainable development and use of natural resources while promoting justifiable economic and social development.

Clearly this contemplates a right to an environment of an acceptable quality and the realisation of this right through appropriate statutory and administrative measures. This right in relation to the environment is clearly linked to the health and well-being of the citizens. Section 27(1)(b) provides specifically that everyone has the right to have access to sufficient food and water. Subsection (2) of section 27 goes on to impose a duty upon the state to take reasonable legislative and other measures, within its available resources, to achieve the progressive realisation of each of these rights. While there is no absolute right to water, there is a right of access to sufficient water and an obligation on the state to take reasonable measures to realise this right.

While constitutional provisions such as these are potentially enforceable, much depends upon the terms of the legislation which seeks to give effect to them. For example, section 3 of the Water Services Act of 1997 of South Africa declares that everyone has a right of access to basic water supply and requires every water services institution to take reasonable measures to realise this right. This set of provisions has been judicially interpreted in South Africa to constitute a qualified rather than an absolute right of access to water as well as to water itself. In a practical context these provisions have been summarised in this way;

> In South Africa, where the right to water is enshrined in the Constitution, every family is entitled to 6,000 litres of water free of cost per month. The wealthy therefore subsidise the supply of water to the poor through its price.[8]

While such a right may be treated in this way as an entitlement, it is in both constitutional and statutory terms a qualified rather than an absolute entitlement.

THE RELEVANCE OF CUSTOM AND PRACTICE IN RELATION TO WATER

A human right to water – similar to rights in relation to water arising *ex iure naturale* or arising incidentally from a right linked to the quality of the environment – exists as a consequence of its own rational strength, whether or not it may have had its existence confirmed or even extended by declaration in a constitution or in a similar statement of fundamental values. Even if a right to water does not constitute an entitlement in the form of an enforceable claim to water, such a right is the basis upon which normative values are attributed formally to water. In this sense these rights are original and inherent. Rights in relation to water, it has already been suggested, may alternatively be contingent and derivative: for example those granted by and hence derived from a person or institution with the power to do so. But rights in relation to water may also derive from continuous activities over the years, and in this way are contingent upon a general acknowledgment of this practice. The existence of customary rights in this sense is just as significant for international as for national systems of water resources governance.

What happens habitually as a matter of practice may in certain circumstances develop as a social norm, and from that as a customary rule of law. There is no single authoritative approach to identifying whether and when a habit or practice has become a customary rule of law. Much depends upon the culture of the society in question and the nature of its legal arrangements and, in an international context, upon the relationships between national states and their behaviour towards each other. Is there, for example, any intrinsic difference between arrangements among states on the one hand and within states on the other hand?

Customary international law is particularly important in the absence of an international legislature. It may be the principal – perhaps even the only – source of international rules of general application. A rule of customary international law has been described as 'a complex mixture of state practice, *opinio juris*, and express or tacit expression of consent'.[9] Thus, what states actually do and their belief that what they do fulfils a legal obligation lead to their acceptance of the custom and the need to comply. Is there, however, any need for formal recognition for a custom to become a rule? The International Court of Justice has said this:

> Where two states agree to incorporate a particular rule in a treaty, their agreement suffices to make that rule a legal one, binding upon them; but in the field of customary international law, the shared view of the parties as to the content of what they regard as a rule is not enough. The court must satisfy itself that the existence of the rule in the *opinio juris* of states is confirmed by practice.[10]

While a rule of customary international law may exist in general, for it to be given effect in this way seems to require judicial recognition and judicial satisfaction of its existence.

It may be asked, for example, whether sustainable development itself is a rule of customary international law. More particularly, is the precautionary principle – as an important theoretical and practical element of sustainable development – a rule of customary international law? It has been suggested that:

> The manner in which this principle has been dealt with in recent times [by the appellate body of the World Trade Organisation and by the International Tribunal for the Law of the Sea] is evidence not merely of the legal issues but also of the political, and, above all, the economic, issues at stake. At the same time, however, the principle's legal status remains uncertain.[11]

Whatever its status, the precautionary principle performs a function within the international environmental legal system either as a legal rule or a paralegal rule in the sense already discussed. In any event its precise function within the system remains unclear to the extent that it depends upon the terminology according to which it is structured.

It has been suggested that it is by no means easy to demonstrate that the precautionary principle is 'a well-defined and established customary rule in positive international law'.[12] More particularly:

> Its formulation, implementation, impact, and implications still raise some difficult issues, whatever we may wish to be the case, not only among prudent diplomats but also within the scientific and legal communities. The varying formulations of the precautionary 'principle' may imply that, in the case of scientific uncertainty, states are allowed or obliged to regulate, to prohibit, or to allow only under certain conditions activities that may cause environmental damage. The question thus remains: even if we agree that the precautionary principle has become customary international law, which of these formulations is binding upon states?[13]

This is a question that may be asked not only in relation to the precautionary principle but also in relation to sustainable development itself. In any event, the existence of a rule and the manner of its implementation are frequently distinguished in both international and national legal regimes. The exercise of a discretion in complying with a rule does not by itself preclude the existence of the rule.

CUSTOMARY RIGHTS IN RELATION TO WATER

Customary law is no less important for the way water resources are governed nationally. And many of the issues are the same nationally as internationally. However the focus of national arrangements tends to be upon rights in relation to water rather than the rights of the state in relation to water. More specifically, what customary arrangements may be in place for the acquisition of interests in water? What is the nature of these interests? Does the existence of these interests lead to the recognition of water rights as such. And what is the nature of these rights?

Customary law, it has been suggested, 'is a popular normative pattern that reflects the common understanding of valid, compulsory rights and obligations. These underlying social norms may become the acknowledged law of the land'.[14] But when does a custom become a customary law? The response in relation to a national system is not so different from the response in relation to the international system. Thus:

> For a custom to acquire the status of law it must carry a popular perception of valid legal obligation (*opinio necessitatis sive obligationis*). The key to determining whether a custom constitutes a customary law is whether the public acts as if the observance of the custom is legally obligated.[15]

It is therefore unclear whether a custom becomes a law only if it is formally recognised in some form by a court. Even if a customary law exists independently of judicial recognition, this is not to suggest that the courts have no particular function to perform.

Clearly the courts do perform functions in relation to custom – for example, settling disputes about compliance with the custom or about its interpretation and application. Indeed the interpretation of a custom by a court may well lead to the development of a customary rule as a rule of the common law and perhaps also of the civil law. It has been suggested that 'if a custom is not unreasonable and meets the prerequisites, a court is bound to apply it as law'.[16] By way of an example:

> An itinerant English judge wrote down the various customs used by the inhabitants of the fens in a document called the Code of Romney Marsh, which then became treated as a 'customal' – what we would today call a 'restatement' of the customary law, which in turn was used by the courts to resolve wetland disputes in other regions.[17]

Custom, then, is important in water resources governance not only for its own sake but also as a foundation upon which new legal arrangements are modelled.

Custom has played and continues to play an important function in many jurisdictions in relation to the use and development of water resources. During the period of monarchy for the Romans, prior to the republic, custom was a major source of arrangements for managing land and consequently water resources.[18] In some parts of Europe during the 20th century local associations and local customs have had a part to play: for example, water users' associations in the Netherlands and Belgium, water protection associations in Germany, the Water Tribunal of Valencia in Spain and customary practices in parts of Italy.[19] It is no surprise that custom and customary laws are equally important elsewhere. In India, for example, a range of legal regimes operates simultaneously. There is, of course, the constitution together with the legislation that impacts upon the use of water resources. Additionally, 'there are customary Hindu, Islamic and tribal laws for water use and distribution which the courts recognise'.[20] The complexity of these arrangements is no doubt a challenge, but also perhaps a benefit for the local community.

Let us look in more detail at the role of customary law in Africa. There, it has been commented, 'customary law is without doubt the most important of the sources of law, and of water law in particular, as it is the one which is most known and respected by the population'.[21] And further, 'customary rules become more detailed where water is scarce'.[22] One example from Tanzania illustrates the point. It is the village of Mkuu close to the border with Kenya and dominated by Mount Kilimanjaro. Mkuu is one of the woodland villages of the Chagga people. Water is either carried from its source or in some cases piped to households. The water arrangements have been described in this way:

> Numerous perennial steams run down from the mountain and the upper zone of the area is cool, humid, and shaded either by residual forest trees or by cultivated bananas and arabica coffee. Cattle in barns receive much of their water from banana stocks rather than by drinking. Further down, the soil is drier. Everything is on a slope, and this has been utilized in the planning of irrigation works.
>
> The streams are dammed and partly diverted almost along the contour lines to bring water very near to many houses under a system of customary law which specifies rights as well as responsibilities for maintaining the channels. Such streams provide the only common type of water source; standing water is rare. As in other sectors of the slopes of Kilimanjaro to the south and west, a pipeline delivers water by gravity to standpipes at intervals along one of the ridges between streams.
>
> The Mkuu study area extends far enough down the mountain to leave these damp plantations in its lower part and enter a more arid zone, stretching towards the Kenya border, resembling the other seasonally arid areas studied. Its contrasts in water availability and climate pose the problem of whether these are reflected in household use. And its long traditions of community management of water call for inquiry into the extent to which such customary action or a formal organisation is needed in organising new water improvements.[23]

This poses an important question: the relationship between formal and informal arrangements or between law and custom. In practical terms, however, the choice is perhaps between communal or shared sources of water on the one hand and private or exclusive sources of water on the other hand. What was the perception at that time? Clearly access to water was both an important and also a sensitive issue:

> It was found that users in East Africa prefer private to communal sources when they have the opportunity. They avoid situations where there is risk of irritating confrontation. They may enjoy sharing communal sources when this is required, but when presented with a genuinely viable choice between a communal source and one affording privacy, they seem to choose the latter except in certain special circumstances. These include water sources and containers which make it necessary for a woman to have the aid of another woman in hoisting the water to her head before carrying it home. It was found that where the assignment of rights to land conveys with it control of access to water sources, even though the water sources are regarded under the customary law as being common property, individuals tend to avoid using sources that would imply a need for trespass or would interfere with what may be newly conveyed rights of ownership in the sources. Thus the trend toward assignment of property rights in areas where formerly there had been no cadastral lines and no deeded rights to land reduces the range of choice in drawing upon water.[24]

This description of cultural attitudes in East Africa points to a range of paradigmatic challenges not only for Africa but also for the global community. It raises the very important issue of the relationships between custom and law and between land and water. Not only that: it raises the issue of how rights and entitlements in relation to water are acquired, transferred and terminated.

CONCLUSION

The idea of a right is as inherently ambiguous as the word itself. Rights might be human or natural in nature; they may be stated in constitutions, in legislation or in rules of the common law or the civil law; they may be inherent or derivative; they may exist as a matter of law; they may be acquired through use; or they may be granted by way of some form of allocation. The substance of these rights varies from freedoms to engage in particular activities or not to be interfered with by the activities of others, through statements in the form of basic needs, to specific entitlements in particular sets of circumstances. Individual water rights are thus a reflection not only of the culture of the society in question but also of the extent to which the arrangements for the management of its water resources have been formalised. Custom plays a significant role in international law and just as significant a role in national legal systems. Local practices and customs, for example, have often been based upon the notion of water as a common resource in which each member of the community has an interest. As we have seen, in some parts of the world – for example in parts of eastern Africa – custom continues to play a very significant role. On the other hand, in many parts of the world more formalised and in many respects more complex arrangements have emerged through the creation of constitutional or statutory normative frameworks.

What is emerging both internationally and nationally at one end of the spectrum is the idea of water as a basic human need and consequently as an inchoate human right and at the other end of the spectrum as an entitlement to a particular share of a specific water resource or to a particular quantity of water able to be extracted from a specific location and used for a particular purpose. The techniques of ensuring compliance with these arrangements depend upon the ways disputes are settled within the society in question. In many jurisdictions disputes are settled locally and often informally. This approach is apparent even in jurisdictions with complex formal legal structures. Ancient civilisations often embraced local practices and customs as well as sophisticated central administration. It has been noted that a consequence of such arrangements has been the sustainable use and development of these water resources. The 19th and 20th centuries probably saw a polarisation of these two extremes. The 21st century, however, seems to be recognising that sustainable water resources governance involves a recognition not only of the importance of individual water rights but also of the need for a system that brings together human rights, customary rights and individual entitlements within one system which is unified, consistent and coherent structurally and substantively.

NOTES

1. The same idea was acknowledged by the English jurist Bracton and accepted by the courts in England in *Calvin's case* (1608, p. 392).
2. Schlatter (1951, p. 28).
3. *Chasemore's case* (1859, p. 379).
4. *Chasemore's case* (1859, p. 388).

5. World Health Organisation (2003, p. 6).
6. Bar and Kurschner-Pelkmann (2007, p. 346).
7. Beintema et al. (2006, p. 7).
8. Bar and Kurschner-Pelkmann (2007, p. 347).
9. Dupuy (2007, p. 454).
10. *Nicaragua case* (1986, p. 1065).
11. Dupuy (2007, p. 451).
12. Dupuy (2007, p. 452).
13. Dupuy (2007, pp. 452 and 453).
14. Ørebach and Bosselman (2005, p. 17).
15. Ørebach and Bosselman (2005, p. 17).
16. Ørebach and Bosselman (2005, pp. 17 and 18).
17. Ørebach and Bosselman (2005, p. 19).
18. Caponera (1992, p. 30).
19. Food and Agriculture Organisation of the United Nations (1975, p. 152) and Caponera (1992, p. 68).
20. Singh (2000, p. 50).
21. Caponera (1992, p. 98).
22. Caponera (1992, p. 98).
23. White (1972, pp. 39 and 41).
24. White (1972, p. 264).

GENERAL REFERENCES

Bar, Rosmarie and Frank Kurschner-Pelkmann (2007), 'How do we guarantee the human right: sufficient water for all?', in José L. Lozán et al. (eds), *Global Change: Enough Water for All?* (Hamburg:Wissenschaftliche Auswertungen), 346–349.

Beintema, Nienke et al. (2006), *World Water Forum Bulletin* (New York: International Institute for Sustainable Development).

Caponera, D.A. (1992), *Principles of Water Law and Administration* (Rotterdam: A.A Balkema Publishers).

Dupuy, Pierre-Marie (2007), 'Formation of customary international law and general principles', in Daniel Bodansky et al. (eds), *The Oxford Handbook of International Environmental Law* (Oxford: Oxford University Press), pp. 449–466.

Food and Agriculture Organisation of the United Nations (1975), *Water Laws in Selected European Countries* (Rome: Food and Agriculture Organisation).

Ørebech, Peter and Fred Bosselman (2005), 'The linkage between sustainable development and customary law', in Peter Ørebech et al. (eds), *The Role of Customary Law in Sustainable Development* (Cambridge: Cambridge University Press).

Salman, Salman M.A. and Siobhán McInerney-Lankford (2004), *The Human Right to Water: Legal and Policy Dimensions* (Washington DC: World Bank).

Schlatter, R. (1951), *Private Property: The History of an Idea* (London: Allen and Unwin).

Singh, Chhatrapati (2000), 'Water rights and the principles of water management in India', in D.E. Fisher and Noeleen McNamara (eds), *Integrated Water and Land Management* (Brisbane: Queensland University of Technology), pp. 49–79.

White, Gilbert F. et al. (1972), *Drawers of Water: Domestic Water Use in East Africa* (Chicago, Ill.: University of Chicago Press).

World Health Organisation (2003), *The Right to Water* (Geneva: World Health Organisation).

LEGAL INSTRUMENTS

Constitution of South Africa.

Constitution of the Philippines.

Constitution of the Union of India.

Draft Declaration of Human Rights and the Environment (1994): United Nations Draft Declaration of Human Rights and the Environment 1994, at University of Minnesota Human Rights Library, www1.umn.edu/humanrts/instree/1994-dec.htm.

International Covenant on Economic, Social and Cultural Rights (1966), in Salman M.A. Salman, and Siobhán McInerney-Lankford (2004) *The Human Right to Water: Legal and Policy Dimensions* (Washington, DC: World Bank), pp. 113, 114, 153 and 154.

Water Services Act 1997 of South Africa.

CASES

Calvin's case: *Calvin v Smith* (1608), English Reports 77, 377–411.

Chasemore's case: *Chasemore v Richards* (1859), English Reports, 11, 140–156.

Nicaragua case: *Nicaragua v United States of America* (1986), International Legal Materials, 25, 1023–1091.

7. Water entitlements

INTRODUCTION

Whether or not a legal system recognises or creates a right to water, much clearly depends upon how a legal entitlement to water may be either acquired or granted. In other words, it is a function of the legal system, particularly a national legal system, to enable rights in relation to water – or, more precisely, enforceable water entitlements – to be acquired or to be granted or, more directly, for water itself to be appropriated or allocated. A legal system may permit rights in relation to water to be acquired or water appropriated simply as a matter of law without recourse to any bureaucratic procedures. Alternatively, a legal system may prohibit either generally or in particular sets of circumstances either the acquisition of rights in relation to water or the appropriation of water without the involvement of the state or an agency of the state in one form or another.

These arrangements are, in other words, a rules based system for acquiring rights or a regulatory system for granting rights: a simple but critical distinction that goes to the very doctrinal foundations of any system of water resources governance. Historically, water entitlements have been based originally on a set of rules and then on an evolving system of bureaucratic grants. But this is an oversimplification. Any system of water resources governance is a combination of both rules and administrative regulation. This is because a system of enforceable water entitlements cannot ignore the nature of water and of the hydrological cycle and the range of diverse and potentially conflicting interests in water.

The physical, biological, ecological and cultural connections between water, land and the environment have probably always been recognised. These connections or relationships have over the millennia influenced the customary and legal arrangements for the use and development of water resources. Frequently – but perhaps somewhat unhelpfully – these arrangements have been analysed doctrinally in the context of rights of ownership or even more unhelpfully as rights of property. From a legal perspective – as distinct from an emotional or sentimental perspective – rights based upon ownership or property are, as we have suggested, intrinsically no more than rights of control. What is significant is the scope of these rights of control, the criteria according to which these rights are exerciseable and ultimately the extent to which the exercise of these rights themselves is controlled by the law.

Let us take this approach further. A right of access to water may arise by way of *ius publicum* or public right simply because the person is a member of the relevant public or community so benefited. A right of access to water may arise as an incident of the ownership of the land connected with the water or as an incident of the ownership of the water itself. A right to use water may arise in consequence of the exercise of the right of access. A right to use water may alternatively arise simply because the water is taken into possession or brought under some form of occupation or control, and consequently used for a particular purpose. This may mature into a legal right immediately or perhaps after a period of time. Once rights in relation to water have been acquired in any of these ways, the right may in appropriate circumstances be transferred to someone else or another right of a subsidiary nature created out of the original and prior right then in existence.

The evolution of rights in relation to water over the millennia has taken place against these theoretical and doctrinal ambiguities. In a metaphorical sense, therefore, the ways in which rights in relation to water have developed have ebbed and flowed as much as the waters themselves. Let us at this stage consider no more than three examples: the very influential principles of Roman law; the doctrinal ambiguities of the English common law; and the congruence of customary law and Islamic law in a contemporary society. Our focus in this chapter is the source and structure of these rights rather than their substantive or normative content.

ROMAN LAW

The principles of Roman law were based fundamentally upon a distinction between public waters and private waters. Although not directly described as such, these categories are – using the singular rather than the plural noun – *aqua publica* and *aqua privata* or *propria* which would correspond in relation to water with *res publica* and *res propria*. This is the use of the adjective *privata* or *propria* in the sense of exclusive. Although these descriptors derived from concepts of ownership, this approach in effect described the subject matter controlled by the holder of these rights. In this sense ownership could be seen as the mechanism by which positive or negative control was exercised over the water resources in question.

The relationship between land and water was close. In this respect, as has been stated in relation to classical Roman law during the republic, 'the legal status of water followed that of land'.[1] Thus water physically connected with public land was *aqua publica* and water physically connected with private land was *aqua privata* or *propria*. Public land was land not allocated. Thus:

> All rivers and some streams, the springs feeding urban aqueducts, mountain lakes and such rain water as was collected by natural mountain pools or artificial tanks were public. Springs, wells and pools located within private property were private.[2]

The distinction between public and private waters reflected not ownership but access. Perennial rivers – *flumina* – were *res publicae*. As such these rivers were controlled by the state. The agencies of the state had access to them. And so had the public. Thus:

> Rivers were regarded as public things if they were perennial, i.e. normally flowing all the year round. Although such a river could not be owned, its beds and banks could be; but it seems that the riparian owners (i.e. the owners of the river banks) could not prevent the public from using the banks to enjoy the river. Remedies were normally available to protect the enjoyment or use of public things.[3]

So the public had a right of access to public waters. How these waters might be used in the exercise of this right of access was another issue.

Rights in relation to water were based upon this classification of water. Apart from the public right of access, public waters were made available by means of appropriate infrastructure – normally aqueducts – for use in public facilities for washing, bathing and recreation.

Accordingly, 'private connections with public aqueducts or reservoirs were strictly forbidden; only excess water (*aqua caduca*) could be diverted exceptionally from public mains, under a perpetual concession from the administrative authorities, and subject to payment of a rent (*vectigal*)'.[4] Thus the state – subject only to the public right of access – had exclusive control of public waters.

Private waters were controlled by the owners of the private land in question. The holder of rights in relation to private waters was able to create complementary or secondary rights out of the principal or original rights by means of servitudes. Commonly these servitudes were *aquae ductus* and *aquae haustus*. In effect:

> *Aquae ductus* conferred the right to divert water from a private water source and to convey it to one's own land by means of suitable water pipes and fittings; *aquae haustus* conferred only the right to draw water from a private source and the additional right of access thereto.[5]

While the right to channel water across another's land was clear enough, were there any rights of ownership arising during or after the activity permitted by the servitude? Interestingly, 'the prevailing view appears to have been that since flowing water was in the category of *res communes*, it could not be privately owned while it was in flow'.[6] In any event, these servitudes were real rights or rights *in rem* and were a burden upon the land in question. In addition, personal rights in relation to water – rights *in personam* – were able to be acquired by usufruct – that is by use of the water or of the benefit of the water.[7] Much later, this was taken further by the recognition of acquisition of rights in relation to water by prescription – the use of the water over a long period of time.[8]

During the later years of the Roman empire, this system of water resources governance based upon public waters controlled by the state and private waters controlled by enforceable legal rules was increasingly overtaken by a system of regulation based upon the administrative grant of rights in relation to water.[9] A phenomenon, of course, not uncommon in other jurisdictions in response to increasing uses of water leading to increasing scarcity of water resources. In any event, the rules based system which constituted the formal structure of water resources governance under Roman law laid the foundations upon which the regimes not only of the civil law but also of the common law were themselves to some extent based.

ENGLISH LAW

(a) Water and Land during the Middle Ages

The doctrinal foundations of the common law of England are even more difficult to identify. Rights in relation to water were originally recognised and protected as incidents of the possession and use of land through land based remedies such as the possessory assizes, the assize of nuisance, trespass and trespass on the case. Consistently with this approach, the common law recognised neither a right of common in relation to water nor a right in relation to water as a *profit à prendre*. To do so would have recognised rights to use water independently of land. Although the use of water may well have been linked in practice to the use of land, during the medieval period the doctrinal foundations of rights in relation to water remained unclear. Thus:

> The law could not decide whether the right of a riparian owner to water flow was to be characterised as a natural right inherent in ownership, or else as an easement resting on grant, prescription, or custom – an issue left blurred by Bracton's eclectic use of the Roman categories of servitude and *usucapio* in describing the rights of amenity incident to land-ownership.[10]

During the 15th and 16th centuries rights in relation to water continued to be considered largely as servitudes or burdens upon the possession and ownership of land. But what was their doctrinal basis? Was it a natural right to the flow of water or a right based on use. Consistently with the development of the common law, much depended upon the circumstances of the case and on the nature and extent of the formulary remedies debated before the court. In 1625, however, the Chief Justice said this:

> The question here is of *aqua profluens*, and I conceive that there needs no prescription or custom in this case, for water hath its natural course, and *natura sua descendit,* and it may be called *usu captio* or usage.[11]

The issue in the case was whether a right in relation to water could be extinguished by unity of possession: that is, when the relevant lands comprising the servient and dominant servitudes were possessed by the same person. In this context the nature of the right in relation to water was critical. Thus:

> Where the thing hath its being by prescription, unity will extinguish it; but where the thing hath its being *ex jure naturae*, it shall not be extinguished.[12]

In this case the right had its being *ex jure naturae* and therefore was not extinguished. This is not to suggest that in all cases a right in relation to water did not have its being by prescription, but not in this case. The common law was thus moving towards a conception of rights in relation to water as a set of natural rights.

(b) The Appropriation Doctrine

During the 18th century there emerged the doctrine of appropriation promulgated by Blackstone but perhaps influenced by the earlier writings of John Locke. According to Locke:

> [The natural resources] where I have a right to them in common with others, become my property without the assignation or consent of anybody. The labour that was mine, removing them out of that common state they were in, hath fixed my property in them.[13]

This applied as much to water as to other natural resources:

> Though the water running in the fountain be every one's, yet who can doubt but that in the pitcher is his only who drew it out? His labour hath taken it out of the hands of Nature where it was common, and belonged equally to all her children, and hath thereby appropriated it to himself.[14]

This emerging philosophy – whether or not consistent with the Roman conception of *aqua profluens* as *res communis* – may well have influenced Blackstone's commitment to the doctrine of appropriation. But how was appropriation achieved? By occupancy. Thus:

> If a stream be unoccupied, I may erect a mill thereon, and detain the water; yet not so as to injure my neighbour's prior mill, or his meadow: for he hath by the first occupancy acquired a property in the current.[15]

The right in relation to water so acquired became, according to Blackstone, qualified property, qualified in the sense that the right remained in existence only in so far as it – in this case the water – was actually used and occupied. And qualified also in the sense that the holder of the right was under a duty not to injure or damage the interests of the holder of the prior right.

The doctrine of appropriation was favoured by the courts in England until the second quarter of the 19th century. Water had for long been a valuable economic and commercial resource and its value had no doubt increased during periods of industrial and commercial expansion. Perhaps the doctrine of appropriation received its judicial apotheosis in 1831 when the Chief Justice said this:

> By the law of England, the person who first appropriates any part of the water flowing through his land to his own use, has the right to the use of so much as he thus appropriates, against any other. And it seems consistent with the same principle, that the water, after it has been so made subservient to private uses by appropriation, should again become *publici juris* by the mere act of relinquishment. There is nothing unreasonable in holding that a right which is gained by occupancy should be lost by abandonment. Suppose a person, who formerly had a mill upon a stream, should pull it down, and remove the works, with the intention never to return. Could it be held, that the owner of other land adjoining the stream, might not erect a mill and employ the water so relinquished? Or that he could be compellable to pull down his mill, if the former mill-owner should afterwards change his determination, and wish to rebuild his own?[16]

A difficult question to resolve in practice, no doubt, but nevertheless a clear statement of principle.

(c) The Riparian Doctrine

It was not long before the doctrine of appropriation began to give way to the doctrine of riparian rights in relation to water as a set of controlled entitlements available to the community comprising the relevant riparians. The Chief Justice in 1833 was faced with a choice between the appropriation doctrine and the riparian doctrine. The riparian doctrine was preferred. In effect this recognised a community of interest rather than an exclusive interest in the water resource in question. But what was the rationalisation for this approach?

The Chief Justice emphasised that land and water were interconnected resources of actual or potential economic and commercial value. Two sets of circumstances were distinguished: injury to a mill using water from a stream running through the land on which the mill was built on the one hand and injury to or loss of the natural advantages of the land on the other hand. The doctrine of appropriation precluded a remedy in the second but not in the first set of circumstances. This was because in the second case no water had been appropriated so that no protectable entitlement existed, while in the first case a protectable right existed as a result of appropriation.

The Chief Justice preferred a legal regime that protected potential lawful uses as well as actual lawful uses of water. Thus:

> But it is a very different question, whether he can take away from the owner of the land below, one of its natural advantages, which is capable of being applied to profitable purposes, and generally increases the fertility of the soil, even when unapplied; and deprive him of it altogether by anticipating him in its application to a useful purpose. If this be so, a considerable part of the value of an estate, which, in manufacturing districts particularly, is much enhanced by the existence of an unappropriated stream of water with a fall, within its limits, might at any time be taken away; and by parity of reasoning, a valuable mineral or brine spring might be abstracted from the proprietor in whose land it arises, and converted to the profit of another.[17]

This represented a utilitarian view of the value of water resources. The riparian doctrine reverted to the relationship between land and water as its foundational element. But it introduced a notion of community rather than exclusivity in the use of the resource because of the mutuality or commonality of interest of all riparians in the one resource.

The mature common law doctrine of riparian rights emerged soon after in the form of the formulation provided by Baron Parke:

> The right to have the stream to flow in its natural state without diminution or alteration is an incident to the property in the land through which it passes; but flowing water is *publici juris*, not in the sense that it is a *bonum vacans*, to which the first occupant may acquire an exclusive right, but that it is public and common in this sense only, that all may reasonably use it who have right of access to it, that none can have any property in the water itself, except in the particular portion which he may choose to abstract from the stream and take into his possession, and that during the time of his possession only. But each proprietor of the adjacent land has the right to the usufruct of the stream which flows through it.[18]

The essential characteristics of the riparian doctrine were clear:

- a right incidental to the possession or ownership of land;
- a right common to all riparians;
- a right of access;
- a right of reasonable use;
- ownership of the water arising only on taking the water into possession;
- a usufructuary right.

The doctrinal foundations of the riparian right were simple. It was a rules based set of arrangements. Compliance with the rules established the right. There was no grant or allocation of the right by a bureaucratic or executive institution. It was a function of the judiciary to adjudicate upon any issues in dispute.

(d) Intervention by the Legislature

The rules of the riparian doctrine have been complemented, or in some cases superseded, by legislation in most common law jurisdictions. During the second half of the 19th century it became common for the legislature in England to pass ad hoc measures providing for the management of a particular water resource. During the second half of the 20th century more comprehensive statutory regimes for the management of water resources were increasingly introduced. The courts in England have given effect – as indeed they must – to the terms of the legislation. If this restricts the application of the common law, then this is a necessary consequence of the enactment of the legislation.

In one case the court was asked whether the exercise of statutory powers for the purpose of making a river and its streams navigable interfered with the common law rights of riparians. In responding positively the court looked at the objects of the legislation and commented:

> We cannot construe the statute as giving the [statutory body] only such a limited right in the river as a private grant of the 'said river and stream' might have conveyed, but as creating a new species of statutory property and interest in the water, which in our opinion was interfered with by the abstraction of it for the purposes to which it was applied by the [riparians]; which purposes were more extensive than those for which a riparian proprietor, as such, could insist upon appropriating the stream as it passed by his land.[19]

So the legislation was applied according to its terms and its objects. In this case the exercise of statutory powers in relation to water was given precedence over the exercise by riparians of their common law rights.

ISLAMIC LAW

Rights in relation to water in Islamic law have evolved from custom, the application of Shari'ah law and, more recently, the codification of the principles informing these rights. Indeed, 'in south-west Arabia Shāfi'ī jurists have for centuries upheld customs which existed long before Islam emerged'.[20] In Islamic law rights in relation to water and in relation to land have been closely linked. 'The connection between water and land law', it has been suggested, 'is both natural and logical'.[21] In the arid countries of the Middle East water has always been critical: so critical that Shari'ah law was water law before it assumed a wider connotation.[22] The basic proposition was that water was not – at least in the first instance – owned by anyone. The focus was upon rights and responsibilities. Much, however, depended upon the source of the water in question, its capacity to be controlled and who was able to control it. This in turn led to practices and customs that responded to local circumstances rather than one common approach.

According to one school of juristic thought, water was classified 'under three headings, according to its source: as river water, well water, and spring water'.[23] While river water was in most cases available to anyone, access to well water and spring water was much more restricted. Well water and spring water were particularly important sources of water in many of the arid Moslem countries. Rights in relation to well water and spring water were generally held in common for use by members of the community or held exclusively by the owner of the land.[24] Ownership in this sense was a means of control and the right in relation to the water matured into a right of ownership of the water when it was exercised and the water was taken into possession.

Islamic law comprised and continues to comprise a subtle and sophisticated set of interests and rights in relation to water. Some arose because of membership of the community and others because of ownership of land. The fundamental doctrine in Shari'ah law was that land was either land which was used or owned by someone or alternatively land which was not used or owned by anyone. Similarly there were two categories in relation to water: water which was owned and water which was not owned.[25] This had clear implications for rights in relation to water. Water on land not owned by anyone was similarly not owned by anyone. But the flow of water to and from land in different classifications of ownership introduced some complexity into these arrangements. For example:

> The provisions for public ownership of *mubah* water [that is, water that is not owned by anyone such as the water of one of the great rivers] appeared clearly when a piece of land is held in private ownership. 'Imrani explains that, if the owner of a piece of land draws water for its irrigation from a river or from rain water, the owner is not regarded by law as the owner of the water, even if the water he takes is already in his privately owned channel. He points out the owner of land has preferential rights to use it, because it is on his land. In addition, 'Imrani points out that no one has the right to trespass upon the owner's land. Yet he admits that, if a person enters upon another person's land without permission and takes some of the *mubah* water he finds on that land, he is guilty of trespass but the law regards him as the legal owner of the *mubah* water he acquired in this manner.[26]

This describes a set of rules for the governance of water resources that accommodated both the public interest and the private interest in their use in ways which were regarded by the community as reasonable and equitable in the circumstances. Significantly this was a rules based system, and not a system of bureaucratic regulation. But, as in other jurisdictions, in many instances legislation has modified these arrangements.

CONCLUSION

The doctrinal foundations of rights in relation to water in Roman law, in the common law of England and in Islamic law disclose remarkable similarities. Common to each of them is the legal recognition of the actual physical connection between land and water and the overarching need to accommodate the interests of members of the relevant community, the interests of land owners and the public interest in ensuring an adequate supply of water for a range of purposes.

While ownership in one form or another forms part of the doctrinal foundations of each of these systems, ownership is essentially a technique or mechanism for either permitting or restricting access to and use of water. Ownership of water is realised only when the holder of these rights in relation to water exercises them by actually taking the water into possession. These rights are acquired in a number of different ways:

- as an incident to the ownership of the land;
- by occupation or occupancy;
- by prescription;
- by usufruct;
- by transfer from the original holder of the rights.

Each of these amounts to a form of appropriation in the sense that the person seeking these rights has the capacity to engage in and actually engages in the activities which enable the rights to be acquired.

Such a system of appropriation is to be distinguished from a system of allocation. Rights in relation to water are allocated when the person seeking the beneficial use of water is granted them by a person or institution with the capacity to do so. Such a person or institution may be the 'owner' of the water resource, but it is not necessarily or perhaps at all in that capacity that the rights are granted through an allocation system. In an allocation system, it is the act or decision of the allocator which is critical, whereas in an appropriation system it is the act or decision of the appropriator which is critical. In an allocation system the holder of rights in relation to water receives these rights, but in an appropriation system the holder of these rights assumes or takes these rights.

It is a function of the state in exercise of its rights of sovereignty to determine which system of water resources governance is appropriate for the water resources located within its territory but always in accordance with international law. Similarly it is a function of the holder of rights in relation to water located within a state to exercise these rights in accordance with national law. The norms and rules of international and national law acknowledging or creating these rights constitute the formal structure of water resources governance. But they tell us little about how the rights should or must be exercised. That is a matter for the normative structure of water resources governance.

NOTES

1. Caponera (1992, p. 31).
2. Caponera (1992, p. 32).
3. Borkowski and du Plessis (2005, p. 154).
4. Caponera (1992, p. 32).
5. Caponera (1992, p. 32).
6. Borkowski and du Plessis (2005, p. 172).
7. Caponera (1992, p. 37) and Borkowski and du Plessis (2005, pp. 172 and 173).
8. Caponera (1992, p. 42) and Borkowski and du Plessis (2005, pp. 177).
9. Caponera (1992, pp. 40–43).
10. Getzler (2006, p. 116).
11. *Sury's case* (1625, p. 1266).
12. *Sury's case* (1625, p. 1267).
13. Locke (1690, edn of 1924, p. 130).
14. Locke (1690, edn of 1924, p. 131).
15. Blackstone (1765–1769, Book II, p. 403).
16. *Liggins's case* (1831, p. 268).
17. *Mason's case* (1833, pp. 698 and 699).
18. *Embrey's case* (1851, pp. 585 and 586).
19. *Medway case* (1861, p. 232).
20. Maktari (1971, pp. 29 and 30).
21. Maktari (1971, p. 9).
22. Mallat (1995, p. 128).
23. Maktari (1971, p. 14).
24. Maktari (1971, p. 14–16).
25. Maktari (1971, p. 13).
26. Maktari (1971, p. 16 and 17).

GENERAL REFERENCES

Blackstone, William (1765–1769), *Commentaries on the Laws of England* (edn of 1809, Abingdon: Professional Texts Ltd).

Borkowski, Andrew and Paul du Plessis (2005), *Textbook on Roman Law* (3rd edn, Oxford: Oxford University Press).

Caponera, D.A. (1992), *Principles of Water Law and Administration* (Rotterdam: A.A Balkema Publishers).

Getzler, Joshua (2006), *A History of Water Rights at Common Law* (Oxford: Oxford University Press).

Locke, John (1690), *Two Treatises of Civil Government* (edn of 1924, London: JM Dent & Sons Ltd).

Maktari, A.M.A. (1971), *Water Rights and Irrigation Practices in Lahj* (Cambridge: Cambridge University Press).

Mallat, Hyam (1995), 'Water laws in Lebanon', in J.A. Allan and Chibli Mallat (eds), *Water in the Middle East: Legal, Political and Commercial Implications* (London: Tauris Academic Studies), pp. 151–176.

CASES

Embrey's case: *Embrey v Owen* (1851), English Reports, 155, 579–587.
Liggins's case: *Liggins v Inge* (1831) English Reports, 131, 263–268.
Mason's case: *Mason v Hill* (1833), English Reports, 110, 692–702.
Medway case: *Medway Company v Earl of Romney* (1861), English Reports, 142, 226–233.
Sury's case: *Sury v Pigot* (1625), *English Reports*, 79, 1263–1269.

PART III

The normative structure of international arrangements

The concepts of sovereignty and property lie at the foundations of the formal structure of water resources governance. A state exercises its rights of sovereignty by complying with its obligations under international law and by performing its legislative, executive and judicial functions accordingly. In relation to water resources these include recognising, creating and enforcing rules about the appropriation, the allocation and the exercise of rights in relation to water. Owners exercise their rights of property similarly by complying with their obligations under the laws of the state. In Part II we considered how rights of sovereignty, rights of property and the associated rights of control are formally structured. This did little more than review the sources and the nature of these institutional rights. In other words it answered questions such as who controls water resources; who has access to water resources; who may use water resources; and who is excluded from these functions.

Let us now consider the substance and content of these rights. This answers questions such as what is controlled; what may the holder of these rights do with them; and what must or must not be done with them. The focus becomes more negative than positive if we analyse the prohibitions and restrictions upon the exercise of these rights. The two are interdependent. Rights and duties are correlatives. It is, however, the existence of duties – particularly enforceable duties – that gives to these rights their substance. In this part we look at the normative structure of international arrangements for the governance of water resources and in Part IV the normative structure of national arrangements.

Rights in relation to water – whatever their nature and source – perform a range of functions within the legal system. Rights of sovereignty, rights of property and customary rights, for example, empower the undertaking of activities or the grant of authorisation to undertake activities. Human rights, on the other hand, such as the right to development, the right to life, the right to an environment of a particular quality are statements of value or of principle which may guide the exercise of these empowering rights. It is in this sense that the right to water is a human right. While this is an expression of a value, it is no more than that until it is realised through the creation of a reciprocal duty which is enforceable. In this way a right to water may constitute a legal entitlement to take water or to receive water in accordance with an enforceable rule.

Water may be a common-pool resource, a product capable of direct and immediate consumption, or something in between. The existence of a right – including the right of sovereignty inhering in a nation state – in relation to water and however acquired is no more than a statement that it exists and who is its holder. What its holder may or may not do with it is a different question. The mere existence of a right does not answer questions such as:

- what is the best use of the water resource;
- for what purposes may it be used;
- how are different uses to be prioritised;
- how to plan for the best use of it;
- what qualitative standards are appropriate for the various uses to which it may be put;
- how should the water resource and its environment be protected from degradation.

Even more significantly, it does not answer the question whether the water resource and the water in question will be used sustainably. In other words, what is the range of norms according to which water resources are governed? The formal existence of rights in relation to water is thus enlivened by the range of legal and paralegal rules which govern their exercise. This applies alike to rights of sovereignty and rights of property. Let us now concentrate on the former.

The sovereign right of a nation state to use and develop its natural resources carries with it the right to enter into arrangements by way of agreements with other nation states about how shared resources are used and developed. It is expected but probably not required that these agreements apply the general principles of international law as they relate to the management of natural resources including water resources. There is in practice a relationship between the general principles of international law and the conduct of states in relation to each other. Chapter 8 looks at the normative structure of the general principles of international law as they relate to water resources. This includes how water resources are used as well as the limitations placed upon their use in order to protect the ecosystems of which they are a part and the environment generally. In effect the evolution of the principle of reasonable and equitable utilisation of water resources and the principle of prevention of harm to the interests of other states and to the environment generally. Chapter 9 looks at the normative structure of interstate agreements and in particular those agreements that deal with how shared water resources are used, developed and generally managed.

8. Norms of international law

INTERNATIONAL NORMS CONTROLLING NATIONAL WATER RESOURCES

(a) The Issues

The *Grundnorm* of international law in relation to natural resources is the sovereign right of a nation state to use, develop and control its natural resources including the water resources within its territory. The exercise of this right has become increasingly constrained over recent years. This applies to water resources confined within the territorial boundaries of the state. It also applies to water resources that are common-pool resources in the sense of waters flowing across the surface of the land in watercourses or located in underground aquifers but not confined within the territorial boundaries of the state – in other words, international water resources. International watercourses able to be navigated have traditionally been available to be used for that purpose. International watercourses are also the source of energy for industrial and commercial purposes and the destination for the deposit of waste from agricultural, industrial and domestic activities. Thus the activities of upper riparian states impact upon lower riparian states and vice versa. This is not a new phenomenon and it has been debated by international jurists for many years.

Not all water resources located within a nation state are common-pool resources in this sense. First, a nation state may be an island or a group of islands. Secondly, some or all of the water resources of a nation state which borders another nation state may be confined within its territorial boundaries. In each case these are national rather than international water resources. How these water resources are used and the impact of their use are increasingly a matter of international concern. There is in this situation no upper or lower riparian state whose interests in the water resource may be directly or even indirectly affected. But how water resources are used and managed in one location may have an impact upon the quality and condition of natural resources located elsewhere within the global environment: for example, as a habitat for migratory species or as an aspect of climate change. In this sense the international community shares a common concern about the quality of the environment in any location and for its own sake. The international legal regime as it relates to the environment may thus be seen to be both anthropocentric and ecocentric and its conservation and protection justified for both human and ecological reasons.

The concern of the international community about the quality of freshwater resources manifested itself in Agenda 21 – the action plan formulated in the context of the United Nations Conference on Environment and Development 1992. It noted in paragraph 35 of chapter 18 'the interconnectedness of the elements related to freshwater and freshwater quality' and that 'there are few regions of the world that are still exempt from problems of loss of potential sources of freshwater supply, degraded water quality and pollution of surface and groundwater sources'. Issues of quantity and of quality were thus inextricably linked. Not only that, water and water resources were seen as part of the wider environment. Accordingly there was a need to protect water as a natural resource and as a social and economic good, but also a need to protect the ecosystems of which water is a part. This approach not only recognised the multiple functions of water resources. It also stated the priorities to be given to the purposes for which water is used, namely, to the satisfaction of basic needs and to the safeguarding of ecosystems. These principles relate as much to international water resources as they do to national water resources. Let us turn to the normative structure of the international arrangements for protecting the relevant sectors of the environment which indirectly impact upon how water resources are used and managed.

(b) The Contribution of Paralegal Rules

There is no international multilateral agreement of general application for the protection of freshwater resources. It is therefore necessary to consider whether and to what extent the range of legal and paralegal rules designed to protect elements of the environment may apply to water resources within a nation state. It is generally recognised that a state is liable for environmental damage arising outside the state and caused by activities within the state. The issue, however, is the extent to which international law controls activities taking place within the state and causing environmental damage also within the state. And to what extent are water resources protected in this way?

Principle 2 of the Stockholm Declaration of 1972 states that the natural resources of the earth including water must be safeguarded for the benefit of present and future generations through careful planning or management. Principle 3 requires the capacity of the earth to produce vital renewable resources to be maintained, and principle 5 the non-renewable resources of the earth to be employed in such a way as to guard against the danger of their future exhaustion. Whether water resources are renewable or non-renewable, these three principles make it clear that water resources are expected to be used and developed by nation states sustainably. And for this purpose sustainability in general terms includes – at least by implication – protection of the environment of these resources.

The World Charter for Nature of 1982 states in general that all areas of the earth are subject to the principles of conservation set out in the charter. This includes the duty in Article 4 that ecosystems – no doubt including water resources as ecosystems – shall be managed to achieve and maintain optimum sustainable productivity, but not in such a way as to endanger the integrity of those other ecosystems or species with which they coexist. More particularly, it is stated in Article 10 that natural resources shall not be wasted but used with a restraint appropriate to the principles in the charter and in accordance with a number of rules.

Two of these rules are relevant:

- resources, including water, which are not consumed as they are used shall be reused or recycled;
- non-renewable resources which are consumed as they are used shall be exploited with restraint, taking into account their abundance, the rational possibilities of converting them for consumption, and the compatibility of their exploitation with the functioning of natural systems.

Each of these statements is of general application. They are not cast as duties imposed upon the states or persons. Whatever their description – as principles, functions or rules – they are not binding rules of international law. The charter was adopted by resolution of the General Assembly of the United Nations by a large majority. It is clearly a view representative of the international community about how natural resources, and in this case water resources, should be managed for the future.

The Draft International Covenant on Environment and Development was formulated by a group of international jurists on behalf of the International Union for the Conservation of Nature and is regarded by many as a statement of the principal rules of international law relating to environment and development. The draft covenant assumes the form of an integrated set of rights and obligations. Article 11, for example, states conjunctively the sovereign right of a state to use its resources and to protect its environment. This right correlates with obligations which include one to protect its environment and one to ensure harm is not caused to the environment of other states. This implies an obligation to protect water resources as ecosystems which are part of the environment of a state. This is expanded in relation to water by the obligation in Article 19:

> The parties shall take all appropriate measures to maintain and restore the quality of water, including atmospheric, marine, ground and surface freshwater, to meet basic human needs and as an essential component of aquatic systems. The parties also shall take all appropriate measures, in particular through conservation and management of water resources, to ensure the availability of a sufficient quantity of water to satisfy basic human needs and to maintain aquatic systems.

This is a reflection in more specific legal terminology of the values and approaches stated in Agenda 21 for the management of freshwater resources.

Water resources are natural systems; hence the relevance of the obligations in Article 20 of the draft covenant about natural systems. There is a duty to take appropriate measures to conserve life-supporting natural systems and, more particularly, to maintain and restore the ecological functions of these systems as an essential basis for sustainable development. Two systems are relevantly identified for this purpose. These are:

- forests as natural means to control erosion and floods;
- freshwater wetlands and floodplains as recharge areas for groundwaters, floodwater buffers, filters and oxidising areas for contaminants.

Then there is the obligation to manage natural systems as single ecological units: more particularly to manage aquatic systems as entire units covering the full extent of the catchment area.

None of these three instruments – the Stockholm Declaration, the World Charter for Nature or the Draft International Covenant on Environment and Development – contains binding rules of international law. These instruments describe rather than prescribe the approaches expected to be taken in relation to the use and development of water resources. These approaches may be formulated comprehensively as a set of potential duties:

- a duty to safeguard water as a natural resource for the benefit of present and future generations;
- a duty to manage water resources as ecosystems to achieve and maintain optimum sustainable productivity without endangering the integrity of other ecosystems;

- a duty not to waste water resources as natural resources but to use them with restraint appropriate to the principles of conservation;
- a duty to reuse or recycle water not consumed as it is used;
- a duty to exploit water resources with restraint;
- a duty to protect and preserve the environment within the state;
- a duty to take all appropriate measures to maintain and restore the quality of water to meet basic human needs and as an essential component of aquatic systems;
- a duty to take all appropriate measures to ensure the availability of a sufficient quantity of water to satisfy basic human needs and to maintain aquatic systems;
- a duty to take all appropriate measures to maintain the ecological function of forests as natural means to control erosion and floods;
- a duty to take appropriate measures to maintain and restore the ecological function of freshwater wetlands and flood plains as recharge areas for groundwaters, floodwater, buffers, filters and oxidising areas for contaminants.

While these are paralegal rules and not enforceable legal rules, their incorporation in these instruments gives increasingly detailed guidance about how those exercising rights of control in relation to water resources should exercise these rights.

(c) The Contribution of Legal Rules

So much for paralegal rules. Let us now turn to a range of legal rules that give substance and content to the restrictions upon the exercise of sovereign rights by nation states; in other words, the rules of international law that bind nation states in relation to their governance of water resources. Some of these multi-lateral agreements apply directly to water resources, while most apply indirectly or by implication. The Ramsar Convention on Wetlands of International Importance of 1971 necessarily applies to wetlands within the territorial boundaries of the state in question. Parties to the convention are required to designate suitable wetlands within their territory for inclusion in the relevant list of wetlands of international importance. International significance is determined in terms of ecology, botany, zoology, limnology or hydrology.

The principal obligation does not depend upon whether the wetland in question is listed or not. According to Article 4:

> Each contracting party shall promote the conservation of wetlands and water fowl by establishing nature reserves on wetlands, whether they are included in the list or not, and provide adequately for their wardening.

It seems clear enough that wetlands are to be conserved and nature reserves to be established within the territory of the state. However, it is provided that the inclusion of a wetland in the list does not prejudice the exclusive sovereign rights of the contracting party in whose territory the wetland is situated. Presumably this means that it is up to the state in question to decide how to discharge the obligation related to conservation. Otherwise the acceptance of the obligation by entering into the agreement becomes rather pointless. In any event the convention is an early example of restrictions imposed by an international legal instrument on how certain water resources are used and managed.

While the Paris Convention for the Protection of the World Cultural and Natural Heritage of 1972 does not refer to water or water resources as such, these resources may well fall within the ambit of the protective provisions in the convention. Article 5(d) places an obligation on parties to endeavour, in so far as possible and as appropriate for each country, to take the appropriate legal, scientific, technical, administrative and financial measures necessary for the identification, protection, conservation, presentation and rehabilitation of the heritage protected by the convention. This is not an absolute obligation; it is qualified in a number of respects, and it leaves considerable discretion to the state in question. In any event, while water and water resources fall within its scope, water is more likely to form part of the natural heritage rather than the cultural heritage. The natural heritage includes natural features consisting of physical and biological formations or groups of such formations and natural sites or precisely delineated natural areas. Water gathering naturally on the surface of land or indeed underground may fall within the scope of one of these descriptions. It is, however, unlikely to be the water resource by itself: rather the water resource in association with the area of land or landscape with which it is associated.

The scope of the protection afforded by the United Nations Convention on Biological Diversity of 1992 much more clearly includes water resources. The objectives of the convention include the conservation of biological diversity. This is complemented by the imposition on parties of a number of general and some more specific obligations. No obligation is absolute. Some are qualified by reference to the particular conditions and capabilities of the state in question while others are qualified by reference to what is possible and appropriate. Subject to this, Articles 6(b) and 8(d) and (f) impose a duty:

- to integrate the conservation and sustainable use of biological diversity into relevant sectoral or cross-sectoral plans, programmes and policies;
- to promote the protection of ecosystems;
- to habilitate and restore degraded ecosystems.

An ecosystem is a dynamic complex of plant, animal and micro-organism communities and their non-living environment interacting as a functional unit. Biological diversity is the variability among living organisms from all sources including terrestrial, marine and other aquatic ecosystems and the ecological complexes of which they are part. If water resources are not themselves ecosystems, then they are clearly a part of ecosystems. In any event the obligations stated in the convention – particularly those identified – apply to water resources in relevant sets of circumstances.

The ultimate objective of the United Nations Framework Convention on Climate Change of 1992 is to achieve stabilisation of greenhouse gas concentrations in the atmosphere at a level that would prevent dangerous anthropogenic interference with the climate system. For this purpose the climate system is the totality of the atmosphere, hydrosphere, biosphere and geosphere and their interactions. Protection of the hydrosphere from dangerous anthropogenic interference falls within the scope of the convention. More particularly one of the commitments stated in Article 4(e) – again qualified rather than absolute – is to develop and elaborate appropriate and integrated plans for coastal zone management, water resources and agriculture, and for the protection and rehabilitation of areas affected by drought and desertification as well as by floods. How water resources are managed, not only in themselves but also in association with other natural resources, has been included as an important element in achieving the objective of the convention.

The objective of the United Nations Convention to Combat Desertification of 1994 is to combat desertification and mitigate the effects of drought in countries experiencing these problems. Desertification is land degradation in arid, semi-arid and dry sub-humid areas. Drought is the naturally occurring phenomenon that exists when precipitation has been significantly below normal recorded levels, causing serious hydrological imbalances which adversely affect land resource production systems. The integration of land and water management is critical to the achievement of the objective of the convention. According to Article 2(2):

> Achieving this objective will involve long-term integrated strategies that focus simultaneously, in affected areas, on improved productivity of land, and the rehabilitation, conservation and sustainable management of land and water resources, leading to improved living conditions, in particular at the community level.

This is supported by a series of obligations in Articles 4–7. One is to adopt an integrated approach addressing the physical, biological and socio-economic aspects of the processes of desertification and drought. Another is to give due priority to combating desertification and mitigating the effects of drought and to establish strategies and priorities within the framework of sustainable development plans and policies to combat desertification and mitigate the effects of drought. Some countries in Africa are the most affected by desertification and drought, and there is a specific obligation to give priority to them. While the outcome of the convention is clear enough, the mechanisms for achieving it relate for the most part to process rather than substance: for example the promotion of cooperation, the establishment of strategies and priorities, and the formulation of relevant plans. Again, there is little doubt that how water resources are managed is an integral part of these arrangements.

(d) Conclusion

While none of these five conventions informs how water resources should be managed as such, the way water resources have been, are being and will be used and managed is a factor likely to assist or impede the achievement of the objective of each of these conventions. Consequently water resources are within the scope of the convention. How they are relevant and what function their consideration performs are matters that depend upon the objective of the conventions, the principles described in them and the obligations imposed by them. Their relevance and function can, therefore, be determined only in more specific sets of circumstances. National water resources are thus governed by the norms and rules of international law that relate to the environment and the development of natural resources rather than by any particular set of norms and rules that apply only or principally to water resources.

INTERNATIONAL NORMS CONTROLLING INTERNATIONAL WATER RESOURCES

(a) The Issue

While the international community has concerns about how water resources, including national water resources, are managed and has responded according to the norms and rules discussed in the preceding paragraphs, the international community has focused its particular attention upon how international water resources, and in particular international watercourses, are managed. It is thus the relationship between and among riparian states and the impacts that their activities have upon each other that have attracted the particular attention of the international community. The rights and duties of nation states that share common-pool resources, and in particular water resources, is not a new issue. The response of international law has been evolving for at least 100 years.

(b) The Emergence of a Community of Interest

One of the earliest attempts to address this issue resulted in the Madrid Declaration of 1911 of the Institute of International Law. It noted that:

> Riparian states with a common stream are in a position of permanent physical dependence on each other which precludes the idea of the complete autonomy of each state in the section of the natural watercourse under its sovereignty.

The declaration then proposed – but could not authoritatively state – a set of rules to deal with this. Since these rules constrained the exercise of sovereign rights, they almost necessarily assumed the form of restrictive or prohibitive obligations.

What form did they take? Some took the form of a requirement for consent while others took the form of a prohibited activity. For example:

- each riparian state must obtain the consent of other riparian states to making or allowing to be made alterations detrimental to the river bank of the other state;
- a state must obtain the consent of other states to a change of the point where a stream crosses the frontiers of two states;
- a riparian state must not on its own territory utilise or allow the utilisation of water so as seriously to interfere with its utilisation by other states or members of other states;
- there must be no alterations injurious to the water (for example by emptying into the water injurious matter from factories);
- water must not be taken so that the utilisable or essential character of the stream downstream is seriously modified;
- a downstream riparian state must not erect or allow to be erected within its territory constructions which would subject the upper riparian state to the danger of inundation.

What is particularly significant about these prohibitions is this. It is not the activity itself that is prohibited. It is the impact or the consequence of an activity that is prohibited. While there were no such rules of international law in 1911, the creation of norms linked either to the need for consent or to the prohibition of impacts of activities was a clear indication of how the normative structure of water resources governance would develop in later years.

The significance of consensual arrangements between nation states became apparent when the Permanent Court of International Justice in 1937 gave its judgment in the case concerning the diversion of water from the river Meuse in a dispute between Belgium and the Netherlands. A treaty concluded in 1863 between these two states provided for the construction of an intake that diverted water from the river Meuse – an international river – for the purpose of 'feeding' navigation canals and irrigation channels in the Netherlands. In the interim both states had undertaken works impacting upon the waters of the river Meuse. The approach adopted by the court was to permit each riparian state to engage in activities within its territory as it saw fit provided that there was compliance with the terms of the treaty. Thus:

> As regard such canals, each of the two states is at liberty, in its own territory, to modify them, to enlarge them, to transform them, to fill them in and even to increase the volume of water in them from new sources, provided that the diversion of water at the treaty feeder and the volume of water to be discharged therefrom to maintain the normal level and flow in the Zuid-Willemsvaart is not affected.[1]

The two states, in other words, had agreed on what the likely impacts or results of the diversion were. Unless the impacts or results were otherwise, neither state could complain.

The Madrid Declaration, it will be recalled, noted the physical interdependence of riparian states 'with a common stream'. In 1929 in a judgment of the Permanent Court of International Justice determining jurisdictional issues in relation to the river Oder in Poland, this physical interdependence emerged as the legal doctrine of the mutuality of interests of riparian states in international watercourses. While the court directed its comment to the right of navigation in international watercourses, it may be and indeed has been extrapolated to interests in international watercourses generally. Thus:

> But when consideration is given to the manner in which states have regarded the concrete situations arising out of the fact that a single waterway traverses or separates the territory of more than one state, and the possibility of fulfilling the requirements of justice and the considerations of utility which this fact places in relief, it is at once seen that a solution of the problem has been sought not in the idea of a right of passage in favour of upstream states, but in that of a community of interest of riparian states. This community of interest in a navigable river becomes the basis of a common legal right, the essential features of which are the perfect equality of all riparian states in the user of the whole course of the river and the exclusion of any preferential privilege of any one riparian state in relation to the others.[2]

The requirements of justice and the consideration of utility justify this approach. But how is this community or mutuality of interests to be protected and what are the restrictions placed upon riparian states in order to protect it?

One response to this question was the duties to collaborate and to negotiate in the Convention Relating to the Development of Hydraulic Power 1923. While there was no reference to community of interest or mutuality of interest in the convention, it was certainly implicit in the approach adopted. Article 2 imposed an obligation to agree to investigations if the reasonable development of hydraulic power involved such investigations. Article 3 imposed a duty to enter into negotiations with a view to the conclusion of agreements allowing operations to be undertaken partly on the territory of one state and partly on the territory of another state. And Article 4 imposed an obligation to enter into negotiations similarly with a view to the conclusion of agreements allowing the development of hydraulic power which might cause serious prejudice to another state. The norms in this case were procedural rather than substantive. Significantly, however, there were obligations in relation to negotiations with a view to agreement rather than the requirement for agreement itself.

This question was addressed by the tribunal in the Lake Lanoux Arbitration in 1957. The tribunal in this arbitration acknowledged that one of the rules proposed by the Madrid Declaration of 1911 might be part of international law, namely:

> A principle which prohibits the upstream State from altering the waters of a river in such a fashion as seriously to prejudice the downstream State.[3]

There was no rule requiring the consent of riparian states to the use of international watercourses. Thus:

> The rule that states may utilize the hydraulic power of international watercourses only on condition of a prior agreement between the interested states cannot be established as a custom, even less as a general principle of law.[4]

However the tribunal went on to acknowledge or perhaps even to create an obligation – a procedural norm – designed to protect the community of interest in international waterways. Thus:

> The second question is to determine the method by which these interests can be safeguarded. If that method necessarily involves communications, it cannot be confined to purely formal requirements, such as taking note of complaints, protests or representations made by the downstream state. The tribunal is of the opinion that, according to the rules of good faith, the upstream state is under the obligation to take into consideration the various interests involved, to seek to give them every satisfaction compatible with the pursuit of its own interests, and to show that in this regard it is genuinely concerned to reconcile the interests of the other riparian state with its own.[5]

This points to a much more significant norm than one merely of collaboration or negotiation; in effect a norm or standard of methodology of decision-making. Significantly:

- it is not purely formal;
- it is based upon good faith;
- it is a specific obligation placed upon the upstream state;
- it is a duty to consider the relevant interests;
- it is a duty to seek to give them satisfaction compatible with the interests of the upstream state;
- it is a duty of genuine concern to do so.

It falls short of requiring agreement. But it goes well beyond negotiating with a view to agreement. Most importantly, there is a duty on the upper riparian state to demonstrate how all the interests comprising the community of interest in the watercourse have been mutually accommodated.

The community of interest – as it may now perhaps be called – was taken further by the Dubrovnik Declaration of 1956 of the International Law Association. A state, it was declared, was responsible for public or private acts producing change in the existing regime of a river to the injury of another state which it could have prevented by reasonable diligence. While the fundamental normative approach was agreement, the declaration went further by placing upon riparian states an obligation – or at least an expectation – that in reaching agreements states should weigh the benefit to one state against the injury done to another through a particular use of water. A similar obligation or expectation was expected of those settling disputes.

The relevant factors to be taken into consideration in doing so were specified in some detail. Two procedural norms emerged from this declaration. The first was weighing benefits against injuries and the second was considering a range of specified factors in doing so: once again standards of methodology of decision making.

(c) Towards Reasonable and Equitable Use

Not long after the recognition of a community of interest in shared water resources, the focus moved from procedural and methodological norms to include norms of substance. The New York Declaration of 1958 of the International Law Association went one step further in this direction. It resolved:

> Except as otherwise provided by treaty or other instruments or customs binding upon the parties, each co-riparian state is entitled to a reasonable and equitable share in the beneficial uses of the waters of the drainage basin. What amounts to a reasonable and equitable share is a question to be determined in the light of all the relevant factors in each particular case.

Significantly this created a substantive norm. An entitlement to a reasonable and equitable share in the beneficial uses of the waters in question. Not only that: this substantive norm was linked to the procedural norm of considering all the relevant factors in each particular case.

The Institute of International Law confirmed this emerging set of norms in the Salzburg Declaration of 1961. Article II confirmed the right of a state to make use of the waters flowing across or bordering its territory but restricted its exercise in four ways:

- any disagreement to be settled on the basis of equity taking into consideration in particular the respective needs of the states;
- activities in relation to water likely to affect the use of the same water by other states to be conditional upon preserving for those states the benefit of the advantages derived from the norm of equity;
- a duty to enter into negotiations in the event of objections;
- a duty to refrain from activities in dispute pending negotiations.

This approach comprised two substantive and two procedural norms. One was the primary substantive norm of equity protected by the duty – the second substantive norm – imposed effectively on all states to preserve the benefit of this norm of equity for each of them. The norm of equity was further protected by the two procedural requirements. The foundations had thus been set over a period of 50 years from 1911 to 1961 for a more formal and comprehensive set of arrangements for the governance of international watercourses.

The International Law Association took the first step by formulating rules on the uses of the waters of international rivers at its conference in Helsinki in 1966 leading to the informal but influential arrangements known as the Helsinki Rules 1966. These rules provided for five matters in some detail:

- equitable utilisation of the waters of an international drainage basin;
- pollution;
- navigation;

- timber floating;
- procedures for the prevention of settlement and disputes.

Let us concentrate for present purposes upon the substantive norms formulated in the rules: in particular Articles IV, V and X.

The fundamental doctrine supporting the rules was stated in Article IV. Namely, 'each basin state is entitled, within its territory, to a reasonable and equitable share in the beneficial uses of the waters of an international drainage basin'. An entitlement to a share comes close to the ownership of a usufructuary interest in the relevant waters. It is not an open ended share or an equal share: it is a reasonable and equitable share. According to Article V what this means was to be determined in the light of all the relevant factors in each particular case and a range of potential relevant factors was stated in Article V.

While the meaning of what is a reasonable and equitable share was not in any way restricted, the factors described were legally relevant but not necessarily circumstantially relevant. The factors included those that were geographical, hydrological, climatic, demographic, economic and social. Significantly there was included the degree to which the needs of a basin state may be satisfied without causing substantial injury to a co-basin state. The reference to needs and to injury in the same context is interesting. Are the needs economic? Is it injury to these economic needs that is contemplated? Or is it injury in a wider sense?

In any event injury in the wider sense was contemplated by Article X. Article X stated:

> Consistent with the principle of equitable utilization of the waters of an international drainage basin, a state:
> (a) must prevent any new form of water pollution or any increase in the degree of existing water pollution in an international drainage basin which would cause substantial injury in the territory of a co-basin state, and
> (b) should take all reasonable measures to abate existing water pollution in an international drainage basin to such an extent that no substantial damage is caused in the territory of a co-basin state.

This applied whether the pollution originated within the territory of a state or whether it was caused by the conduct of the state. Paragraph (a) created a norm linked to the impact of an activity rather than the activity itself. While such an approach is common in the 21st century, the juxtaposition of norms linked to use and norms linked to impact in relation to the use of water was perhaps an example of or even an extension of the approach adopted by the tribunal in the *Trail Smelter* Arbitration in 1941. Whatever its provenance, the acknowledgement of water pollution prevention as a requirement for the equitable utilisation of international watercourses constituted a significant doctrinal development in the normative structure of water resources governance. In effect, it brought together issues of quality and quantity in how water resources should be managed.

(d) A Set of Coherent Principles

While doctrinal approaches such as these have manifested themselves in other instruments, the United Nations itself has responded by promoting the Convention on the Protection and Use of Transboundary Watercourses and International Lakes of 1992 and the Convention on the Law of the Non-navigational Uses of International Watercourses of 1997. The 1992 convention is open to states that are members of the Economic Commission for Europe, while the 1997 convention is open to any state. The 1992 convention applies to all states that are parties, including but not restricted to riparian states.

Let us consider first of all the normative structure of the 1992 convention. While its principal focus is the prevention and control of pollution of transboundary water resources, this is complemented by the recognition that this is linked in practice and doctrinally with the way water resources are used. Article 2(1) imposes an obligation upon states to take all appropriate measures to prevent, control and reduce any significant adverse impact on the environment resulting from a change in the conditions of transboundary waters caused by human activity within a state. This is supported by a duty to take all appropriate measures linked to more specific outcomes. One is related directly to the prevention of pollution – the principal focus of the convention. Two others are directed at how water resources are used. The first of these is to ensure that transboundary waters are used with the aim of ecologically sound and rational water management and the second is to ensure that they are used in a reasonable and equitable way. In taking these measures there is an obligation to be guided by a number of principles that have become those associated with sustainable development: for example, the precautionary principle, the polluter-pays principle and the principle of intergenerational equity.

Article 3 requires states to develop and implement measures – including legal measures – to ensure a number of more specific outcomes. Most of these are directed at the control of pollution. However one such outcome is to ensure that sustainable water resources management, including the application of an ecosystems approach, is promoted. This is one of the first – if not the first – references in formal instruments to sustainable water resources management.

Importantly, the norms created by the 1992 convention are structured somewhat differently from those in the Helsinki Rules and its doctrinal predecessors. Essentially the convention requires states to take measures to achieve outcomes. The subject matter of the outcomes varies: for example, rational water management, use of waters in a reasonable and equitable way, and sustainable water resources management. The functions of management and use are in themselves neutral. But if adjectives such as rational, reasonable, equitable and sustainable qualify the neutrality of the nouns management and use, then a value or a type of norm is implicitly attached to the concept. There are, therefore, two normative elements in this approach. There is, first, a norm in the form of an outcome: for example, the aim of rational water management or use in a reasonable and equitable way. In addition, a normative quality is attached to this norm of outcome by the use of these adjectives. The convention thus sets a standard of decision-making and of behaviour in relation to the water resources to which it applies.

The 1992 convention, finally, recognises the special responsibilities of riparian states in the context of international watercourses. While riparian states are subject to the general obligations in the convention, they are also under a specific obligation to enter into bilateral or multilateral agreements in order to fulfil their obligations under the convention about the prevention, control and reduction of transboundary impact. Article 9(1) states that this obligation is based upon 'equality and reciprocity'. This is an implicit recognition of the community of interest in these watercourses – a point recognised many years earlier. It is, however, equality in the context of negotiation rather than equality in the context of outcome. Outcome is governed by the norm of equity.

The United Nations Convention on the Law of the Non-navigational Uses of International Watercourses of 1997, although not in force, is generally considered to express many of the norms of the international legal system in relation to international watercourses. The convention contemplates watercourse agreements as the principal – perhaps the only – mechanism according to which international watercourses are managed. However such agreements are formulated on the basis of the principles and rules – in effect substantive and procedural norms – set out in the convention. While the focus of the 1992 convention was protection of watercourses from harm linked to the equitable and reasonable use of international watercourses, the focus of the 1997 convention is the equitable and reasonable use of international watercourses linked to their protection from pollution.

The 1997 convention incorporates a set of procedural norms in the context of strategic planning for the best use of international watercourses together with a set of arrangements for protecting not only international watercourses from pollution but also for ensuring the protection and preservation of the ecosystems of international watercourses. While the functions contemplated by the 1997 convention are more extensive than those of the 1992 convention, there is a consistency of doctrinal approach to the recognition and formulation of the relevant norms. In other words the normative structure of the two conventions is intrinsically consistent.

The normative foundations of the 1997 convention lie in Articles 5, 6, 7 and 8. Article 5(1) provides:

> Watercourse states shall in their respective territories utilize an international watercourse in an equitable and reasonable manner. In particular, an international watercourse shall be used and developed by watercourse states with a view to obtaining optimal and sustainable utilization thereof and benefits therefrom, taking into account the interests of the watercourse states concerned, consistent with adequate protection of the watercourse.

There are four substantive norms in this set of provisions:

- a duty to use in an equitable and reasonable manner – a standard of conduct or behaviour;
- a duty to use and develop with a view to obtaining optimal and sustainable utilisation – a standard of outcome;
- a duty to take into account the interests of watercourse states – a standard for the methodology of decision-making;
- a duty for the utilisation of a watercourse to be consistent with its adequate protection – a standard for the methodology of decision-making.

The first two clearly address the operational substance of how international watercourses are used and developed. The second two state the substantive methodological standards of decision-making by prescribing how decisions about the use of watercourses are made. Each of these tests or standards is intrinsically different. What is important is that strategic decisions about use are integrated with decisions about operational use. So there is inherently an element of planning for the future – at least implicitly – when an outcome is prescribed as a relevant norm.

Article 6 states the range of factors and circumstances that must be taken into account in determining whether an international watercourse is being utilised in an equitable and reasonable manner. Interestingly, this duty is constructed passively rather than actively. In other words, the duty is not imposed upon a watercourse state. Rather it is the activity of utilisation that attracts the duty. The duty thus attaches to the use and not to any particular state or person. However, this is expanded by the obligation placed upon watercourse states to enter into consultations in a spirit of cooperation in the context of these duties. What emerges, therefore, is a trilogy of fundamental duties:

- a duty on watercourse states to utilise in an equitable and reasonable manner;
- a duty for utilisation to require the prescribed factors to be taken into account;
- a duty on watercourse states to enter into consultations in the performance of these two functions.

Paragraph 1 of Article 7 places a duty on watercourse states in utilising an international watercourse in their territory to take all appropriate measures to prevent the causing of significant harm to other watercourse states. This is consistent with Article 5, and in this sense it amounts to a duty on watercourse states to protect the watercourses of other watercourse states. In the event of such harm, issues are resolved either by agreement or by consultation. There is in Article 8 a general obligation on watercourse states to cooperate. The outcome of the duty to cooperate is the optimal utilisation and adequate protection of an international watercourse. This takes place on the basis of sovereign equality, territorial integrity, mutual benefit and good faith. While these four presumptions are clear enough in their own terms, how they operate in practice may be more equivocal. What is important, however, is the range of duties imposed upon watercourse states by Articles 7 and 8. These are:

- a duty to prevent significant harm – a substantive norm of outcome;
- an implied duty to resolve issues by agreement – a procedural norm;
- a duty to resolve issues by consultation – a procedural norm;
- a duty to cooperate – a procedural norm;
- a duty to attain optimal utilisation and adequate protection – a substantive norm of outcome.

While the 1997 convention expects international watercourses to be managed by agreement among the watercourse states, any such negotiated arrangements are expected to comply with the general principles described in Articles 5, 6, 7 and 8. These principles themselves describe a series of norms of different types. Some are procedural: for example, negotiation to agreement, consultation and cooperation. Some are substantive: for example, how to use, outcomes of use, outcomes of cooperation, how to make decisions about use and development. These norms disclose different characteristics and they perform different functions within the overall international legal system.

(e) The Emergence of Strategic and Operational Norms

The range of functions performed by these norms invites closer scrutiny. To what extent are these norms translated into enforceable rules of law which impose obligations or duties that are on the one hand strategic and on the other hand operational? What is the relationship between strategic and operational norms? Are they part of a continuum of legal relationships or are they disconnected? Is their relationship informal rather than formal? There are no easy answers to these questions. However, in its judgment in the dispute between Hungary and Slovakia about the development of water resources in sections of the river Danube – itself an international watercourse – the International Court of Justice included analysis of some of these issues.

The arrangements between the two states took the form of a treaty which had three principal objectives:

- production of hydroelectricity;
- improvement of navigation;
- protection against flooding.

It was agreed that the operational arrangements for undertaking the project would take the form of a joint contractual plan. The treaty provided for the achievement of additional objectives. Article 15 stated that the parties:

> Shall ensure, by the means specified in the joint contractual plan, that the quality of the water in the Danube is not impaired as a result of the construction and operation of the System of Locks.

And Article 19 that the parties:

> Shall ensure compliance with the obligations for protection of nature arising in connection with the construction and operation of the System of Locks.

Neither of these requirements is cast in direct and peremptory terms. Article 15 leaves the detail of how the duty is to be discharged to the joint contractual plan. But the outcome is clear: namely protection of the quality of the water in the river. Article 19 is essentially tautologous. There are two duties: a duty to ensure compliance and a duty to protect nature. Together there is the duty to ensure compliance with the duty to protect nature. Similarly the substance of the obligation is an outcome – the protection of nature.

The analysis by the court of the nature of the responsibilities arising from these provisions is particularly interestingly. The validity of the treaty attracted much of the attention of the court. However, in a rather indirect way the requirements of international law in relation to the protection of the environment became relevant. Articles 15 and 19 were quite clearly about the protection of the environment. The issue was whether the terms of the treaty – assuming its validity – should be interpreted in the light of what were arguably 'intervening developments in environmental law'. Judge Weeramantry in his separate opinion discussed in detail these developments including the principle of sustainable development. While the court itself did not review these developments in detail, it impliedly acknowledged their existence.

It was in this context that the Court specifically commented upon the nature of the obligations in Articles 15 and 19. Thus:

> The court wishes to point out that newly developed norms of environmental law are relevant for the implementation of the treaty and that the parties could, by agreement, incorporate them through the application of articles 15, 19 and 20 of the treaty. These articles do not contain specific obligations of performance but require the parties, in carrying out their obligations to ensure that the quality of

> water in the Danube is not impaired and that nature is protected, to take new environmental norms into consideration when agreeing upon the means to be specified in the Joint Contractual Plan.[6]

There are three points to note. First, Articles 15 and 19 do not contain 'specific obligations of performance'. Second, there is a requirement to take new environmental norms into consideration when negotiating the terms of the joint contractual plan. Third, Articles 15 and 19 contain 'obligations to ensure' the non-impairment of the quality of water and the protection of nature. It would appear, therefore, that Articles 15 and 19 contain obligations linked to an outcome – the protection the environment. These are not, however, specific obligations of performance. In other words, they do not state specifically how this outcome is to be achieved. But the court has itself indicated how these obligations are to be discharged: namely by taking new environmental norms into consideration in formulating the joint contractual plan. If this analysis is justified, then these provisions of the treaty comprise an obligation of outcome and a non-specific obligation of performance made more specific by the obligation implied by the court to consider new environmental norms. This implied obligation of consideration may be described as an obligation of decision-making methodology.

The Court was acutely aware that the protection of the environment was neither the only nor indeed the ultimate objective of the treaty. Thus:

> The 1977 treaty was not only a joint investment project for the production of energy, but it was designed to serve other objectives as well: the improvement of the navigability of the Danube, flood control and regulation of ice-discharge, and the protection of the natural environment. None of these objectives has been given absolute priority over the other, in spite of the emphasis which is given in the treaty to the construction of a System of Locks for the production of energy. None of them has lost its importance. In order to achieve these objectives the parties accepted obligations of conduct, obligations of performance, and obligations of result.[7]

The matrix of norms thus includes the objectives of the instrument – in this case the treaty. The achievement of these multiple objectives is an integral part of the arrangements. Whatever might be the legal status of these objectives as norms part of the overall system, the Court specifically recognised that in order to achieve these objectives there had been acceptance of obligations of three kinds: of conduct, of performance and of result. These three descriptors are complex and controversial. Earlier in its judgment, the Court had already indicated that, while Articles 15 and 19 did not contain specific obligations of performance, these Articles did contain obligations and it has been suggested that these obligations are obligations of outcome. This may correspond with the Court's reference to obligations of result. What is important, however, is that the Court appeared to acknowledge that these obligations – in this case accepted voluntarily by the parties to the treaty – were designed to achieve the objectives of the treaty.

It is particularly significant that the court went on to recognise or perhaps even create a further obligation in these circumstances. Thus:

> The court is of the opinion that the parties are under a legal obligation, during the negotiations to be held by virtue of article 5 of the Special Agreement, to consider, within the context of the 1977 treaty, in what way the multiple objectives of the treaty can best be served, keeping in mind that all of them should be fulfilled.[8]

There is thus a duty to negotiate to reach an agreement about how to implement the judgment of the Court in accordance with Article 5 of the Special Agreement. This implies a duty – described by the court as a legal obligation – to consider how all of the multiple objectives of the treaty can be achieved. And this duty to consider – an obligation of decision-making methodology – arises in the context of the 1977 treaty at large. The 1977 treaty, of course, included Articles 15 and 19. These contained, according to the Court itself, obligations of various kinds. What emerges from this set of comments by the court? There is a matrix of three norms:

- an obligation to negotiate with a view to an agreement – a procedural norm;
- an obligation to consider how best to achieve the objectives of a treaty – an obligation of decision-making methodology linked to the achievement of an objective;
- an implied obligation to comply with the obligations in the treaty – an obligation of decision-making methodology linked to substantive obligations in the treaty.

The judgment of the International Court of Justice in this case – including the separate opinion of Judge Weeramantry – goes in many respects to the doctrinal foundations upon which the international regime for the governance of water resources is based. There emerges a set of arrangements that impact upon strategic and operational decision-making and behaviour. These are, in the sense already discussed, in the form of legal rules and paralegal rules. Underlying the system is the doctrine of the sovereignty of a nation state, but this is clearly limited not only by the obligations undertaken voluntarily by a nation state but also by the obligations which arise from the rules of international law that relate to international watercourses and the environment. The arrangements contain a variety of norms of different kinds, both procedural and substantive. These norms perform different functions within the legal system. There are obligations of varying degrees of specificity, and these obligations prescribe what decisions are to be made, how decisions are to be made, what operational activities are to be undertaken, how operational activities must be undertaken and the overall context of the system including in particular its objectives.

CONCLUSION

Ultimately the question is whether this set of arrangements encapsulates all the elements of sustainable development. We have looked at the evolution of the system over a period of about 100 years. The fundamental *Grundnorm* underlying the management of international watercourses appears to be their equitable and reasonable utilisation linked to the prevention of pollution and the protection of the environment. If this is so, it is moving towards the notion of sustainable development. But sustainable development may not – at least yet – have emerged formally as the principle around which the specific rules for the management of international watercourses have been gathered. We have looked at substantive and procedural norms separately, but they are elements of one system. What does the system look like when the elements are synthesised? It may be instructive for this purpose to take note of the opinions expressed by a number of jurists about the normative structure of international water resources governance.

One conclusion is that: ‘equitable utilization is the fundamental rule governing the use of international watercourses. It is much more than a rule, however. It is a dynamic process, which depends heavily upon active cooperation between states sharing fresh water resources.'[9] In this sense the system based upon an arrangement of norms is a process. While it is a process aimed at the equitable utilisation of the water of watercourses, 'achieving equity is not an exercise in unguided discretion'. Further, 'an equitable result is a requirement of general international law'.[10] Accordingly, there is a rule, a process and a result: in other words, three different types of norms that are all part of an interdependent system.

The justification for such an approach is the equitable apportionment of shared water resources. This has led to a second conclusion that 'the rights of states in the use of shared resources are not unlimited'.[11] In reaching this conclusion the equitable apportionment and hence use of water resources is linked to the prevention of their pollution. Thus:

> Water resources which are the subject of a 'common legal right', including rivers or lakes or groundwaters, may not be used by states in such a manner as to prevent or otherwise limit other 'riparian' states from making full use of their equitable and reasonable entitlements in relation to that shared resource. Although international law does not necessarily prohibit all pollution, it is clear that the quality of freshwaters should not be altered in such a way as to result in significant or substantial damage to the point that the resource may no longer be used, or that its potential for use is materially diminished.[12]

This is a reflection to some extent of the doctrine of community of interest in shared water resources including international watercourses. Community of interest thus extends not only to the quantity of water available for use but also the quality of water so available.

Whether the doctrine of equitable use by itself contains any restrictions on use causing pollution is a matter of some controversy. While the focus of Article 5(1) of the Convention on the Law of the Non-navigational Uses of International Watercourses of 1997 is, as we have seen, upon the equitable and reasonable use of water in international watercourses, it is linked to the need for adequate protection of the watercourse. This is supported by the obligation in Article 7 not to cause significant harm to watercourse states, and this no doubt includes their watercourses. This perhaps leads to the incorporation within the doctrine of community of interest in shared watercourses of a need for their protection from harm.

What has emerged from the analysis based on these two conclusions is a combination of concepts:

- community of interest in shared water resources;
- equitable apportionment;
- equitable use;
- dynamic process;
- protection from harm;
- prevention of pollution.

How are these to be accommodated? This leads to the third conclusion: namely, the doctrine of common management. The use of these two words is instructive. Management is a neutral term that describes a process. Common connotes the involvement of all interested parties. Despite the community of interest in shared water resources, there may well be a conflict of interest. This is to be resolved by reference to equity and prevention of harm. These are themselves dynamic standards set in a dynamic process.

In a sense, therefore, a notion of common management brings together all of these various elements of the system. Thus:

> Common management is the logical combination of the idea that watercourse basins are most efficiently managed as an integrated whole, and the need to find effective institutional machinery to secure co-operation on environmental, social, and economic objectives. It represents a community-of-interest approach which goes beyond the allocation of equitable rights, however, and opens up the possibility of integrating development and international regulation of the watercourse environment.[13]

This brings together in an integrated fashion on an institutionalised basis the equitable use and development of water resources, the protection of the water and the watercourse from pollution, and the protection of the environment of which the water and the watercourse are an element. This takes the normative structure of water resources governance in the direction of sustainable development.

NOTES

1. *River Meuse Case* (1937, p. 186).
2. *River Oder Case* (1929, p. 154).
3. *Lake Lanoux Arbitration* (1957, p. 373).
4. *Lake Lanoux Arbitration* (1957, p. 373).
5. *Lake Lanoux Arbitration* (1957, p. 381).
6. *River Danube Case* (1997, p. 196).
7. *River Danube Case* (1997, p. 200).
8. *River Danube Case* (1997, p. 200).
9. McCaffrey (2007, pp. 404 and 405).
10. McCaffrey (2007, p. 405).
11. Sands (2004, p. 462).
12. Sands (2004, p. 462).
13. Birnie et al. (2009, p. 544).

GENERAL REFERENCES

Birnie, Patricia, Alan E. Boyle and Catherine Redgwell (2009), *International Law and the Environment* (2nd edn, Oxford: Oxford University Press).

Food and Agriculture Organisation of the United Nations (1998), *Sources of International Water Law* (Rome: Food and Agriculture Organisation of the United Nations).

McCaffrey, Stephen C. (2007), *The Law of International Watercourses: Non-Navigational Uses* (2nd edn, Oxford: Oxford University Press).

Sands, Phillipe (2004), *Principles of International Environmental Law* (2nd edn, Cambridge: Cambridge University Press).

LEGAL INSTRUMENTS

Agenda 21 (1992): Action Plan arising from the United Nations Conference on Environment and Development 1992, in Johnson, S.P. (1993), *The Earth Summit: the United Nations Conference on Environment and Development (UNCED)* (London: Graham Trotman), pp. 125–503.

Convention for the Protection and Use of Transboundary Watercourses and International Lakes 1992, (1992) *International Legal Materials* 31, 1312–1329.

Convention on the Law of the Non-navigational Uses of International Watercourses 1997, (1998) *International Legal Materials* 36, 700–718.

Convention relating to the Development of Hydraulic Power 1923, in Food and Agriculture Organisation of the United Nations (1998), *Sources of International Water Law* (Rome: Food and Agriculture Organisation), pp. 21–26.

International Union for the Conservation of Nature (2004), *Draft International Covenant on Environment and Development* (3rd edn, Gland: International Union for the Conservation of Nature).

Dubrovnik Declaration 1956: Statement of Principles in the Exercise of Rights over the Waters of an International River, in Food and Agriculture Organisation of the United Nations (1998), *Sources of International Water Law* (Rome: Food and Agriculture Organisation), pp. 283–284.

Helsinki Rules 1966: Rules on the Uses of the Waters of International Rivers, in Food and Agriculture Organisation of the United Nations (1998), *Sources of International Water Law* (Rome: Food and Agriculture Organisation), pp. 290–298.

Madrid Declaration 1911: International Regulation regarding the Use of International Watercourses for Purposes other than Navigation, in Food and Agriculture Organisation of the United Nations (1998), *Sources of International Water Law* (Rome: Food and Agriculture Organisation), pp. 269–270.

New York Declaration 1958: Resolution on the Use of the Waters of International Rivers, in Food and Agriculture Organisation of the United Nations (1998), *Sources of International Water Law* (Rome: Food and Agriculture Organisation), pp. 285–286.

Paris Convention for the Protection of the World Cultural and National Heritage 1972, (1972) *International Legal Materials* 11, 1358–1366.

Ramsar Convention on Wetlands of International Importance 1971, (1972) *International Legal Materials* 11, 969–976.

Salzburg Declaration 1961: Resolution on the Use of International Non-maritime Waters, in Food and Agriculture Organisation of the United Nations (1998), *Sources of International Water Law* (Rome: Food and Agriculture Organisation), pp. 275–276.

Stockholm Declaration (1972): Declaration of the United Nations Conference on the Human Environment (1972), *International Legal Materials* 11, 1416–1421.

United Nations Convention on Biological Diversity 1992, (1992) *International Legal Materials* 31, 818–841.

United Nations Convention to Combat Desertification 1994, (1994) *International Legal Materials* 33, 1328–1382.

United Nations Framework Convention on Climate Change 1992, (1992) *International Legal Materials* 31, 849–873.

World Charter for Nature 1982, (1983) *International Legal Materials* 22, 455–460.

CASES

Lake Lanoux Arbitration: (France and Spain) (1957), International Environmental Law Reports, 1, 332–385.

River Danube case: Judgment in the case concerning the Gabcikovo-Nagymaros Project (Hungary and Slovakia) (1997), *International Legal Materials* 37, 162–242.

River Meuse case: Diversion of Water from the Meuse Case (Belgium and The Netherlands) (1937), International Environmental Law Reports 1, 157–230.

River Oder case: Case relating to the Territorial Jurisdiction of the International Commission of the River Oder (1929), International Environmental Law Reports 1, 146–156.

Trail Smelter Arbitration: (Canada and the United States of America) (1941), International Environmental Law Reports 1, 231–331.

9. Norms in interstate agreements

INTRODUCTION

The normative structure of arrangements between states with an interest in a common or shared water resource depends to some extent upon the relationship between two of the fundamental principles of international law. On the one hand there is the principle of the permanent sovereignty of states over their natural resources and on the other hand the principle that *pacta sunt servanda*. The obligation to comply with consensual arrangements may be seen in two different ways. The power to enter into agreements, and hence to be bound by them, may be taken to be an exercise of the right of sovereignty. Alternatively it may be seen as a restriction upon the exercise of the right of sovereignty. The first is positive and the second is negative. The acceptance of restrictions in this negative sense can be justified in the context of sovereignty as the outcome of the exercise of a right of sovereignty rather than the existence of the right in the first instance.

Perhaps even more important for present purposes is the principle of the equitable and reasonable use of international watercourses. This principle is and can only be a substantive restriction, not only upon an exercise of the right of sovereignty but also upon the normative substance of consensual arrangements entered into in exercise of this right of sovereignty. If this a correct analysis, then the validity of consensual arrangements entered into between states depends upon their compliance, or at least concordance, with this principle. Consensual arrangements must therefore comply not only with the procedural norm of good faith but also with the substantive norm of equitable and reasonable use.

The approach adopted by the International Court of Justice in the *River Danube case* impliedly indicates a move in this doctrinal direction. The Court in this case decided that the agreement entered into between the states in question remained in force. Hungary and Slovakia were then instructed to put together a regime to achieve the outcomes intended by the agreement, but taking into account the normative developments in international law since 1977 when the agreement was put in place. As the Court itself said:

> Re-establishment of the joint regime will also reflect in an optimal way the concept of common utilization of shared water resources for the achievement of the several objectives mentioned in the treaty, in concordance with article 5, paragraph 2, of the Convention on the Law of the Non-navigational Uses of International Watercourses.[1]

The Court took particular notice of the provisions of paragraph 2 which, as we have seen, refer specifically to the normative standard of what is equitable and reasonable. It was thus the expectation, but perhaps not the requirement, of the Court that the common management regime for the relevant parts of the river Danube would be guided by and perhaps even observe the principle of the equitable and reasonable use of shared watercourses. Hence, the power to enter into agreements – in exercise of a right of sovereignty – is restricted among others by the expectation, and perhaps even the requirement of, compliance with this normative standard.

It is against this background that we now consider the normative structure of a number of examples of agreements between states. The structure, nature and purpose of these agreements vary enormously. Some are statements of general application about the management of the water resources or of the natural resources of the region. A number create a regime of common management for a particular project or a series of projects. In some cases the norms described in the agreement are procedural; in other cases they are substantive; in a growing number they are both. An increasing number of arrangements include an organisational structure for the common management of the resources in question. Let us begin by considering norms of strategy.

NORMS OF STRATEGY

(a) African, American and Central Asian Arrangements

The African Convention on the Conservation of Nature and Natural Resources of 1968 stated a fundamental principle in the form of an obligation 'to undertake to adopt' the measures necessary to ensure conservation, utilisation and development of – among others – water resources in accordance with scientific principles and with due regard to the best interests of the people. Despite the tautology in the formulation of the duty, it is a duty the implementation of which was left to the discretion of the state, but restricted methodologically by reference to scientific principles and the best interests of the people. This was supported by a more particular responsibility in relation to water: namely to establish relevant policies and to endeavour to guarantee a sufficient and continuous supply of suitable water. This was complemented in relation to shared water resources – whether surface or underground – by a duty to consult and to resolve problems through inter-state commissions. These are, therefore, norms of outcome, norms of methodology and norms of procedure.

A few years earlier in 1965, the Organisation of American States prepared the draft Convention on the Industrial and Agricultural Use of International Rivers and Lakes. It 'establishes the general standards concerning the utilization of the waters of international rivers and lakes' for these purposes. The standard created by article 5 was that the use of these waters for these purposes 'must not cause substantial injury, according to international law, to the riparian states or alterations to their boundaries'. Significantly the obligation related to the use of the waters rather than a duty imposed directly upon a state. It probably did no more than confirm the existence of an obligation in international law not to cause substantial injury to the environment beyond the jurisdiction of the state in question. This was supported by Article 6 which required, first, the consent of the state sustaining or likely to sustain damage or injury in this way as well as, second, payment or indemnification for any such damage. In any event the construction of works for the use of the waters of international rivers or lakes had to be notified to interested states. There was therefore a combination of procedural and substantive norms, and in this case the substantive norms were prescriptive in nature. An approach of this nature was confirmed by the Asuncion Declaration of 1971 at a meeting of foreign ministers of the countries of the river Plate basin.

The trilateral agreement of 1998 between a group of central Asian states – Kazakhstan, the

Republic of Kyrgyz and Uzbekistan – disclosed a different normative structure. Essentially it comprised procedural duties of cooperation and notification. The preamble and Article 6 provided a substantive context within which these procedures were to take place. The preamble attached great significance 'to environmental protection and rational use of the natural resources' of the area and recognised – among others – the 'prudent and zealous use of natural resources for effectuation of economic and social development with due regard to the interests of the living and future generations'. The reference to intergenerational equity in this context was perhaps restricted to economic and social sustainability without reference to ecological sustainability. Nevertheless Article 6 attached 'priority importance' to providing environmental protection.

How was this to be achieved? By cooperation. In particular the states were required to cooperate:

> To harmonize the ecological laws, interstate normative and legal regulations in the area of environmental protection and use of natural resources providing the mechanism for compensation of the damages and losses incurred by one party to the other party.

While the duty was to cooperate in the harmonisation of the relevant laws, the expectation was that the substance of these laws would be normative and regulatory on an interstate basis and that these laws would provide remedies for relevant damage and loss. The duty to cooperate, finally, extended to the protection, rational use and prevention against pollution of the transboundary water resources. The substantive norms in this agreement had no independent existence but were linked in all cases to the procedural duty to cooperate.

(b) Arrangements in the European Union

The arrangements for the governance of water resources within the European Union are not formally agreements between the states of the European Union but legislation in the form of directives enacted in accordance with the treaties setting up the union. One of the most recent directives in this regard has been the Directive 2000/60/EC Establishing a Framework for Community Action in the Field of Water Policy of 2000. Its normative structure is particularly significant.

The directive contains:

- a lengthy preamble which states a number of significant fundamental values and principles;
- a statement of purpose which prescribes an overarching outcome supported by a set of additional outcomes directed at more specific sets of circumstances;
- a duty on states to identify individual river basins within their territory;
- a duty on states to ensure that an interstate river basin is assigned to an international river basin and district;
- a duty on states to endeavour to establish appropriate coordination with other states where a river basin district extends beyond the territory of the European Union;
- a duty on states to ensure the production of river basin management plans for river basin districts and international river basin districts;
- a duty on states to ensure the establishment for each river basin district of a programme of measures comprising a set of prescribed basic measures that must be included;

- a statement of environmental objectives to be achieved in making operational the programmes of measures specified in the river basin management plans;
- a range of procedural duties on states in relation to environmental impact analysis, economic analysis, monitoring, provision of information, consultation and reporting;
- a statement of normative definitions of ecological status classifications.

There are, therefore, norms of procedure and norms of substance; obligations of performance and obligations of outcome or result; and standards of ecological quality. The directive applies to all waters, including groundwater and surface waters, whether located inland, at the landward coastal margin or at the seaward coastal margin. It thus discloses an integrated concept of water and water resources within an integrated framework of procedural and substantive norms performing a range of functions.

Despite its complexity, the fundamental focus of the directive seems clear enough. The preamble, for example, makes a number of points. These include:

- water is not a commercial product like any other but, rather, a heritage which must be protected, defended and treated as such;
- the diversity of conditions and needs indicates that decisions should be taken as close as possible to the locations where water is affected or used.

The first acknowledges that water may be a commercial product in certain circumstances but, unlike other commercial products, it is 'a heritage' to be treated as such. In this sense, water is a natural resource the various attributes and functions of which must be maintained for future generations. Additionally, whatever may be the strategic framework for managing water resources, the ultimate effectiveness of water resources governance is likely to be determined at the point of storage, abstraction, consumption or use.

The overarching rationale for the directive may be found in this statement in the preamble:

> Common principles are needed in order to coordinate member states' efforts to improve the protection of community waters in terms of quantity and quality, to promote sustainable water use, to contribute to the control of transboundary water problems, to protect aquatic ecosystems, and terrestrial ecosystems and wetlands directly depending on them, and to safeguard and develop the potential uses of community waters.

The directive proceeds to formulate a set of 'common principles'. Yet many of these principles appear as norms of various kinds, and some of these are in the form of prescriptive rules capable of enforcement. For example, the duty to ensure the establishment of a programme of measures in order to achieve the environmental objectives prescribed by the directive.

The preambular statement about the need for common principles incorporates most – probably all – of the elements of sustainable development. While there is the function of coordination, it is directed at a number of outcomes. These include the protection of water in terms of quantity and quality, the promotion of sustainable water use, the control of transboundary water problems, the protection of aquatic ecosystems and the wider ecosystems of which they are a part, and the safeguarding and development of the potential uses of the relevant waters.

These values are reflected – as one would expect – in the purpose stated in Article 1. In terms of structure, it creates a general normative outcome and a more specific set of outcomes designed to achieve the primary result. The primary outcome is the establishment of a framework for the protection of inland surface waters, transitional waters, coastal waters and groundwater. While these waters are to be protected, the set of secondary outcomes includes not only the protection of the waters themselves from pollution, the ecosystems of these waters and the terrestrial ecosystems of which they are a part, but also the promotion of sustainable water use based on long-term protection of available water resources. The statement of purpose in Article 1 together with the statement in the preamble emphasises the integration of the functions of use and protection and the substances of water, aquatic ecosystems and terrestrial ecosystems. This lies at the heart of the normative structure of the directive. While the statement of purpose in Article 1 does not constitute by itself an enforceable rule of law, it is 'interstitially' the paralegal rule which informs, guides and inspires the procedural and substantive norms included in the directive.

NORMS OF OPERATION

(a) Introduction

The sets of principles contained in the European directive, the American draft convention and the African convention prescribe at a strategic level the conceptual and normative foundations for the management of water resources in particular sets of circumstances. In other words, the exercise of rights of sovereignty in an open and unfettered way has been modified in the circumstances in which these instruments apply. Let us now turn from the normative structure of instruments of this kind to the normative structure of arrangements for the management of international watercourses negotiated on an ad hoc basis between states. A review of arrangements in Africa and Asia discloses an interesting pattern of normative approaches to the use and development of international water resources. Some agreements, for example, authorise the undertaking of a specific project, while others provide the normative context within which decisions are made about actual or potential projects.

(b) Africa

In 1959 the Republic of the Sudan and the United Arab Republic entered into an agreement for the full utilisation of the waters of the river Nile. It authorised the construction of the Sudd el Aali at Aswan as the first of a series of projects on the Nile to provide for long term storage of the waters of the Nile. The construction of this dam was the responsibility of the United Arab Republic. In addition the agreement authorised the construction of the Roseires Dam on the Blue Nile. This was the responsibility of the Republic of the Sudan. The focus of the agreement was the distribution of the net benefit of the waters stored in the Sudd el Aali Reservoir. The agreement prescribed how to calculate the net benefit and how to divide it between the two states. The agreement provided an indicative example of how this arrangement would work:

> This means that, if the average yield remains the same as the average of the previous years of this century which is estimated at 84 Milliards, and if the losses of over-year storage remain equal to the present estimate of 10 Milliards, the net benefit of the Sudd el Aali Reservoir shall be 22 Milliards

> of which the share of the Republic of the Sudan shall be $14^{1}/_{2}$ Milliards and the share of the United Arab Republic shall be $7^{1}/_{2}$ Milliards.

It was further agreed that the United Arab Republic would pay to the Republic of the Sudan compensation for the damage resulting to existing Sudanese properties as a result of the storage of water in the reservoir. The Republic of the Sudan undertook to arrange final transfer of the population of Halfa and all other Sudanese inhabitants whose lands would be submerged by the stored water. There was no reference to responsibility for any other impacts that the project might have. The agreement was thus concerned almost entirely with the construction of the dam, the storage of water in the reservoir and the distribution of the net benefit from the waters so stored.

In 1993 Ethiopia and the Arab Republic of Egypt agreed to cooperate to achieve the conservation and protection of the waters of the river Nile. The essence of the arrangements was thus a duty to undertake to consult and cooperate with each other in mutually advantageous projects. However Articles 4 and 5 of the agreement placed this duty in a wider context. According to Article 4 the use of the waters of the river Nile was to be determined on the basis of the rules and principles of international law. These rules and principles are no doubt ambulatory in nature. Although the principle of the equitable and reasonable use of international watercourses was still evolving in 1993, arguably the principle would have been relevant in any contemporary negotiations between the parties. Article 5 required each party to refrain from engaging in any activity related to the waters of the river Nile which might cause appreciable harm to the interests of the other party. This duty was no doubt based upon the liability of a state for harm caused to the environment of another state. This general obligation was well recognised in 1993.

During the second half of the 1980s arrangements were put in place for the use and development of international watercourses in southern Africa principally for the purpose of the generation of hydroelectric power. The preamble to the treaty of 1986 between Lesotho and South Africa noted the mutual benefits for each party 'to be derived from the enhancement, conservation and equitable sharing of the water resources of the Senqu/Orange River and its affluents'. The purpose of the treaty was to enable the project to go ahead and the purpose of the project was to use the water resources in question to deliver water to the parties, and in particular for the generation of hydroelectric power. Minimum water deliveries were specified in detail and a complex set of financial and institutional arrangements incorporated in the treaty. The preamble, it will be recalled, referred to the equitable sharing of the water resources in question. It must be assumed, therefore, that the arrangements in the treaty were taken by each of the parties to constitute an equitable sharing of the water resources.

The agreement of 1987 between Zimbabwe and Zambia was designed to secure for them 'the greatest possible benefit from the natural advantages offered by the waters of the Zambezi River'. In particular it was their use for the production of energy that was the focus of the agreement. It involved the operation and maintenance of the Kariba complex which comprised the Kariba Dam and reservoir and related infrastructure. These activities were the responsibility of the Zambezi River Authority. Article 18 imposed an obligation upon the states to undertake to keep each other informed, to consult the authority and to consult with each other. The purpose of this set of duties was to ensure the efficient and equitable use of the waters of the Zambezi River. Article 23 placed an obligation upon the states to share all the available energy from the Kariba Dam equally in accordance with the specified procedures. It must therefore be presumed that in these circumstances equal shares constituted equitable shares.

While these agreements were concerned essentially with the implementation of specific projects described in the agreement and operated in accordance with the agreement, it was not uncommon for states in Africa to put in place consensual arrangements for the use and development of water resources in general and independently of any specific project. For example, as early as 1964 the riparian states of the river Niger and its tributaries agreed to set up the River Niger Commission (later the Niger Basin Authority) with functions related to the use and development of the resources of the River Niger Basin. According to this agreement, the resources of the River Niger Basin were managed on behalf of the riparian states by the Niger Basin Authority. Cooperation was the essence of these arrangements and the achievement of cooperation was assisted by the imposition of two more specific duties upon the riparian states, namely:

- to inform the authority of all studies and works upon which they proposed to embark;
- to abstain from carrying out on their portion of the river any works likely to pollute the waters or to affect biological characteristics of its fauna and flora without adequate notice to and prior consultation with the authority.

Although these duties were procedural and not substantive, it is interesting that the duty to consult related not to the use of the water but to the potential pollution of the water.

The Convention relating to the Status of the River Gambia of 1978 contemplated that the use and development of the water resources of the river Gambia would be undertaken in accordance with a special agreement between the parties. While Article 4 required prior approval to a project likely to have certain consequences, this requirement terminated after six months. The wording is particularly interesting:

> No project which is likely to bring about serious modifications on the characteristics of the river's regime, on its navigation conditions, the agricultural and industrial exploitation of the river, the sanitary state of the waters, the biological characteristics of its fauna and its flora, as well as its water level, will be implemented without the prior approval of the contracting states. Such approval should be received within six months. Beyond this deadline the state concerned may, as of right, proceed with the execution of its project.

Significantly, the exercise of this right appears not to have been limited or restricted in any way.

About the same time an organisation was established for the management and development of the River Kagera Basin. The commission established for this purpose was given the power to decide which activities should be undertaken. Thus:

> *La Commission a le pouvoir de décider quels projets, travaux ou programmes d'ordre international intéressant le developpement du bassin de la riviere Kagéra seront poursuivis.*

This reflects – not surprisingly – the French doctrine of these public domain or the exercise of *la puissance publique.*

The agreement of 1992 between Swaziland and South Africa concerning the water resources of the Komati River Basin enabled the development of these water resources in accordance with a comprehensive development plan involving the construction of dams and related works. The agreement in Article 12(2) provided for annual entitlements to water from the project in accordance with specified net consumptions, namely:

- for Swaziland:
  15.1 cubic hectometres at high assurance
  260.2 cubic hectometres at low assurance
- for South Africa:
  157.8 cubic hectometres at high assurance
  381.0 cubic hectometres at low assurance.

Significantly the parties agreed to take all reasonable measures to ensure that the design, construction, operation and maintenance of the project were compatible with the protection of the existing quality of the environment and to pay due regard to the maintenance of the welfare of persons and communities immediately affected by it. The water entitlements were thus quite specific, while the environmental and social norms were essentially obligations of outcome or result.

It may be assumed, once again, that these arrangements reflected the perception of each party of its reasonable and equitable share in the use of the waters in question. This is implicitly supported by paragraph 5 of Article 3. The Komati River Basin is an integral part of the Inkomati River Basin. Mozambique had an interest in the waters of the Inkomati River Basin. Accordingly Swaziland and South Africa specifically recognised the right of Mozambique to a reasonable and equitable share in the use of the waters of the Inkomati River Basin. This would be determined by negotiation. Thus this agreement – dated 1992 – quite specifically acknowledged the existence of a right to a reasonable and equitable share in the use of the waters of an international watercourse.

Two years earlier a concept of equity underpinned the agreement between Nigeria and Niger for managing their common water resources. This agreement contained procedural obligations in relation to the collection of information, the provision of information and consultation. In terms of its normative structure, however, Articles 2 and 5 are particularly significant. According to Article 2:

> Each contracting party is entitled, within its territory, to an equitable share in the development, conservation and use of the water resources in the shared river basins.

Subject to the reference of disputes to the Commission of Mediation, Conciliation and Arbitration of the Organisation of African Unity for binding determination, the determination of equitable shares was a matter for the two states concerned.

Article 5 imposed an obligation in making such a determination. Paragraph 1 stated the factors to be taken into account in making such a determination. Paragraph 2 required each factor to be given the weight warranted by the circumstances peculiar to each individual river basin, all factors so weighted to be considered together; and a determination to be arrived at on the basis of the whole. The factors to be taken into account included climatic, meteorological, hydrological, economic, social and environmental issues. One in particular was the maintenance of an acceptable environmental balance in and around a particular body of water. The right to an equitable share in the water resources was thus tempered by the duty to consider what might be described as the principle of intragenerational equity. Clearly not, at this stage, the principle of intergenerational equity.

In 1987 a number of states in southern Africa entered into an agreement about an action plan for the environmentally sound management of the common Zambezi river system. While the preamble to the agreement referred to cooperation for sustainable development, the agreement did no more than adopt the action plan. According to paragraph 15 of the plan, it was expected to contribute to the incorporation of environmental considerations in water resources management while increasing long-term sustainable development in the river basin. The key to sustainable, environmentally sound development was stated by paragraph 29 to be the proper management of the resource base. It was suggested that a number of activities might assist in doing so. These included:

- improvement of drinking water supply, sanitation and human health;
- environmentally sound development of water resources to meet the demand for water for industries, mines, irrigation, hydropower, navigation and drinking water supply.

While the focus was again upon intragenerational equity, the agreement and the action plan embraced to some extent the notion of sustainable development in linking social, economic and environmental values.

Almost a decade later a larger group of states in the Southern African Development Community Region developed further the normative structure of their arrangements for the management of their shared water resources in accordance with the Protocol on Shared Watercourse Systems 1995. The Protocol was itself revised in 2000. These arrangements were based upon three fundamental *Grundnorms*:

- each riparian or basin state is entitled to use shared water resources;
- this is without prejudice to the sovereign rights of each state;
- the use of shared water resources is in accordance with the principles in the protocol.

Each state nevertheless undertook specifically to respect and apply the existing rules of general or customary international law relating to the use of shared water resources, and in particular to respect and abide by the principle of community of interests in the equitable use of these resources.

While the arrangements included obligations in relation to cooperation and the exchange of information, a number of substantive obligations were included. For example:

- to utilise a shared watercourse in an equitable and reasonable manner;
- to use and develop a shared watercourse with a view to attaining optimal and sustainable utilisation thereof and benefits therefrom;
- to do so by taking into account the interests of the watercourse states concerned consistent with adequate protection of the watercourse for the benefit of current and future generations;
- in determining equitable and reasonable utilisation, to take all relevant factors and circumstances into account including a range of technical, social, economic and environmental issues;
- to do so by considering all relevant factors together and reaching a conclusion on the basis of the whole.

By 2000, the normative structure of these arrangements for southern Africa rested upon the emerging concept of sustainable development and an acknowledgement of a number of its supporting principles.

(c) Asia

In 1996 India entered into two agreements in relation to shared watercourses: one with Bangladesh on the sharing of the waters of the river Ganges at Farakka and one with Nepal on the integrated development of the river Mahakali. The agreement with Bangladesh did little more than determine the distribution of waters between India and Bangladesh, while the agreement with Nepal not only distributed the shared waters from the Sarada barrage but also created the framework for a proposed multi-purpose project at Pancheshwar.

The purpose of the agreement with Bangladesh was clear from the preamble: namely to share by mutual agreement the waters of the international rivers flowing through the territories of the two countries and to make the optimum utilisation of the water resources of their region in the fields of flood management, irrigation, river basin development and generation of hydro-power for the mutual benefit of the peoples of the two countries. Article II shared the waters of the river Ganges at Farakka with reference to the formula in annexure 1 and an indication of how it would be implemented in annexure 2. The formula was based upon the availability of water at Farakka. Three sets of circumstances were contemplated and a percentage share stated. Thus:

- 70,000 cusecs or less – 50% to India and 50% to Bangladesh;
- 70,000 to 75,000 cusecs – 35,000 cusecs to Bangladesh and the balance to India;
- 75,000 cusecs or more – 40,000 cusecs to India and the balance to Bangladesh.

If the flow fell below 50,000 cusecs in any ten-day period, there was an obligation to consult on emergency arrangements 'in accordance with the principles of equity, fair play and no harm to either party'. There was an obligation to review the arrangements every five years at least and to make any adjustments based on the same principles. The agreement contained no indication of how these principles should be applied and no reference to the quality of the waters or of their environment.

The agreement with Nepal stated not only rights but also obligations, including obligations of a social and environmental nature. Paragraph 1 of Article 1 stated that Nepal shall have a right to a supply of 28.35 m^3/5 (1,000 cusecs) of water from the Sarada barrage in the wet season (ie from 15 May to 15 October) and 4.25 m^3/5 (150 cusecs) in the dry season (ie from 16 October to 14 May). Paragraph 2 placed an obligation on India to maintain a flow of not less than 10 m^3/5 (350 cusecs) downstream of the Sarada barrage in the Mahakali River to maintain and preserve the river eco-system. While Article 7 prevented any use or interference with the waters of the river that adversely affected its natural flow otherwise than by agreement, this did not prevent the use of the waters by the local communities living along both sides of the river where this did not exceed 5% of the average annual flow at Pancheshwar.

A multi-purpose project was proposed for Pancheshwar, and this included the construction of power stations of equal capacity on each side of the river. Significantly, in relation to this project, it was agreed that each of the parties would have equal entitlement to the use of the waters and to the benefits from the development. In particular it was stated that 'total energy generated shall be shared equally between the parties'. Assuming that the states reached this agreement in accordance with the emerging principle of the equitable and reasonable use of shared water resources, in this case equality was equitable and reasonable.

In 1995 Cambodia, Laos, Thailand and Vietnam agreed to cooperate for the sustainable development of the Mekong river basin. The preamble to the agreement made it clear that these arrangements were all about the use and conservation of the water and related resources of the Mekong river basin, the social and economic development of the states and the protection of the environment of the river basin, but all in the context of sustainable development. The basis of the arrangements for cooperation was stated by Article 4 to be 'sovereign equality and territorial integrity in the utilisation and protection of the water resources of the Mekong river basin'. Consistently with this, Articles 1 and 2 imposed obligations to cooperate in achieving the outcomes contemplated by the agreement and to promote the development of the full potential of sustainable benefits to the states. In this sense the rights of sovereignty of the states were in a technical sense retained. Article 3, however, prescribed an obligation to protect the environment and the ecological balance of the Mekong river basin from pollution or other harmful effects. This is positive and constitutes a restriction on the exercise of rights of sovereignty, but probably in a way consistent with the general principles of international law in 1995.

Even more importantly, Article 5 required the parties to utilise the waters of the Mekong river system in a reasonable and equitable manner in their respective territories. The obligation to do so in a reasonable and equitable manner is consistent with the emerging principle's of international law at that time. Article 5, however, qualified this by reference to all relevant factors and circumstances, the set of rules for water use contemplated by the agreement, and the procedural obligations in Article 5 itself about notification and consultation. The rules are particularly significant. These rules were concerned quite specifically with the day to day operations involved in the use and development of the water resources of the Mekong river basin: for example, the location of hydrological stations, flow level requirements, and the criteria for determining surplus quantities of water. Issues such as these are critical for the international waters of the river basin to be managed effectively on an integrated basis. While the agreement did not say so in terms, these rules would be enforceable. The protection of the environment was highlighted not only by the obligation in Article 3 but also by the recognition of state responsibility for environmental harm in Article 8. Where substantial damage was caused by the use of the waters of the river Mekong or by the discharge of matter into the waters of the river, responsibility would be determined in conformity with the principles of international law relating to state responsibility and any disputes settled in accordance with the arrangements in the agreement. The agreement in many respects reflects a number of the principles for the management of international watercourses that were emerging at the time these arrangements were negotiated.

CONCLUSION

The legal and paralegal rules that constitute the normative structure of international water resources governance derive from a number of sources:

- the norms of international law that relate to the use and development of natural resources generally;

- the norms of international law that relate to the protection of the environment;
- the norms of international law that relate to the sustainable development of natural resources;
- the norms of international law that focus specifically upon the use, development and sustainable development of international watercourses, the protection of their ecosystems and the protection of their environment;
- the consensual arrangements put together for these purposes by two or more nation states for the common management of shared water resources;
- the consensual arrangements put together for these purposes by two or more nation states in the context of specific projects linked to the achievement of these purposes.

In essence these norms represent restrictions upon the exercise of the right of sovereignty of nation states over their natural resources, and these restrictions are either imposed by the rules of international law or accepted voluntarily by the nation states in question.

The nature of the rules for the governance of water resources reflects their purpose and context. Many are procedural. A growing number are substantive. The distinction between procedural and substantive rules is not always clear and a number no doubt fall uneasily between the two. Similarly some rules relate to the formulation of strategies and policies, while others regulate more directly the actual or potential operational activities within nation states. In other words the rules do not always fit comfortably within either of these categories.

The relevant norms of procedure include obligations to inform, to negotiate or to agree. Information may be required about the proposed use and development of water resources or about the impact of the proposed use and development of water resources. Negotiations are often required to be conducted in good faith. The expectation is an agreement satisfactory to all parties. Consensual arrangements are essentially a meeting of minds on the issue in question. This effectively rules out an obligation to agree. In any event, the expectation, and on occasion the rule, is that consensual arrangements reflect the general principles of international law.

It is these general principles that give substance not only to consensual arrangements but also to the framework within which water resources are used and developed. Some of these relate to natural resources and the environment generally while others are more specifically directed at water resources. There is, for example, a general obligation not to harm the environment either inside or outside the territorial boundaries of the nation state in question. This applies to water as an element of the environment just as it applies to any other elements of the environment. The normative structure of international water resources governance, however, incorporates both quantitative and qualitative criteria. This is justified in these ways:

- the nature of water resources as common-pool resources;
- this includes international water resources as shared resources;
- the community of interest of nation states in these shared resources;
- the requirements of justice generally.

While the use and development of its water resources by one state cannot generally in the absence of agreement be prevented by another state, the way in which these activities are undertaken is governed by a number of rules. These include:

- a duty to prevent or at least reduce any prejudicial or harmful impact of these activities;
- an entitlement to a reasonable and equitable share of the water resources;
- a duty to consider specific factors in determining what is a reasonable and equitable share;
- an expectation rather than a rule that water resources will be used and developed sustainably.

While these norms are no doubt of general relevance, they are more likely to be significant at the strategic rather than the operational level. However, the principle of adaptive management contemplates ongoing responsibilities in relation to how water resources are managed. Accordingly, what may be reasonable, equitable or sustainable may change as circumstances themselves change. It is for reasons such as these that the legal and paralegal rules that constitute the normative structure of international water resources governance have reflected, and will no doubt continue to reflect, this flexible and responsive structure. There is nothing new about such an approach. Standards such as what is reasonable or equitable and what amounts to *culpa* or *iniuria* have underpinned the common law and the civil law for centuries.

The normative structure of consensual arrangements is often much more specific. Particularly so in relation to the uses of water for specific purposes and the construction of related infrastructure. For example, the creation of rules according to which water resources are allocated among riparian states; the specific quantity of water available to a specific riparian state; or the grant of approval for the construction of a dam or a hydroelectric power station of a specific size and in a particular location. While the specificity of such arrangements contrasts with the flexibility of the principles according to which water resources are managed, one is no easier of enforcement than the other. It is trite that the more specific the rule the easier it is to recognise and enforce. While it is no doubt possible to measure and monitor water use, it is likely to be difficult in practice to be assured that there is compliance with the quantitative and no doubt also the qualitative limitations on extraction and discharge.

There are – at least according to this analysis – a range of procedural and substantive norms for international water resources governance that comprise a matrix of legal and paralegal rules. While some are rules and potentially enforceable rules of international law, others are the supporting context or framework within which they have effect. What is particularly important is the nature of the obligations or duties that emerge from this structure of procedural and substantive norms. It is not only the source of the norm – a declaration of an international conference or a provision in a multilateral agreement – but also the nature and the substance of the norm that influences its capacity to be recognised as a rule and enforced as such. It has been suggested by the International Court of Justice that obligations may be obligations of conduct, of performance or of result. Although there is no consensus about the usefulness of this classification, it is at least a starting point for an analysis of the nature of obligations as elements of the normative structure of international water resources governance.

What are these different types of obligation? An obligation of conduct explains what must be done, what must not be done and the standard according to which it may or must be done. It is essentially a standard according to which people are expected to behave in doing what they do and in undertaking their activities. An obligation of result indicates what is to be achieved as a consequence of doing something or even of doing nothing. This obligation may also relate to the future behaviour of a person. An obligation of performance indicates how a result may or perhaps even must be achieved. In other words the trilogy is:

- what;
- for what purpose;
- how.

If the subject matter of these obligations is human behaviour – which it traditionally is – then it represents a helpful paradigm against which obligations may be analysed. This is true even if human behaviour is institutionalised through corporate or other juristic arrangements. In relation to water resources governance, this classification of obligations is particularly relevant for activities of an operational kind. That is, the actual use of water, the impacts of this use, for what purposes the water is used and how it is used.

So far so good. But the international arrangements for water resources governance are not directly focused upon human behaviour in this sense. These arrangements are much more concerned with what decisions may be made and how decisions may be made about water resources. Unquestionably decision-making processes and activities related to these processes are linked. Indeed decision-making processes set the parameters for these activities. But the processes are different. Does the trilogy of obligations of conduct, of performance and of result make sense in this context? Yes – but the analysis is different. In this respect obligations of conduct are more procedural, in the sense of obligations to inform and to negotiate. An obligation of result is to ensure that the outcome is achieved rather than to achieve the outcome.

Consider the example of sustainable development. A decision-making process may contain an obligation to ensure that the development is sustainable. The implementation of this decision is that the development will be sustainable. In other words a decision to approve a development cannot achieve its intended result. Only the undertaking of the activity so authorised can achieve the intended result. An obligation of performance in relation to decision-making indicates how the decision may or must be made so that it will achieve its intended outcome or result. In earlier paragraphs this has been described as a norm of methodology in decision-making.

The normative structure of international water resources governance is concerned principally with decision-making and only indirectly with behaviour. There are examples of obligations of conduct, of performance and of result. For the most part the duties are obligations of performance in relation to decision-making – the methodology of decision-making. Obligations of result are beginning to emerge more directly through the matrix of legal and paralegal rules in question: for example, reasonable and equitable use. However, the critical issue is whether and to what extent sustainable development is incorporated as an obligation of result in relation to either behaviour or decision-making. There does not appear to be any specific example of such an obligation. We have examined a range of provisions in international instruments of various kinds which refer – often indirectly – to the concept of sustainable development, its principles and the ways in which it may be achieved. There is no consistency or coherence of approach, of structure, of terminology or of language. But the idea of sustainable development as an obligation of result is perhaps beginning to emerge through the interstices of these international water resources governance arrangements. This classification of obligations and ultimately of norms is important not only for the international regime but also for national regimes that are attempting to give effect to it.

NOTE

1. *River Danube Case* (1997, p. 201).

LEGAL INSTRUMENTS

African Convention on the Conservation of Nature and Natural Resources 1968, in Food and Agriculture Organisation of the United Nations (1998), *Sources of International Water Law* (Rome: Food and Agriculture Organisation), p. 47.

Agreement between the Federal Republic of Nigeria and the Republic of Niger concerning the Equitable Sharing in the Development, Conservation and Use of Their Common Water Resources 1990, in Food and Agriculture Organisation of the United Nations (1997), *Treaties Concerning the Non-navigational Uses of International Watercourses Africa* (Rome: Food and Agriculture Organisation), pp. 219–224.

Agreement between the Government of the Republic of Kazakhstan, the Government of the Kyrgyz Republic and the Government of the Republic of Uzbekistan on Cooperation in the Area of Environment and Rational Nature Use 1998, at ocid.nacse.org/qml/research/tfdd/toTFDDdocs/196ENG.htm.

Agreement between the Republic of the Sudan and the United Arab Republic for the Full Utilization of the Nile Waters, in Food and Agriculture Organisation of the United Nations (1997), *Treaties Concerning the Non-navigational Uses of International Watercourses Africa* (Rome: Food and Agriculture Organisation), pp. 236–241.

Agreement between the Republic of Zimbabwe and the Republic of Zambia Concerning the Utilization of the Zambezi River 1987, in Food and Agriculture Organisation of the United Nations (1997), *Treaties Concerning the Non-navigational Uses of International Watercourses Africa* (Rome: Food and Agriculture Organisation), pp. 273–291.

Agreement Concerning the River Niger Commission and the Navigation and Transport in the River Niger 1964, in Food and Agriculture Organisation of the United Nations (1997), *Treaties Concerning the Non-navigational Uses of International Watercourses Africa* (Rome: Food and Agriculture Organisation), pp. 16–18.

Agreement on the Action Plan for the Environmentally Sound Management of the Common Zambezi River System, in Food and Agriculture Organisation of the United Nations (1997), *Treaties Concerning the Non-navigational Uses of International Watercourses Africa* (Rome: Food and Agriculture Organisation), pp. 84–112.

Agreement on the Cooperation for the Sustainable Development of the Mekong River Basin (1995), *International Legal Materials* 34, 864–880.

Asuncion Declaration 1971: Act of Asuncion on the Use of Internatonal Rivers, in Food and Agriculture Organisation of the United Nations (FAO) (1998), *Sources of International Water Law* (Rome: Food and Agriculture Organisation), pp. 51–52.

Convention relating to the Status of the River Gambia 1978, in Food and Agriculture Organisation of the United Nations (1997), *Treaties Concerning the Non-navigational Uses of International Watercourses Africa* (Rome: Food and Agriculture Organisation), pp. 43–46.

Directive 2000/60/EC Establishing a Framework for Community Action in the Field of Water Policy 2000 [2000] OJ L327/1–69.

Draft Rio de Janeiro Convention on the Industrial and Agricultural Use of International Rivers and Lakes 1965, in Food and Agriculture Organisation of the United Nations (FAO) (1998), *Sources of International Water Law* (Rome: Food and Agriculture Organisation), pp. 48–50.

Framework for General Cooperation between the Arab Republic of Egypt and Ethiopia 1993, in Food and Agriculture Organisation of the United Nations (1997), *Treaties Concerning the Non-navigational Uses of International Watercourses Africa* (Rome: Food and Agriculture Organisation), pp. 159–160.

Protocol on Shared Watercourse Systems in the Southern African Development Community (SADC) Region 1995, in Food and Agriculture Organisation of the United Nations (1997), *Treaties*

Concerning the Non-navigational Uses of International Watercourses Africa (Rome: Food and Agriculture Organisation), pp. 146–156.

Revised Protocol on Shared Watercourse Systems in the Southern African Development Community (SADC) Region 2000, (2000) *International Legal Materials* 40, 321–333.

Treaty between His Majesty's Government of Nepal and the Government of India Concerning the Integrated Development of the Mahakali River (1996), *International Legal Materials* 36, 531–546.

Treaty between the Government of the Republic of India and the Government of the People's Republic of Bangladesh on Sharing of the Ganga/Ganges Waters at Farakka 1996, (1997), *International Legal Materials* 36, 519–528.

Treaty on the Development and Utilization of the Water Resources of the Komati River Basin between the Kingdom of Swaziland and the Government of the Republic of South Africa 1992, in Food and Agriculture Organisation of the United Nations (1997), *Treaties Concerning the Non-navigational Uses of International Watercourses Africa* (Rome: Food and Agriculture Organisation), pp. 242–265.

Treaty on the Lesotho Highlands Water Project between the Government of the Kingdom of Lesotho and the Government of the Republic of South Africa 1986, in Food and Agriculture Organisation of the United Nations (1997), *Treaties Concerning the Non-navigational Uses of International Watercourses Africa* (Rome: Food and Agriculture Organisation), pp. 172–212.

CASES

River Danube case: Judgment in the case concerning the Gabcikovo-Nagymaros Project (Hungary and Slovakia (1997), *International Legal Materials* 37, 162–242.

PART IV

The normative structure of arrangements within states

The international arrangements for water resources governance have developed from the recognition, then the protection and subsequently the enforcement of a community of interest in water resources. The international community is concerned about water resources generally – wherever they happen to be located – and more particularly about international water resources including international watercourses. The community of interest is protected first by the imposition of procedural norms and then by the imposition of substantive norms. In many instances these have been linked with each other, and in some cases with the potential of leading to the concept of sustainable development. The principal focus of the normative structure of international water resources governance is strategic, but it incorporates operational rules in the context of some sets of arrangements.

The arrangements for water resources governance within nation states disclose similar characteristics. A community of interest may be recognised in a number of ways:

- by the recognition of water resources as common-pool resources;
- by enabling the state to control directly the use and development of water resources;
- by providing for a regime of common property;
- by delegating the use and development of water resources to local communities.

It has frequently been recognised that the management of water resources in particular sets of circumstances, in particular locations and for particular purposes is best left in the hands of local communities, not, of course, in the hands of individual users and developers. While this is a feature of customary arrangements, it is also a feature of more formal sets of arrangements.

How water resources are managed within nation states depends to a great extent upon the fundamental *Grundnorms* that either underlie the system or are stated by it. As we have already seen, rights in relation to water may be original or derived. Rights may exist simply as a matter of law without further involvement of the state. They may arise through a system of appropriation. They may be granted through a system of allocation. Their sources may be custom, rules of the common law or of the civil law, rules deriving inspiration from religious sources or legislation. In many contexts it is a combination of some or all of these. In this part we turn our attention from the formal structure of rights in relation to water in nation states to the normative structure of these rights. In other words, what the holders of these rights may do with the water; for what purposes the water may be used; for what purposes the water resources may be developed; and what restraints there are upon the exercise of these rights. We shall consider the arrangements within federal jurisdictions before turning our attention to those in unitary jurisdictions.

The management of shared water resources in a federation discloses features similar to the management of water resources shared by members of the international community: in other words transboundary watercourses or aquifers. The way in which these water resources are managed may be determined either in accordance with principles recognised or created by the courts or by agreement between or among the states within the federation. Chapter 10 considers in this context the approaches adopted in the United States of America, Canada and Australia. While these arrangements affect legal relationships between and among states within the federation, the rights and duties of persons and entities other than the states are governed by the law of the state. This applies to the rights and duties of all persons – whether individual or corporate – within a state with a unitary system of government. Chapter 11 considers the normative arrangements in these jurisdictions. First of all there is consideration of the fundamental principles of Roman law, of Islamic law and of English law. These principles – of major historical interest – remain intrinsically relevant as the foundations for later codified and statutory systems in many parts of the world. Chapter 11 consequently considers the codified arrangements in Islamic law and civil law jurisdictions.

10. Normative arrangements in federal jurisdictions

INTRODUCTION

The essential characteristic of a federal system is the division or distribution of sovereignty among a number of entities within the federation. The arrangements between these entities are quasi-international in character. Each such entity – often called a state or a province – may exercise its legislative, executive and judicial powers within its territory as it sees fit, but subject to the constraints in the constitution setting up the federation. Water resources in a federation may in some circumstances be shared resources in much the same way as in an international context. To what extent, therefore, is a community of interest recognised in a federal structure and to what extent is there a capacity within a federal structure to integrate the governance of water resources? For this purpose we shall consider three federal jurisdictions: the United States of America, Canada and Australia.

UNITED STATES OF AMERICA

(a) The Constitutional Context

When the United States of America was established towards the end of the 18th century, the common law of England was in a state of doctrinal uncertainty: were water rights acquired according to the appropriation doctrine or were water rights recognised in accordance with the riparian doctrine. In the event the eastern states within the United States preferred the riparian doctrine and the western states the appropriation doctrine. But there are many rivers located within the north American continent which flow across the boundaries of two or more states. The Supreme Court of the United States has recognised that admission to the union carries with it 'the full powers of local sovereignty'[1] which belong to all states of the union. The legislative capacity – or local sovereignty – of each state includes the power to legislate with the respect to water. Accordingly, 'it is well recognised that a state, under its power to provide for the general welfare, has the power to regulate water flowing within its boundaries. However, a state's power does not extend to regulation of water flowing in other states or water rights arising under the jurisdiction of other states'.[2]

Inter-state water disputes have accordingly been settled either by inter-state compact – contractual arrangements – or by litigation before the Supreme Court of the United States – judicially imposed arrangements. According to Article I(10)(3) of the Constitution of the United States such consensual arrangements require the consent of the Congress of the United States. The Supreme Court of the United States has acknowledged its jurisdiction to settle these disputes. Thus, 'the question of the extent and the limitations of the rights of the two states becomes a matter of justiciable dispute between them, and this court is called upon to settle that dispute in such a way as will recognise the equal rights of both and at the same time establish justice between them'.[3]

(b) Judicially Imposed Arrangements

But how is justice established in the context of the equal rights of the disputants? The dispute in question was between the states of Kansas and Colorado in relation to the flow of water through the Arkansas river. The issue was whether Kansas had a right to the flow of the waters of the river as the waters had always flowed or whether Colorado had a right to appropriate the waters of the river for the irrigation of otherwise arid land. In the event the decision was reached by balancing the benefits and the detriments of interfering with the natural flow of the waters of the river. Thus:

> The diminution of the flow of water in the river by the irrigation of Colorado has worked some detriment to the south western part of Kansas, and yet when we compare the amount of this detriment with the great benefit which has obviously resulted to the counties in Colorado, it would seem that equality of right and equity between the two states forbids any interference with the present withdrawal of water in Colorado for purposes of irrigation.[4]

While the rights of each state were equal, it was equity as between the two states that justified interference with the natural flow of the water.

Some years later the Supreme Court developed the riparian doctrine rather than the appropriation doctrine as the doctrinal basis for resolving inter-state water resources disputes. The state of New York proposed to increase the water supply of the city of New York by diverting large amounts of water from some of the tributaries of the Delaware river and from the watershed of the Delaware river, to the watershed of the Hudson river. This would diminish the flow of water in the Delaware river and hence affect the interests of the states of Pennsylvania and New Jersey. The Supreme Court sought 'to secure an equitable apportionment'[5] of the water resources in question. Thus:

> A river is more than an amenity, it is a treasure. It offers a necessity of life that must be rationed among those who have power over it. New York has the physical power to cut off all the water within its jurisdiction. But clearly the exercise of such a power to the destruction of the interest of lower states could not be tolerated. And on the other hand equally little could New Jersey be permitted to require New York to give up its power altogether in order that the river might come down to it undiminished. Both states have real and substantial interests in the river that must be reconciled as best they may be. The different traditions and practices in different parts of the country may lead to varying results, but the effort always is to secure an equitable apportionment without quibbling over formulas.[6]

Significantly, therefore, an equitable rather than a formulary approach.

On the face of it, the norm of equitable apportionment leaves considerable discretion to the court. In the decision concerning the Arkansas river the appropriation doctrine was moderated by equity, while in the decision concerning the Delaware river it was the riparian doctrine that was moderated by equity. Subsequently, the Supreme Court adopted what has been described as the 'multifactor standard of equitable apportionment'.[7] So equity itself is tempered by a set of methodological norms. This is nothing new. While the terminology varies between the common law and the civil law, the application of a rule is often tempered by a principle of equity, while the exercise of an equitable discretion is often itself tempered by a consideration of relevant matters or – as in this case – factors. Thus, according to the Supreme Court:

> Apportionment calls for the exercise for an informed judgement on a consideration of many factors. Priority of appropriation is the guiding principle. But physical and climatic conditions, a consumptive use of water in several sections of the river, the character and rate of return flows, the extent of established uses, the availability of storage water, the practical effect of wasteful uses on downstream areas, the damage to upstream areas as compared to the benefits to downstream areas if a limitation is imposed on the former – these are all relevant factors.[8]

This approach incorporates three elements: a rule, a principle, and a set of factors to consider in reaching an informed judgement or decision. It is the interrelationship between these that is important. It needs to be borne in mind that the institution applying this approach is the Supreme Court – an adjudicative institution. But in exercising this jurisdiction the Supreme Court is not adjudicating upon the rights, duties and liabilities arising out of activities that have taken place. Rather it is deciding what uses and the impacts of these uses are acceptable for the future. Although decisions for the future are based – at least to some extent – upon what has happened in the past, the process is essentially predictive in function and executive in nature. In this sense the function of the Supreme Court is to seek an outcome of a process in accordance with defined criteria. In this case it is to secure the equitable apportionment of shared water resources. It is, in effect, an imposed solution, and perhaps contrary to the preferences of one or more of the parties. It is, however, a final determination until it is adapted or modified by a subsequent lawful determination.

(c) Consensual Arrangements

Then there are consensual arrangements among the interested states. On the face of it, consensual arrangements contain rights and duties acceptable to all parties. It may perhaps be assumed, therefore, that the arrangements are equitable as between the parties. While equity or at least consensus is the outcome, there is no requirement that equity must be achieved. It has been suggested that 'apportioning the rights of states to use interstate waters by compact offers flexibility that may not be available in judicial apportionment'.[9] On the other hand, 'compacts often lack the flexibility to deal with changing conditions and they cannot be easily amended'.[10] In other words, flexibility characterises the creation of rights and duties by agreement but inflexibility is their hallmark at the later stages of enforcement and change. In any event the purpose and structure of these consensual arrangements vary enormously.[11] Many deal with water allocation; some with planning and flood control; several with pollution control; and a few are multi-purpose. Structurally the arrangements range through statements of management principles, strategic apportionment of water resources, provision of a formula for the apportionment of water resources, a very detailed set of rules for the apportionment, diversion and use of water resources to a comprehensive institutional structure for their management on an ongoing basis. Let us consider some examples.

The Great Lakes Charter of 1985 sets out a set of principles for the management of the Great Lakes water resources. The parties are two provinces in Canada and eight states in the

United States. It is some extent international as well as interstate in effect. Its substantive purposes include the conservation of the levels and flows of the Great Lakes and their tributary and connecting waters, and the protection and conservation of the environmental balance of the Great Lakes Basin ecosystem. The purposes are achieved in accordance with the five principles for the management of the water resources of the Great Lakes and the arrangements for their implementation.

Three of the principles are procedural and two substantive. The procedural principles involve cooperation in general, cooperative programmes and practices, and arrangements for prior notice and consultation in relation to proposed major or increased diversion or consumptive uses of the relevant waters. The substantive principles are the integrity of the Great Lakes Basin and the protection of the water resources of the Great Lakes. Essentially the natural resources and ecosystem of the Great Lakes Basin are to be considered – and by implication managed – as a unified whole, and this necessarily involves a recognition in principle III of the 'shared responsibility' of all the relevant provinces and states 'to conserve and protect the water resources of the Great Lakes Basin for the use, benefit, and enjoyment of all their citizens'. The charter is, however, a statement of principles rather than a set of legal rights and obligations, for in the section on reservation of rights it reserves specifically to each of the provinces and states the right 'to represent and protect the rights and interests of their respective jurisdictions and citizens in the shared water and other natural resources of the Great Lakes region'.

The Colorado River Compact of 1928 divides the Colorado river basin into an upper basin and a lower basin so far as it is located within the United States. The uses of the waters of the Colorado river system are apportioned strategically between the upper basin and the lower basin. Thus, as stated in Article III (a):

> There is hereby apportionment from the Colorado River system in perpetuity to the upper basin and to the lower basin respectively the exclusive beneficial consumptive use of 7.5 million acre-feet of water per annum which shall include all water necessary for the supply of any rights which may now exist.

In addition the upper basin states are under a duty not to cause the flow of the river at the boundary point between the two basins to be depleted below an aggregate of 75 million acre-feet for any period of ten consecutive years. The upper basin states are under a further duty not to withhold water and the lower basin states are under a duty not to require the delivery of water which cannot reasonably be applied to domestic and agricultural uses.

Significantly Article IV states the priorities of use of the waters. Thus the use of the waters for the purposes of navigation are subservient to the use of the waters for domestic, agricultural and power purposes. But the use of the waters for the generation of electrical power is subservient to the use and consumption of the waters for agricultural and domestic purposes and must not interfere with or prevent use of the waters for these dominant purposes. It is however provided in Article IV(c) that these rules of priority do not restrict the power of a state to regulate and control the appropriation, use and distribution of water within the boundaries of the state.

A different approach is taken by the Alabama–Coosa–Tallapoosa River Basin Compact of 1997. It applies to the waters within the drainage basin of these three rivers in Alabama and Georgia. Its purpose according to Article VII is to develop an allocation formula for equitably apportioning the surface waters of the river basin among the states while protecting the water quality, ecology and biodiversity of the basin in accordance with the relevant legislation of the United States. The allocation formula requires the unanimous approval of the states and the concurrence of the United States. The allocation formula is the methodology in whatever form by which an equitable apportionment of surface waters within the basin among the states is determined. It may be represented by a table, a chart, a mathematical calculation or any other method of expressing the apportionment of the waters. Each state – in whatever capacity it functions – is responsible for using its best efforts to achieve compliance with the allocation formula and each state takes such actions as may be necessary to achieve compliance with the allocation formula.

While a provision of the compact is enforceable, it is specifically provided in Article XIV that proceedings may not be brought in relation to the adoption of the allocation formula, enforcement of it, or compliance with it. In addition, the provisions of the compact do not interfere with the right of a state to regulate the use and control of water within the boundaries of the state but provide that such action by the state is not inconsistent with the allocation formula. The allocation formula thus represents the methodology for apportionment of the water resources of the basin, but the methodology as such is not enforceable as a matter of law.

The South Platte River Compact of 1926 prescribes in detail the rights and obligations of Colorado and Nebraska in relation to the diversion and use of the waters of the South Platte river and its tributaries. These include:

- the maintenance of a stream-gauging station on the river;
- the exchange and publication of records and data;
- the location of the point of diversion of the waters of a tributary;
- the right of Nebraska to the full and unmolested use and benefit of all waters flowing in this tributary above the point of diversion;
- the right of Colorado to the exclusive use and benefit of all waters flowing at or below the point of diversion;
- the right of Nebraska to use the channels associated with the river and its tributaries;
- specification of the times during which Colorado has the right to the full and uninterrupted use and benefit of the waters of the river flowing within the boundaries of Colorado;
- specification of the times during which Colorado must not permit diversions from the lower section of the river;
- an obligation upon Nebraska to use the flow of the river at the interstate station to supply the needs of present perfected rights to the use of water from the river within the state before permitting diversions from the river;
- a power to permit minor or compensating irregularities and fluctuations in the flow at the interstate station because of variable climatic conditions;
- the right of Colorado to maintain, operate and extend specified canals located within Nebraska;
- the right of Nebraska to construct, maintain and operate a canal and to divert water from the river within Colorado for use in Nebraska;
- a duty that the diversion of water by this canal must not diminish the flow necessary to pass the interstate station to satisfy superior claims of users of water from the river in Nebraska;

- the right of Nebraska to regulate diversions of water by this canal for the purpose of protecting other diversions from the river within Nebraska.

The detail of these provisions is compounded by their complexity. There is no specific provision for the settlement of disputes between the states. Presumably, therefore, protection of rights, enforcement of obligations and adjudication of disputes are functions of the courts of either of the states or of the United States.

The Delaware River Basin Compact of 1961 sets up the Delaware River Basin Commission to plan for the use and development of the water resources of the basin, to provide for their management or to manage them directly. Its functions are therefore strategic, policy formation, planning, operational and enforcement. It contemplates a comprehensive multi-purpose plan which will bring the greatest benefits and produce the most efficient service in the public welfare. Accordingly one institution replaces a complex combination of a multitude of state and federal agencies that otherwise would be involved in the management of the water resources of the basin. For this purpose water resources are defined broadly to include water and related natural resources, in, on, under or above the ground, including related uses of land which is subject to beneficial use, ownership or control. What is contemplated, therefore, is an integrated set of arrangements, and it is integration in the sense of legal integration as well as hydrological, political, bureaucratic and institutional integration.

Significantly the commission is stated in Article 2.1 to be a 'body politic and corporate'. Its functions include:

- to develop and effectuate plans, policies and projects relating to the water resources of the basin;
- to adopt and promote uniform and coordinated policies for water conservation, control, use and management in the basin.

Its powers are directed at water supply, pollution control, flood protection, water shed management, recreation and hydroelectric power generation. Significantly, its powers include the allocation of water resources. Thus according to Article 3.3:

> The commission shall have the power from time to time as need appears, in accordance with the doctrine of equitable apportionment, to allocate the waters of the basin to and among the states signatory to this compact and to and among their respective political subdivisions, and to impose conditions, obligations and release requirements related thereto.

While this power is subject to certain limitations, it is significant that it must be exercised in accordance with the doctrine of equitable apportionment – the doctrine created by the Supreme Court of the United States. However, it is specifically provided by Article 14.19 that 'nothing contained in this compact shall be construed as affecting or intending to affect or in any way to interfere with the law of the respective signatory parties relating to riparian rights'. It would seem, therefore, that while rights of access to interstate water resources and to use them are granted in accordance with the doctrine of equitable apportionment, such rights in relation to water resources within a state are governed by the riparian doctrine.

(d) Conclusion

The *Grundnorm* underlying the governance of shared inter-state water resources in the United States is the doctrine of equitable apportionment. It comprises, as we have seen, a rule, a principle and a methodology of decision-making. Although it emerged in the exercise by the Supreme Court of its adjudicative function, it is in many respects executive or administrative in character. The court, in other words, exercises a proactive function in creating rights and duties for the future in accordance with the doctrine. While the Supreme Court appears to be reluctant to perform this function, the doctrine created by it appears to have influenced the way in which the states in conjunction with the United States have negotiated consensual arrangements for the management of inter-state water resources.

These arrangements seek to achieve different purposes, adopt a range of approaches and disclose a variety of structures. Sometimes the normative arrangements are very specific. Sometimes they are strategic. It is not uncommon for these functions to be delegated to a specific institution. Occasionally the arrangements are in the form of a set of overarching principles that guide strategic, planning and operational decision-making. Despite these differences, there are certain common features. One is a recognition of a common interest in shared water resources. Another is a need for a unified or integrated approach to their governance. A third is a sense of common responsibility for the way they are governed. While the approaches in the United States – at least those discussed in the foregoing paragraphs – do not acknowledge sustainable development or its principles, aspects of it manifest themselves from time to time in the arrangements in the United States.

CANADA

The governance of water resources in Canada is in the first instance a matter for the provinces. This arises either from the proprietary rights of the crown in relation to land – and by implication water – or the legislative powers of the provincial legislatures.[12] These powers are restricted to the water resources within the boundaries of the province. It has been suggested that the use of interprovincial water resources may be addressed by adjudication by the Supreme Court of Canada, by the exercise of legislative power by the Parliament of Canada or by intergovernmental agreements.[13] So far in Canada – unlike the United States – the common law has not directly addressed this issue.[14] The Canada Water Act of 1970 – whether for political or constitutional reasons – focuses upon agreements between Canada and the provinces in relation to water management.[15] In any event the extent to which the Parliament of Canada may constitutionally legislate more directly for the management of water resources remains a matter of some controversy.[16] Let us therefore concentrate upon intergovernmental agreements.

Intergovernmental agreements are an aspect of cooperative federalism. And there have been many such agreements relating to water.[17] Many of these provide for federal funding for projects undertaken within the provinces.[18] Then there are agreements that create institutions with a range of different functions. Thus:

> These agreements do not contemplate joint management of the resource in the sense of an integrated bureaucracy engaging in direct control of the resource on an ongoing basis. However, there do exist

> in Canada a number of inter-jurisdictional boards designed to accomplish this function to varying degrees. Some of these boards are very skeletal in nature, and are, in fact, no more than data-gathering centres or instruments for consultation between different jurisdictions. Others, while also ultimately responsible to their political masters, may have much more sophisticated research capabilities and exercise a greater degree of influence in the actual management of the water resource.[19]

One such institution is the Prairie Provinces Water Board.[20] It has been suggested as a model for integrated watershed management.[21] How inter-provincial water resources should be managed led in 1969 to the Master Apportionment Agreement designed for this purpose. According to one interpretation, the general rule in the agreement, with some specified exceptions, 'permits the upstream province to make a net depletion of one-half of the natural flow of water, and requires it to pass on the remaining one-half to the downstream province'.[22] According to another interpretation:

> The parties agreed that each upstream province should allow one-half of the natural flow, measured on an annual basis, to pass to its downstream neighbour. Minimum allowable flows, and the time when this share should be released during the year, are not specified beyond the requirement that the water be delivered 'on an equitable basis'.[23]

The principle of equality thus seems to have been applied in the first instance but extended by the application of the principle of equity where the principle of equality did not apply. In any event the legal status of these agreements is ambiguous. For the most part they are seen to be political arrangements and therefore legally unenforceable. Thus, 'the most important limitation from a legal perspective is that, unless given constitutional status, intergovernmental agreements are not enforceable against a government determined not to be bound'.[24] Their success thus depends upon political will and good faith.

AUSTRALIA

(a) Introduction

In Australia the power to enact laws about water resources lies with the states. However, if the way in which water resources are managed falls within one or more of the areas of legislative competence of the Commonwealth – the federal entity – then it is open to the parliament of the Commonwealth to enact legislation that indirectly yet constitutionally impacts upon how water resources are used and developed.[25] It is not uncommon in Australia for water resources to flow across or underlie the territory of two or more states or territories. In a number of instances the relevant states have entered into agreements for managing these water resources, and these are traditionally supported by legislation of the relevant states.[26] With one major exception, the Commonwealth has historically refrained from involvement in the management of Australia's water resources. The principal exception has been the Murray–Darling Basin which straddles most of that part of south eastern Australia which lies west of the Great Dividing Range. The involvement of the Commonwealth has been motivated for the most part by two considerations: increasing overallocation and overuse of the water resources of the Murray–Darling Basin and the acknowledgement of the need for their sustainable development.

(b) Intergovernmental Agreements

Water resources intergovernmental agreements in Australia are either strategic, operational or both. The Lake Eyre Basin Intergovernmental Agreement of 2000 applies to a large area in central and north eastern Australia located within two states and comprising river systems, catchments, flood plains, overflow channels, lakes, wetlands and sub-artesian waters. The purpose of the agreement is to develop, adopt and implement policies and strategies about the water resources and their associated natural resources in the Lake Eyre basin. The agreement recognises the multiple uses, activities and values in the area, including pastoralism, tourism, oil and gas extraction and mining, and areas of social, cultural and heritage values. The principles guiding the implementation of the agreement include the conservation and promotion of the social, environmental, economic and cultural values of the area: for example, the importance of naturally variable flow regimes, the maintenance of water quality and the maintenance of water requirements for ecological processes, biodiversity and ecologically significant areas. A precautionary approach is specifically incorporated within the guidelines.

The parties to the agreement are the Commonwealth, Queensland and South Australia. The agreement is supported by legislation in each of the jurisdictions. The involvement of the Commonwealth is essentially strategic and procedural except in relation to the management of natural resources on land which it owns or occupies. The involvement of the states is similarly strategic and procedural in relation to each other. Thus the implementation of strategies and policies and the control and regulation of operational activities in relation to water resources are governed by the law of each of the states.

The water resources of the Murray–Darling Basin have always been of interest and concern to the Commonwealth.[27] The original agreement for the management of these waters was concluded in 1914 between the Commonwealth and the three relevant states. It was intended to bring about the economical use of the waters of the river Murray and its tributaries for irrigation and navigation. Each of the four jurisdictions enacted legislation in support of the agreement. In 1987 a new agreement was concluded and later ratified by the Murray–Darling Basin Act of 1993 of the Commonwealth. This was designed to promote and coordinate effective planning and management for the equitable, efficient and sustainable use of the water, land and environmental resources of the Murray–Darling Basin. To some extent the principles of sustainable development and integration underpinned this approach.

The scheme incorporated in the agreement for the management of the water and other resources of the basin was managed by the executive institution set up by the scheme in accordance with the set of prescriptive rules contained in the agreement. Part X dealt with the distribution of waters. It stated, for example, the monthly entitlement of the lower riparian state to receive specific quantities of water from the river Murray. The agreement went on to state the entitlements of the upper riparian states at various points in the waters of the upper river Murray and of the river Darling. The uppermost riparian state – Queensland – became a party to the agreement in 1997 but the agreement has remained silent about the specific rights and entitlements of this state. In any event, the agreement was intrinsically a political arrangement between the parties. It contained, for example, no provision for settling disputes between the Murray–Darling Basin Commission – the executive body responsible for its implementation – and individual members of the community benefitted or disadvantaged by these arrangements. In particular, it seems that the rights and duties of the users of the water resources of the basin remained to be governed not by the agreement but by the law of each of the relevant states.

(c) The Constitutional Foundations of the Water Act of 2007

However in 2007 much of this changed when the parliament of the Commonwealth enacted the Water Act. The Commonwealth decided that it should become more directly involved in the management of the Murray–Darling Basin. But how could this be done constitutionally? One way was for the matter to be referred to the parliament of the Commonwealth by the parliaments of the states. But essentially this required the agreement of the states to transfer their legislative capacity in this respect to the Commonwealth. An alternative was for the Commonwealth to exercise its power to enact legislation about the aspects of water resources and their management that fell within the areas of legislative capacity identified for the Commonwealth in the Constitution of the Commonwealth. These included:

- trade and commerce among the states;
- trading or financial corporations;
- external affairs;
- financial assistance.

While the provision of funds is and always has been a critical function of the Commonwealth, it is the explicit reliance by the Commonwealth upon a number of the areas of its legislative capacity which has supported in a unique way the involvement of the Commonwealth directly in the management of Australia's water resources. Indeed, one of the objects of the 2007 Act is to enable the Commonwealth, in conjunction with the basin states, to manage the basin water resources in the national interest.

Australia is a party to a number of international agreements. These include relevantly to the management of water resources:

- the 1971 Convention on Wetlands of International Importance;
- the 1992 Convention on Biological Diversity;
- the 1992 Convention on Climate Change;
- the 1994 Convention to Combat Desertification;
- a number of bilateral agreements with Asian states for the protection of migratory birds and birds in danger of extinction.

Giving effect to obligations undertaken by Australia under agreements such as these enlivens the power of the parliament of the Commonwealth to make laws with respect to external affairs. While none of these agreements deals directly with water resources, they implicitly impact upon how water resources are used and developed. It is clear that a number of the norms for water resources governance in the Water Act of 2007 are likely to derive their constitutional validity from this source.

The power to make laws about interstate trading is similarly relevant. One of the principles of sustainable development is the use of economic instruments including markets. Markets in water rights have been evolving gradually in Australia, but the capacity to engage in interstate trading has been limited largely for institutional reasons. It is the policy of all governments in Australia – Commonwealth, state and territory – to facilitate trading in water rights, and particularly interstate trading. For example the mandatory content of the Murray–Darling Basin Plan under the 2007 Act includes rules for the trading or transfer of tradeable water rights in relation to basin water resources. These rules must contribute to achieving the basin water market and trading objectives and principles set out in the legislation. In this respect the legislation relies for its constitutional validity upon the interstate trade and commerce power in the constitution.

Consistently with this commercial perspective, many, but clearly not all, activities undertaken in relation to the use and development of water resources are undertaken by corporations. In this respect the regulation by the Commonwealth of the use and development of water resources by corporations finds its constitutional source in the power conferred upon the Commonwealth by the constitution to make laws with the respect to trading corporations. This specifically includes the power to regulate the activities – including the trading activities – of a trading corporation.

If the provisions of the Water Act of 2007 cannot be supported constitutionally in one or another of these ways, then they will be valid by reason of the referral by the states to the Commonwealth of the power to legislate on the matters covered by the Water Act 2007. But ultimately this is a matter of agreement. The application of the rules relating to external affairs, interstate trade and commerce, and trading corporations is a matter of law quite independently of agreement.

(d) The Detail of the Water Act 2007

Let us assume that the Water Act of the Commonwealth is constitutionally valid. It enables the Commonwealth to become involved in the management of the water resources of the Murray–Darling basin. The Act however does not give the Commonwealth total, complete and comprehensive control of the water resources of the basin. It creates a planning regime according to which rights in relation to water granted in accordance with the laws of the state or territory are to be exercised. In this respect – apart from the sets of rules about procedure and enforcement – there are three critical aspects of the legislation:

- the function of the Murray-Darling Basin Plan;
- the function of water resource plans for particular water resource plan areas;
- trading in water rights.

While the objects of the Act do not include in terms the achievement of ecologically sustainable development, this is their effect. For example, one is to promote the use and management of the basin water resources in a way that optimises economic, social and environmental outcomes. Another is to ensure the return to environmentally sustainable levels of extraction for water resources which are over-allocated or over-used. A third is to protect, restore and provide for the ecological values in the ecosystem services of the Murray–Darling basin. The Murray–Darling Basin Plan performs a critical function in achieving these objects. Its purpose is to provide for the integrated management of the basin water resources in a way that promotes the objects of the Act. One of the ways of doing this is the establishment and enforcement of environmentally sustainable limits on the quantities of surface water and groundwater that may be taken from the basin water resources. This includes taking by interception activities.

The preparation of the basin plan is mandatory. It is prepared by the Murray–Darling Basin Authority and adopted by the Minister. It is a legislative instrument. There is an obligation on all Commonwealth agencies, including the authority, to perform their functions consistently with the basin plan and in a manner that gives effect to it. Perhaps even more importantly, a duty is imposed upon those responsible for the operational management of the water resources of the basin. This includes an agency of a state, an operating authority, an infrastructure operator or the holder of a water access right. The duty is two-fold:

- not to do an act in relation to the water resources of the basin if the act is inconsistent with the basin plan;
- not to fail to do an act in relation to the water resources of the basin if the failure to do that act is inconsistent with the basin plan.

The test of lawfulness is thus consistency.

Assuming, then, that these elements of the Act rely upon the external affairs power for their validity, there is effectively a close relationship between the relevant international obligations to which the Commonwealth is giving effect, the provisions of the Act itself and the detail of the basin plan. There must be consistency and coherence between each of these elements. What emerges, therefore, is a hierarchy of norms:

- the obligations undertaken by Australia in accordance with the international agreements to which it is a party;
- the statement of objects in the Act;
- the rules formulated in the basin plan for the basin-wide management of the water resources of the basin.

The relationship between these three elements is critical not only for constitutional but also for normative reasons.

The Murray–Darling Basin Plan is implemented in particular areas within the basin through the formulation or accreditation of water resource plans. These plans are mandatory. A plan prepared by a state may be accredited or a plan is adopted by the Minister under the Act. The water resource plan must be consistent with the basin plan. Rather interestingly, in determining whether the plan is consistent with the basin plan, regard must be had to the legislative framework within which the plan operates. This seems to suggest that the water resource plan must be consistent not only with the basin plan but also with the normative structure of the Act as a whole. A water resource plan – just like the basin plan – is a legislative instrument. Its legal effect is exactly the same as the legal effect of the basin plan. In other words, there is a general duty not to do an act and not to fail to do an act inconsistently with the plan. Although the Act does not say so in terms, the test of consistency presumably applies in the context of the Act as a whole and not just in the context of the water resource plan.

What is the relationship between these plans and the entitlements granted under the laws of a state or territory? A relevant law of the Commonwealth prevails over a law of a state to the extent of any inconsistency. The provisions of the basin plan and of a water resources plan therefore prevail over those of a state to the extent of any inconsistency. Clearly these plans affect directly water entitlements granted under the laws of a state. In effect, therefore, while the water entitlements are granted by the state, the specific rights and duties of the holders of water entitlements in the Murray–Darling Basin are regulated by the Murray–Darling Basin

Plan and the water resource plan for the particular area where these rights are exercisable and the duties to be performed. This is so only to the extent that these plans have effect. The holders of water entitlements in the Murray–Darling Basin are accordingly subject to the laws of the state or territory where the activities in question are taking place. Although the rules of the Commonwealth override the rules of the state or territory to the extent of their inconsistency, the holders of these water entitlements cannot assume that compliance with the Commonwealth rules is sufficient.

Finally, there are water market rules. These rules contribute to the achievement of the water market and trading objectives and principles set out in the Act for the Murray–Darling Basin. The objectives include:

- to facilitate the operation of efficient water markets and the opportunities for trading within and between basin states where water resources are physically shared or hydrologic connections and water supply considerations permit water trading;
- to minimise transaction costs;
- to enable the appropriate mix of water products to develop;
- to recognise and protect the needs of the environment;
- to provide appropriate protection of third party interests.

The complex set of principles is designed to facilitate trading as much as possible and to limit restrictions upon trading activities. The need for limitations is, however, recognised largely as a consequence of the nature of the hydrological cycle generally and in the basin in particular.

CONCLUSION

The United States, Canada and Australia have approached the management of water resources somewhat differently. In each of these jurisdictions how water resources are managed is in principle a matter for the state or province rather than for a federal institution. But in each case the federal institutions have shown an interest in and indeed a concern about how water resources are managed, not only within the state or province but also across the boundaries of the state or province. The Supreme Court of the United States has proactively acknowledged or perhaps even created a set of rules according to which interstate water resources should be managed – essentially the *Grundnorm* of equitable apportionment. Equitable apportionment – it may be assumed – is achieved through consensual arrangements among the states. These agreements require the consent of the congress of the United States. So the management of shared water resources in the United States in a subtle combination of the judicial, executive and legislative functions.

Historically, Canada and Australia have managed water resources in accordance with the doctrine of cooperative federalism. In each of these jurisdictions a judicial function has yet to emerge. In each case, however, federal institutions exercising executive functions have influenced either directly or indirectly how shared water resources are managed on an ongoing basis. The model in each jurisdiction has historically been largely unenforceable agreements of a political nature between the federal and the provincial or state institutions – normally the relevant governments or their agencies. The normative structure of these arrangements has been for the most part strategic and high level operational policy. In particular these intergovernmental consensual arrangements have not impacted directly upon the rights and duties of the users and developers of water resources within the province or the state. The agencies of the provinces and the states have no doubt been influenced by these strategies and policies, but as a matter of law these strategies and policies cannot be given direct effect. In other words paralegal rather than legal rules.

However the Water Act of 2007 of the Commonwealth of Australia has created a novel normative structure for the management of shared water resources in the Murray–Darling Basin. An enforceable normative structure comprising the Murray–Darling Basin Plan, the water resource plan for the area in question and the relevant water market rules has emerged to complement the arrangements for water resources governance in the states. But the plan and these rules depend for their validity upon their compliance with the rules contained in the Water Act. The Act in turn depends for its validity upon its compliance with the relevant provision of the Commonwealth Constitution. The scope of the external affairs power in the Constitution – one of these relevant provisions – depends upon the extent to which Australia has given effect to its international obligations. This matrix of legal and paralegal rules – international and national – is designed ultimately to achieve the ecologically sustainable development of the water resources of the Murray–Darling Basin through a national – or federal – regime of integrated management. The Water Act of the Commonwealth of Australia creates a range of mechanisms to ensure compliance. The nature of the judicial response to this normative structure will no doubt emerge in the fullness of time.

The normative structure of the arrangements for the management of interstate water resources by states or provinces within a federation is in many respects similar to the normative structure which governs states that are members of the international community: in other words a set of fundamental principles according to which decisions will be made for the future or which underlie the arrangements consensually accepted by the states in question. What has emerged in these two contexts has been the principle of equity increasingly modified by reference – mostly obliquely – to aspects of sustainable development. These arrangements at the level of strategy and policy impact upon operational activities, as indeed they are intended to do. But they do not create individual rights and duties in particular sets of circumstances. Such individual rights and duties are part of or are derived from the law of the nation state or the law of the state or province within a federation. Hence it is critical how rights are exercised and duties discharged in the performance of both regulatory and operational functions within a nation state, whether structured as a federation or a single unit. Let us now turn from international and federal arrangements to consider normative arrangements within nation states and within states or provinces in a federal jurisdiction.

NOTES

1. *Kansas case* (1906, p. 95).
2. Corbridge and Rice (1999, p. 527).
3. *Kansas case* (1906, p. 98).
4. *Kansas case* (1906, pp. 113 and 114).
5. *New Jersey case* (1931, p. 343).
6. *New Jersey case* (1931, pp. 342 and 343).
7. Tarlock, Corbridge and Getches (2002, p. 929).
8. *Nebraska case* (1945, p. 618).

9. Tarlock, Corbridge and Getches (2002, p. 948).
10. Tarlock, Corbridge and Getches (2002, p. 949).
11. Sherk (2000, pp. 986–998).
12. Saunders (1988, pp. 8 and 9).
13. Kennett (1991, p. 49).
14. Kennett (1991, chapter 4).
15. Saunders (1988, pp. 27–30).
16. Saunders (1988, p. 29) and Kennett (1991, chapter 5).
17. Kennett (1991, p. 50).
18. Saunders (1988, pp. 48–52) and Kennett (1991, pp. 54 and 55).
19. Saunders (1988, pp. 52 and 53).
20. Saunders (1988, pp. 53–56) and Kennett (1991, pp. 56 and 57).
21. Saunders (1988, p. 53) and Kennett (1991, p. 56).
22. Saunders (1988, p. 55).
23. Kennett (1991, p. 56).
24. Kennett (1991, p. 101).
25. Fisher (2000, pp. 37–42).
26. Fisher (2000, pp. 55–59).
27. Fisher (2000, p. 59–61).

GENERAL REFERENCES

Corbridge, James N. and Teresa A. Rice (eds) (1999), *Vranesh's Colorado Water Law* (Niwot, Colo: University Press of Colorado).

Fisher, D.E. (2000), *Water Law* (Sydney: LBC Information Services).

Kennett, Steven A. (1991), *Managing Interjurisdictional Waters in Canada: a Constitutional Analysis* (Calgary, Alta: Canadian Institute of Resources Law).

Saunders, J. Owen (1988), *Interjurisdictional Issues in Canadian Water Management* (Calgary, Alta : Canadian Institute of Resources Law).

Sherk, George William (2000), *Dividing the Waters: the Resolution of Interstate Water Conflicts in the United States* (London and Boston: Kluwer Law International).

Tarlock, A. Dan, James N. Corbridge Jr and David H. Getches (2002), *Water Resource Management: A Casebook in Law and Public Policy* (5th edn, New York: Foundation Press).

LEGAL INSTRUMENTS

Alabama–Coosa–Tallapoosa River Basin Compact 1997, in G.W. Sherk (2000), *Dividing the Waters: the Resolution of Interstate Water Conflicts in the United States* (London and Boston: Kluwer Law International), pp. 487–500.

Colorado River Compact 1928, in G.W. Sherk (2000), *Dividing the Waters: the Resolution of Interstate Water Conflicts in the United States*, London and Boston: Kluwer Law International, pp. 621-625.

Delaware River Basin Compact 1961, in G.W. Sherk (2000), *Dividing the Waters: the Resolution of Interstate Water Conflicts in the United States* (London and Boston: Kluwer Law International), pp. 653–687.

Great Lakes Charter 1985: Principles for the Management of Great Lakes Water Resources, in G.W. Sherk (2000), *Dividing the Waters: the Resolution of Interstate Water Conflicts in the United States* (London and Boston: Kluwer Law International), pp. 697–704.

Lake Eyre Basin Intergovernmental Agreement Act 2001 of the Commonwealth of Australia.

Murray–Darling Basin Act 1993 of the Commonwealth of Australia.

South Platte River Compact 1926, in G.W. Sherk (2000), *Dividing the Waters: the Resolution of Interstate Water Conflicts in the United States* (London and Boston: Kluwer Law International), pp. 857–864.

CASES

Kansas case: *Kansas v Colorado*, 206 US, 46–118 (1906).
Nebraska case: *Nebraska v Wyoming*, 325 US, 589–672 (1945).
New Jersey case: *New Jersey v New York*, 283 US, 336–348 (1931).

11. Normative arrangements in unitary jurisdictions

INTRODUCTION

The effectiveness of how water resources are governed depends ultimately upon what happens to the water at a particular location and at a particular time. This is a reflection of both the formal and the normative structure for managing water resources. What happens may be guided by a sophisticated set of principles, by a range of centrally dictated administrative edicts, by a collection of well recognised rules or simply by the values consciously absorbed by the members of the community. As we have seen, these values may be religious, philosophical, cultural or utilitarian. In any event, most – probably all – of the earliest arrangements for managing water resources have assumed a formal structure based upon control inherent in an omnicompetent institution. For example, the pharaoh in Egypt, the king in Babylonia or the code of Manu imposing duties on the king in the Indus Valley.

The normative structure of water resources management in these instances for the most part comprised duties imposed upon administrative officials, users of the water or both. And the sanction was often criminal. For example, in Egypt 'damage to and neglect of dikes and canals was an offence punishable by death'.[1] The Hammurabi Code provided for the maintenance of waterworks and canals, and a failure to comply attracted a punishment. Thus 'where a man has opened his trench for irrigation and has been slack and has let the waters carry away the soil on his neighbor's land, he shall pay corn corresponding to his neighbor's loss'.[2] The code of Manu contained an obligation to punish 'him who breaks the dam of a reservoir and causes the loss of water by drowning him in the water or have his head cut off'.[3] While these rules reflected the values of the society in question, significantly they were enforced by some form of punitive sanction. This was a circumstantial response to individual conduct. The norm in these cases was clearly an enforceable obligation of conduct. Let us turn now to consider in more detail four examples of traditional normative structures.

CHINESE LAW

The arrangements in China for the management of water resources have traditionally been based upon the close connection 'between the human order and the natural cosmic order'[4] and their provision and regulation for the public benefit[5] in accordance with a sophisticated administrative system that linked the Emperor with the local community of users. And in the absence of any specific administrative solution, 'ancient usages and precedents'[6] provided a solution. It evolved as a regime of duties and not of rights. Thus 'everyone seems to have had duties, with the rights reduced to the positive manifestation of other people's duties'.[7] While rights may have existed formally as correlatives to duties, the arrangements in China thus focused upon duties without substantive reference to rights.

The fundamental concept upon which this set of duties was based was the principle of 'water equalization'.[8] For this purpose equalisation seems to have amounted to equitable distribution. It was, however, the duties reflecting the principle and not the principle that were enforceable. These duties were imposed upon regulators, distributors and users alike.[9] During the Han administration, for example, one of the duties was the corvee – the duty to provide one's labour for one month of the year at the disposal of the administration.[10] This included the construction, repair and maintenance of hydraulic works – a positive duty. By way of an example of a negative duty that emerged during the T'ang administration 'those living on the upper stream must not block up the sources and monopolise its richness'.[11] Minor activities in relation to water were permitted without administrative approval, but always subject to monitoring and inspection. Duties were similarly imposed upon officials who became personally responsible for compliance. For example, 'Tin Chong, Grand Secretary, was adjudicated and committed suicide in 29 B.C. because the Yellow River had broken through the dikes'.[12]

The principle of water equalisation or equitable distribution became the basis upon which there emerged an order of priorities among water users. These were quite specific:

> Except for domestic uses, transportation had first priority over the others, such as for agriculture (irrigation), and industry (water mills). Irrigation, in turn, had priority over water mills, which had to be closed during larger irrigation periods.[13]

The priorities were thus:

- domestic use;
- transportation;
- agriculture;
- industry.

The sanctions for ensuring compliance with these responsibilities, either strategic or operational, were penal rather than civil. This included the imposition of penalties upon the officials within the administration. In many respects, as we shall see, many of these principles have survived into the modern era. Thus 'the basic principles of water equalization, punishments for offences relating to waters and responsibility of local officials, persist up to modern times'.[14] In other words, a comprehensive set of enforceable obligations of conduct based upon the governing principle.

ROMAN LAW

Rights in relation to water, as we have seen, were the focus of Roman law. These were rights of ownership of various kinds and rights of *usufructus*. These were protected against unlawful interference by the grant of *interdicta* granted by the relevant judicial officer: for example, the *interdictum de aqua castellaria* 'which protected existing rights to divert water from public aqueducts or reservoirs'.[15] The existence of this remedy to protect a right implied a general duty not to interfere with this right. Out of an implied general duty such as this emerged the normative structure of Roman law. It crystallised into a set of rules that restricted activities in relation to water with harmful consequences. Roman law in general provided a remedy for a loss or damage brought about unlawfully – *damnum iniuria datum*. In some instances, however, Roman law recognised the existence of a liability for harm or damage caused by activities in relation to water. The most significant remedy was the *actio aquae fluviae arcendae.*[16] According to the digest, 'this action is appropriate whenever water is likely to cause damage to a field as a result of a man-made construction, that is, whenever someone causes water to flow elsewhere than its normal and natural course'.[17]

Most importantly, however, what rights and duties comprised this action? According to one analysis, the action 'enabled upper riparians to prevent lower riparians from impeding the natural flow of water through their fields and, conversely, allowed for lower riparians to prevent the upper riparians from increasing or limiting that natural flow'.[18] This discloses a related or mutually beneficial set of norms:

- a duty on upper riparians not to increase or limit the natural flow of water;
- a duty on lower riparians not to impede the natural flow of water.

A failure to comply with such a duty attracted a liability to compensate for the damage, remove the structure or both. Such an action, however, could be brought only against the person – the neighbour – 'who had made the alterations or altered the natural flow of water'.[19] It appears, however, that the action was not universally available because 'the damage threatened or caused by the alteration of the natural flow of water for the purposes of agricultural cultivation did not fall under this action'.[20] This immunity – apparently protective of irrigators – was consistent with the strategic approach of Roman law during its earlier period of development. Although it became a correlative arrangement of rights and duties, its underlying value may be described as a freedom – but specifically not an open-ended entitlement – to use water but later restricted by a potential liability for any harmful effects of its use.[21]

The way in which Roman law evolved is particularly important for two reasons: first, the existence of protectable rights in relation to water coupled with enforceable duties not to interfere with the flow of water so as to cause damage to the interests of other beneficiaries of the water; secondly, the recognition of the interests of riparians with a particular interest in water resources. Hundreds of years later this aspect of Roman law influenced the development of the common law in England and in the United States of America. In addition, as we have seen, the relatively open access to water resources was increasingly circumscribed by the inclusion of an increasing range of water resources within the classification of *res publicae*. This led – no doubt as a matter of public policy – to more control of water resources by the state in response to increasing demand for a scarce resource.[22] This emerged hundreds of years later as a feature of arrangements for managing water resources in other communities.

ISLAMIC LAW

(a) The Context

The arrangements for managing water resources in Islamic countries have traditionally been very much influenced by local practice and custom and by the values stated in the Qur'an. Such an approach has thus been inspired by the need for practical solutions to managing water resources in arid parts of the world and by the religiously underpinned moral values of Islam. 'Water lay at the heart of the legal system',[23] and from this there emerged a system of rights and duties in relation to water. The original meaning of the expression *shari'a* was the law of water rather than the law in general. Thus:

> The connection between *shari'a* as a generic term for Islamic law, and *shari'a* as the path as well as the law of water, is not a coincidence, and the centrality of water in Islam is obvious in the economic as well as ritualistic sense. What is more important however, is that the jurists – the exponents and expounders of the *shari'a* – did not fail to develop, in answer to this centrality, a highly sophisticated system of rules.[24]

This system of rules comprised sets of rights and duties in a context of customary practices and moral values: in other words a set of legal and paralegal rules. It was in many respects the relationship between these elements of the system that gave to it its flexibility on the one hand and its capacity to be recognised and implemented on the other hand. It was this centrality of water that ensured that it was ultimately the interest of the community that prevailed. Traditionally Islamic law has recognised rights of individuals in land and in water. But these rights may be seen not as values in themselves but more as a means to an end. And the end was ensuring the availability of water to satisfy not only human needs but also those of nature and the natural environment in the wider sense. Let us now look in more detail at the normative structure of Islamic law. The values that have underpinned it for hundreds of years continue to do so in the 21st century.

(b) The Detail

The *Grundnorm* of the system was that water is the gift of God.[25] It is a sin to misuse or abuse water.[26] There was thus an implied duty not to misuse or abuse it. But this was in the first instance a moral duty rather than a legal duty. It was transformed into a legal duty through the 'highly sophisticated system of rules' which has evolved. The idea of responsibility derived to some extent from the notion of balance contained in the Qur'an. Thus 'man must preserve the environmental balance that God has established'[27] and consequently 'man should act with moderation towards nature'.[28] Perhaps even more positively, 'Islam views protection of the environment as a sacred duty of humankind based upon the exalted position it occupies as a higher authority responsible for the treatment of lesser creatures'.[29] When this is translated into a legal paradigm, it 'requires a person to exercise a higher standard of care where his action adversely affects the interest of the public or may cause injury to it'.[30] This duty of care manifested itself in a number of ways from the very general to the more specific. For example:

- a general duty for humankind to care for the earth;
- a duty not to damage the environment;
- a duty not to overuse water;
- a duty to keep water free from pollution.[31]

While Islamic law did not use the legal language of the 21st century to describe these responsibilities, values such as these were the justification for more detailed arrangements.

Islamic law recognised the need of humankind for water as well as the needs of animals and of nature generally for water. Consequently there were rights in relation to water. First of all, there was the right of thirst. 'This right is juridically the right to take water to quench one's thirst or to water one's animals.'[32] This has been transposed into a duty of general application imposed on no person in particular but on everyone in general. 'Water, whether privately owned or not, must be made available for man and animal to drink.'[33] While there were probably no sanctions in support of these rights,[34] it seems that in practice there was a recognition of a community of interest in water resources, and hence no absolute right and probably no absolute duty: in different parlance, a freedom rather than a protectable right. For example, the approach of the Shafi'i school of jurists has been explained in these words:

- the owner of the water has foremost right to drink it and to use it for watering;
- there is a religious and moral obligation to offer water in the desert to a thirsty human being;
- if there is pasture and the animals are unable to graze without water, the person who has surplus water must offer water for their consumption;
- if the person who requires the water for himself or his animals is able to acquire it near at hand, then the owner of the water is under no obligation to offer that person his surplus water.[35]

So much for basic human and animal needs. What about the use of water for other purposes such as irrigation? Islamic law recognised both public and private land and water in appropriate circumstances. If water was to be used for irrigation, it could irrigate the land by the natural forces of flood, by extraction from a watercourse or a well, or by taking it from a reservoir in which it had been stored. In some cases the activity of irrigation involved earlier or simultaneous activities involving the land or the water. These could affect the interests of other land and water users. Islamic law protected these interests to some extent. For example, 'it was forbidden to dig a new well so as not to damage the quality or lower the quantity of the water in the existing one'.[36] Significantly this constituted a restriction upon the impact of the activity and not directly upon the activity itself. There were also restrictions upon the activity of irrigation itself:

- a duty for low lying areas to be irrigated before the irrigation of high lying areas;
- a duty not to retain a quantity of water on the irrigated land that 'reached over the ankles'.[37]

Water was in many cases a shared resource. Indeed water for irrigation could come from a shared water resource. It was recognised that 'a community of farmers may irrigate their land, either individually, each from his own private well or spring, or by sharing *mubah* water'.[38] Wells or springs could alternatively be common property, and in this sense the resource was shared. How was shared water apportioned among those entitled to use it? The general approach was clear:

> Co-owners of wells, channels and springs are free to share water in the manner best suited to their needs. However, where disputes arise between co-owners regarding the apportionment of water, the law requires that each should receive a quantity of water in proportion to his share. Sequence of watering is established according to the position of the land and the volume of the water.[39]

If there was plenty of water, the position was clear. The lands that were nearest to the water were afforded priority.[40] Otherwise three ways of sharing the water were available:

- according to the specified number of hours during the day during which water could be taken;
- in accordance with the number of shares held by the co-owner identified by holes drilled in a wooden plank set in the bed of the channel;
- by drawing lots.[41]

Each of these methods depended upon the existence of an identifiable number of shares held by each co-owner. Significantly these shares were 'determined by the amount of expense and labour which each individual has contributed'.[42] If this was not possible, the share was probably determined by the size of the land held by each individual. Finally, in the absence of any such norms, it was a matter of custom.

(c) Conclusion

Islamic law has traditionally disclosed a clear pattern of values, customs, principles and rules. Each is linked rationally to the others. This set of arrangements has been succinctly described in this way:

- water is a gift of God;
- it belongs in principle to the community;
- there is a primary right for water to be available to be drunk by individuals, cattle and household animals;
- the construction of infrastructural works, especially for irrigation, may create a qualified right of ownership by adding value to water by labour;
- a range of customary water-sharing principles;
- an acknowledgement of a right of prior appropriation combined with a required distribution of surplus;
- the attachment of liability for the withholding or misuse of water including pollution or degradation of clean water.[43]

Structurally, this comprises an integrated set of values, principles, customs, community rights, individual rights, duties and liabilities. This 'highly sophisticated system of rules' had been developed by groups of influential Islamic jurists in response to the 'centrality of water in Islam'.[44] Some of these rules are paralegal, while others are legal rules. In any event, as we shall see, the responsibility for ensuring compliance for the most part lay with the local community and its institutions.

ENGLISH LAW

(a) The Relationship between Water and Land

Rights in relation to water under the common law of England have had a somewhat chequered history. Even if the doctrinal foundations of these rights can be identified, the normative structure according to which these rights are exercised has been described as 'far more inscrutable'[45] than the doctrines themselves. For the purposes of the common law – as in other systems – land and water have always been closely interrelated. In the first instance the common law protected rights in relation to land, and then in relation to water. Much more problematical were the duties associated with the exercise of these rights. In other words, what were the nature and the extent of any duties not to interfere with the exercise of these rights in relation to water? And upon whom were these duties imposed?

The existence of duties in relation to water evolved from those in relation to land. Those in turn evolved from the protection of rights in relation to land. Direct interference with a right in relation to land was a form of disseisin. Thus the early assize of novel disseisin 'could be used to protect from dispossession not only freehold titles but also a range of appurtenant freehold rights, such as water use, where there had been some invasion of the plaintiff's tenement or other acts sufficient to invoke the jurisdiction'.[46] Out of this grew the assize of nuisance which became available where there was no direct interference with the land and later the water associated with it.[47] For example, in an early case in 1284 it was commented:

> Where a bank has been lawfully raised or a fish pond or watercourse diverted or something like that, a certain remedy is available in the king's court of the nature of a novel disseisin, that is the writ 'why he levied or knocked down a wall, bank or pond to the nuisance' [of the person seeking the remedy].[48]

The assize of nuisance thus focused upon the use of land rather than the possession of land, and indirectly the use of water and the relationship between the two.

The assizes of novel disseisin and of nuisance were essentially concerned with protecting interests in land. The writ of trespass, on the other hand, was concerned with the rights and duties of persons as such independently of any interest in land. The rights of a landholder as a person were protectable by the writ of trespass, but the remedy was personal and not real. Originally trespass involved actual physical violation of rights in relation to land or water and subsequently – as it evolved through the action of trespass on the case – indirect rather than direct violation of rights in relation to land or water. All of these remedies, however, were concerned more with an interference with a right than with a standard of behaviour associated with a breach of a duty.

An interesting example of this distinction and of the ambiguities associated with its doctrinal foundations is the *Prior of Southwark's case*, decided in 1498.[49] Let us translate the circumstances into modern terminology. The owner of land adjacent to a watercourse running into the river Thames in London had leased the premises to a group of tenants. The tenants used the waters of the watercourse 'to dye their clothes, to water their beasts, to bake and to brew'. The water in the watercourse was polluted by the activities undertaken by the occupier of land located upstream of and adjoining the watercourse. The occupier of this land 'had made a lime-pit for calf-skins and sheep-skins so close to the stream that a corruption of the said lime-pit had corrupted the said stream'. In consequence of this the occupiers of the land adjacent to the watercourse downstream gave up their occupation of the land. The upstream land was owned and occupied separately from the downstream land. The activities on the upstream land did not impact directly upon the downstream land. However the capacity to use the water of the watercourse adjacent to the land was detrimentally affected by the impact of the activities undertaken on the upstream land adjacent to the watercourse.

The riparian doctrine was not part of the common law of England in 1498. So a breach of a duty or an infringement of a right in relation to water had to be based upon remedies arising out of nuisance or trespass. The assize of nuisance was inappropriate because it was not the right to the use and enjoyment of the land that was infringed, but rather the right to the use and enjoyment of the water indirectly linked to the use and enjoyment of the land. The action was brought by way of trespass on the case. It was successful. So a right in relation to water was protected from interference attributable to the activities of another person which constituted no more than the medium through which the interference came about. The nature of this right was unclear: perhaps an easement; perhaps a right in common property; perhaps an emerging riparian right; or perhaps an extension of a right in land protected by nuisance. In any event, the relationship between land and water had been loosened to the extent that the focus of the action was the use of water and the impact of activities on the use of water through a doctrine of liability based upon trespass on the case. What emerged in practical terms was a normative structure which recognised a duty imposed upon a person not to engage in activities that could cause harm to the use of water by other persons. It was the relationship between the activity and the harm that was important rather than the relationship between land and water or between persons.

As we have seen, the structural foundations of rights in relation to water under the common law remained unclear until the riparian doctrine matured in the second half of the 19th century. However the normative structure of water resources governance under the common law has remained much the same over that period of some 500 years. The theory of water rights under the common law enunciated by Blackstone some 100 years before the emergence of the mature riparian doctrine focussed upon the doctrine of appropriation as the foundation of water rights. Despite this change – or clarification depending upon one's perspective – it did not affect the doctrinal foundations supporting liability for damage or injury to water. The essence of Blackstone's water doctrine has been synthesised in five propositions. In this context two are critical:

- the first appropriator of water wins title under a natural principle defending occupation;
- actions for wrongs, such as nuisance and trespass, are used to defend such water rights; and this raises the issue of identifying actionable damage.[50]

While one of the purposes of these actions is 'to defend' or to protect rights in relation to water, another purpose – because of the correlation between rights and duties – is to state the circumstances in which there is a common law duty not to injure or damage the capacity of water to be used for lawful purposes.

(b) The Normative Structure of the Riparian Doctrine

The emergence of the riparian doctrine provided an opportunity for the common law to develop further its normative structure. The riparian doctrine effectively restricts the right to take and use water to the group of riparians who have a community of interest in the water. The right to use water is limited by the existence of an identical right in all other riparians. The normative structure according to which the right is exercisable depends as much upon the use to which the water will be put as upon the impact of that use. The test is the reasonable enjoyment of the water by all riparians. Thus the right to the benefit and advantage of the water flowing past the land of the riparian has been described as 'a right only to the flow of the water, and the enjoyment of it, subject to the similar rights of all the proprietors of the banks on each side to the reasonable enjoyment of the same gift of Providence'.[51]

What is reasonable is a matter of circumstances. Thus:

> On the one hand, it could not be permitted that the owner of a tract of many thousand acres of porous soil, abutting on one part of the stream, could be permitted to irrigate them continually by canals and drains, and so cause a serious diminution of the quantity of water, though there was no other loss to the natural stream than that arising from the necessary absorption and evaporation of the water employed for that purpose; on the other hand, one's common sense would be shocked by supposing that a riparian owner could not dip a watering-pot into the stream, in order to water his garden, or allow his family or his cattle to drink it.[52]

The existence of a riparian right thus correlates with a duty not to exercise it so that it impacts unreasonably upon the quantity and quality of the flow of the water in the watercourse as it passes the lands of other riparians.

The existence of such a duty is traditionally described as a condition subject to which the right of the riparian is exercisable. This formulation, however, neither denies nor disguises the existence of the duty. One side of the equation is the purpose for which the water may be used. First, a riparian has a right to the 'ordinary use of the water' – namely 'the reasonable use of the water for his domestic purposes and for his cattle'[53] – without any limitation. Secondly, a riparian has a right to the extraordinary use of the water 'provided that he does not thereby interfere with the rights of other proprietors, either above or below him'.[54] This, in other words, is the other side of the equation: namely the duty owed to all other riparians that is part of the normative structure of the riparian doctrine.

This duty focuses upon the impact of the use of the water when the use of the water is otherwise lawful. It is the duty not to impact upon the lawful use of the water by another riparian that constitutes the condition subject to which the right to use the water for extraordinary purposes is exercisable. By way of example:

> Subject to this condition, he may dam up the stream for the purpose of a mill, or divert the water for the purpose of irrigation. But, he has no right to interrupt the regular flow of the stream, if he thereby interferes with the lawful use of the water by other proprietors, and inflicts upon them a sensible injury.[55]

This translates, therefore, into a duty upon each riparian proprietor in the exercise of the riparian right to use water for an extraordinary purpose not to inflict upon all other riparians a sensible injury. The relationship between the right and the duty has been alternatively described in the language of a remedy – the traditional approach of the common law:

> Any invasion of this right causing actual damage, or calculated to found a claim which may ripen into an adverse right, entitles the party injured to the intervention of the court.[56]

A riparian right in association with the correlative duty inheres only in the community of riparians. The common law recognises also a proprietary right in relation to water. Thus a right in relation to water may arise as a consequence of the ownership of the surface of the land in relation either to water flowing across the surface of the land or to water flowing indiscriminately below the surface of the land.[57] The exercise of each of these rights is limited by the existence of a duty imposed by the common law and owed to anyone who may be affected unreasonably by the exercise of the right. In other words, in traditional terminology, the duty exists because of the availability of an action for nuisance or for trespass on the case.

Whether the right is riparian or proprietary, the effect is the same. Riparian rights incidental to the ownership of the banks of a watercourse and proprietary rights incidental to the ownership of the surface of land are protected as a consequence of a liability arising either from nuisance or negligence – the earlier action on the case – or from unlawful interference with the rights of other riparians. Either way there is an implicit duty not to cause damage or injury to the interests of other persons in the water in question – whether these interests are riparian or proprietary. In England, as elsewhere, duties such as these have been affected by legislation. Has it, for example, deprived injured parties of their common law rights? Does it matter whether these rights are riparian or proprietary? Let us turn to the way legislation has intervened.

(c) Statutory Arrangements in England

Until late in the 20th century the intervention by the legislature was ad hoc. For example, during the middle of the 19th century the parliament of the United Kingdom enacted legislation which applied only to the town of Birmingham in the centre of England. Effectively it required the construction and operation of a new drainage and sewerage system for the town. The operation of the new system had an instantaneous effect. 'Immediately the defendants' sewers were opened the fish were killed in the river, and the cattle would no longer drink of it.'[58] The river into which the sewage flowed was the tributary of a river in respect of which the plaintiff, as the owner of the banks of the river, had riparian rights. The plaintiff sought an injunction to stop the discharge of the sewage. Although the plaintiff was a riparian, the claim was based upon the commission of a nuisance by the town of Birmingham. This was perhaps because a provision in the legislation effectively placed a duty upon the town not to create a nuisance by – in this case – constructing and operating the sewerage system. In effect, the legislation did not render the town immune from liability under the common law for a nuisance.

The court approached the issue from the perspective not so much of a commission of a nuisance but rather the infringement of a right. The rights of the land owner in relation to the water were expressed with absolute clarity:

> He has a clear right to enjoy the river, which, before the defendants' operations flowed unpolluted – or, at all events, so far unpolluted that fish could live in the stream and cattle would drink of it – through his grounds, for three miles and upwards, in exactly the same condition in which it flowed formerly, so that cattle may drink of it without injury, and fish, which were accustomed to frequent it, may not be driven elsewhere. He is entitled to the full use and benefit of the water of the river just as he enjoyed them before the passing of the Municipal Act, unless there be in that Act something which says he is not to enjoy them any longer.[59]

There was nothing in the legislation to that effect. The rights were clear. These rights had been infringed. It was a function of the law to protect these rights. An injunction was granted. And this was so even if it disadvantaged – even temporarily – the some 250,000 inhabitants of the town of Birmingham. Significantly – perhaps because of the references to nuisance – the issue was analysed in the context of an injury to the rights of a land owner protected by the common law. Whether these rights were riparian or proprietary is unclear. In effect, it made no difference.

Some 20 years later a dispute arose between a land owner and the operator of a private rather than a public enterprise.[60] In this case there was no reference to nuisance. The issue turned on the nature and extent of the riparian rights of the owner of a cotton mill located on the bank of a river in relation to the riparian rights of the owners of a colliery located on the bank of the river upstream of the cotton mill. It appears that the owners of the colliery pumped water into the river and this contained sulphuric acid and sulphate of iron. This, it was argued, damaged the machinery in the cotton mill and this damage was repaired at great expense. Much of the judgment in this case was concerned with the appropriate remedy. The alternatives were an injunction, damages or both.

In this case there appears to have been damage to the machinery in the cotton mill for which damages by way of compensation would be payable. If there had been no damage to the machinery, the right of the downstream riparian might nevertheless have been infringed. An injunction would prevent the continuation of any such infringement. The court did not present the issue in this way. A distinction was drawn by the court in this case between injury and damage. Or, in accordance with the terminology in the foregoing paragraphs, between the infringement of a right and the breach of a duty. According to the court, the difference was this:

> I may observe in passing that the case of a stream affords a very clear illustration of the difference between injury and damage; for the pollution of a clear stream is to a riparian proprietor below both injury and damage, whilst the pollution of a stream already made foul and useless by other pollutions is an injury without damage, which would, however, at once become both injury and damage on the cessation of the other pollutions.[61]

To change the terminology, the discharge of polluting material into an unpolluted watercourse by an upper riparian is an unlawful interference with the riparian rights of the lower riparian and a breach of the duty not to harm the interests of the lower riparian by causing damage to them. On the other hand, the discharge of polluting material into an already polluted watercourse by an upper riparian is an unlawful interference with the riparian rights of the lower riparian and nothing more. This is because the interests of the lower riparian have already been harmed by the damage caused to them. In this event there has been an infringement of a right but no breach of a duty. There has been no breach of the duty not to pollute the watercourse because the watercourse has already been polluted.

Whether the distinction drawn by the court is entirely rational or not, a perpetual injunction was granted in these terms:

> To restrain the defendants from discharging water from their mines and colliery into the stream so as to cause an injury to the plaintiff's mill, engine, boilers, and works, or so as to cause the stream to flow to the plaintiffs' mill and premises in a state less pure than that in which it flowed thither previously to the commencement of the defendants' pumping.[62]

The first limb of the injunction seems to be directed towards compliance with a duty not to damage the cotton mill, while the second limb of the injunction seems to have been directed towards prevention of the infringement of the riparian rights of the owner of the mill. Although the two may well be related, an infringement of a right is doctrinally different from a breach of a duty. The evolution of water rights under the common law has pointed to the practical as well as the doctrinal significance of this distinction. The distinction becomes even more important to the extent that the legislature intervenes to remove, modify or expand the rights and duties that exist under the common law.[63]

(d) Conclusion

The management of water resources in the ancient civilisations – Egypt and China for instance – centred upon administration supported by penal sanctions. Roman, Islamic and English systems placed greater emphasis upon the rights and responsibilities of individual members of the community. Despite their differences, the normative arrangements in Roman law, Islamic law and English law for the use of water resources disclose a relatively consistent set of rights and duties. The formal structure of Roman law and Islamic law focused in the first instance upon the rights of the public and the correlative duties owed by individuals to the community before adapting in the case of Roman law to more extensive reliance on administration. Conversely, English law tended to focus upon the rights of individuals and the corresponding duties owed to identifiable members of the community rather than to the community itself.

Each of these three systems has not only evolved over time but also influenced the development of doctrine about how water resources are governed in other jurisdictions. Roman law, for example, was the inspiration for the emergence in civil law jurisdictions of the notion of the public domain. The centrality of water in Islamic culture has continued to be the fundamental principle underpinning the legal arrangements of Islamic communities around the world. While the doctrinal foundations of English law in relation to water have been superseded in many respects by legislation in England, a number of the fundamental principles of the common law remain relevant in relation to the impact of the use of water rather than the right to use water itself.

CODIFIED NORMATIVE ARRANGEMENTS IN ISLAMIC LAW

(a) Introduction

Roman law itself, Islamic law and thereafter the civil law have developed from their original forms into codified arrangements. These have included codes not only regimes of civil law but also of public law. Not so in relation to English law. These codified arrangements present a much more unified approach to water resources governance than the relatively fragmented arrangements of the common law of England. The principles of Roman law and of English law have nevertheless influenced the arrangements for water resources governance in a significant number of jurisdictions in Asia, Africa and Latin America as a result of their colonisation by European states over a period of several hundred years. The codification of Islamic law has been just as influential in a number of jurisdictions. Let us begin by considering the normative arrangements in codified Islamic law.

(b) Codified Islamic Law

The 'essential traits of classical water law', it has been noted, 'were consolidated in the

Ottoman Majalla (the civil code) and in the Ottoman land laws of 1858'.[64] The centrality of water in Islamic culture has ensured that the code 'gives particular attention to water'.[65] The code, in the first instance, created a formal structure according to which water resources were managed. While water could be privately owned in certain circumstances, the code treated it essentially as a community resource. Although a member of the community had access to water as a common resource, the code imposed reciprocal restrictions – or duties – upon the exercise of this community right. The general principle was that personal use was limited where it conflicted with community rights. For example, the use of a community water resource was subject to two conditions:

- its use must not cause damage to another person;
- its use must not prevent another person from using it.[66]

More specifically in relation to irrigation, there was a duty not to undertake any irrigation work 'which causes floods, exhausts the water supply of a river or lowers the water level so as to prevent the passage of boats'.[67]

The code by article 1326 dealt specifically with the maintenance of watercourses – community resources – by distributing liability among the community of users. Thus:

> The responsibility for the maintenance of watercourses begins upstream, and all co-owners must contribute to initial expenses. Provided upstream riparians have met their maintenance obligations, they are no longer bound to contribute to the maintenance of the downstream sections. The principle of this provision is that everyone must contribute to maintenance expenses proportionately to the benefit he derives therefrom.[68]

The code, therefore, was based upon the fundamental principles of customary and Shari'ah law, but modified slightly to reflect contemporary circumstances – particularly those of the 19th century.

In Morocco it is the concept of the public domain that underlies their water resources governance. This has been formalised as the principle of the public ownership of water – *domanialité publique*.[69] Consequently water is 'inalienable and its usage is controlled and regulated by the state'.[70] Access to water requires a permit or a concession. But a permit is not necessary to draw water from a well provided the amount does not exceed 200 cubic metres per day.[71] These arrangements exist side by side with the recognition of traditional water rights. These exist 'in numerous regions of Morocco'.[72] Accordingly:

> Water rights may be recognised as belonging to individuals or ethnic groups based on water sources, *seguias* (rivulets), *oueds* (rivers), *khettaras* (canals) or other types of water flow. Rights so recognized remain to ensure that traditional law applicable to water rights is protected and remains in force.[73]

In traditional areas such as these, the water is managed in accordance with customary rules. The collective arrangements that are part of these rules include:

- the priority of the upstream user over downstream users;
- the significance of the sequence in which rights were established;
- participation in the development of the water supply.[74]

The regime of the public domain enables the state acting through its various agencies to undertake large scale operations involving the management of water resources. 'It produces external equipment such as dams, irrigation and drainage canals, as well as internal equipment for clearing areas for cultivation, levelling, internal irrigation and drainage networks.'[75] The rights and duties of irrigators benefited by such schemes are determined by contracts entered into between the state and farmers. In practice it appears that 'the farmer's choices are limited and the weight of the state heavy'.[76] In addition these farmers are required to pay tax 'in return for the use conditions of water distribution and utilization'.[77] The arrangements in Morocco disclose on the one hand direct control and use of water resources – particularly in the context of large scale operations – and on the other hand a diversity of arrangements at the local level in accordance with traditional customary practices: a collocation of individual rights and duties and of public rights and duties.

The *Grundnorm* of water resources governance in Syria and in Lebanon is similarly the public domain. It has been stated judicially that 'the public domain concerns all the elements which, by their nature, are destined for the use of all citizens or for public needs'.[78] Consistently with this – but differently based – much of the Ottoman code – the Majalla – remains in effect in Lebanon.[79] In addition local practices and customs continue to play a significant part in how water is managed and used. In Syria these arrangements have been complemented by a liabilities code, and in Lebanon by an irrigation code.

The irrigation code in Lebanon provides that the irrigation system is public property.[80] This includes all the buildings, installations, aqueducts and other infrastructure which are part of the system. The code places a duty upon the public administration to repair the system regularly. On the other hand the drainage system is private property and the users are responsible for the costs associated with it.[81] In addition the owners of lower plots in the irrigation system must accept water drained from upper plots. The owners of upper plots are under a duty not to damage lower plots, and if lower plots are damaged the owners are entitled to compensation. If the system is blocked in any way, the person who created the blockage must pay compensation for any damage suffered as a result.[82] This comprises again the juxtaposition of private interests and the public interest in codified form.

The liabilities code in Syria creates a number offences for which penalties are imposed. For example, in the absence of special authorisation, it is unlawful:

- to exploit underground water or artesian wells (except on private property to a depth of less than 150 metres);
- to engage in excavations closer than three metres to water canal borders or drainage systems;
- to take material from water resources;
- to prevent the free flowing of public water;
- to take any permanent or temporary action affecting the quantity and flowing capacity of water.[83]

In addition there are a number of offences which comprise generically a duty not to pollute water resources. Some of these activities may be inconsistent with the principles of Shari'ah law or the Ottoman code. They are nevertheless incorporated specifically in this code. In this sense they support the public domain regime by modifying the principles and rules otherwise applicable.

(c) The Code in the Sultanate of Lahj

The Sultanate of Lahj is located in south west Arabia. Its water resources had for many years been managed in accordance with customary law and Shari'ah law. While it was under British protection an administrative regime was created. Ultimately in 1950 a law was enacted to ensure that 'justice be observed in irrigation and other agricultural affairs'.[84] The law established an agricultural council and an agricultural court. One of the functions of the council was this:

> Protection of the *aqnã'* (the right proportions of water established by custom for the irrigation of individual parcels of land), and the *radãyi'* (the sequence of allotting irrigation water to channels and parcels of land established by custom), and the allotting to each channel, barrage, sub-channel and marginal channel, the amount of water to which it is entitled according to the established system (ie the custom).[85]

In other words the function of the council was to protect customary rights in relation to the use of water for irrigation.

The function of the agricultural court was to hear 'those cases which the *shari'ah* court considers itself not competent to decide according to '*urf* (custom)'. The court was expected to function on a temporary basis:

> Until the issue of a law resolving that the *shari'ah* courts shall take account of '*urf* in respect of agricultural questions; it being desired to preserve the unity of the courts and to recognise Islamic laws as the sole law of the country.[86]

In other words it was the function of the council to protect customary rights and the function of the court to adjudicate upon customary rights.

The law created penalties in respect of a range of criminal activities. If these are interpreted as duties breach of which constitutes a criminal offence, these are some examples:

- a duty not intentionally to draw off water for a second time;
- a duty not to break down a small barrage, a large barrage or a bank of a channel in such a way as to deprive an entitled person of a share of water;
- a duty imposed upon the supervisor of a channel or barrage to maintain strict honesty in the division of water and the protection of the people's entitlements and a duty to ensure that such an entitlement is not lost as a result of neglect or carelessness in the performance of this duty;
- a duty not to obstruct the flow of spring water by the construction of a medium size barrage or the erection of a small barrage so that a piece of land not entitled to this water is irrigated.[87]

These duties reflect the substance but not the structure of the arrangements for the management of water resources according to customary law and Shari'ah law.

This law created a judicial body in Lahj for the first time. The law was 'an attempt at rationalising an already existing system'.[88] Indeed customary courts were dealing with legal disputes about water on a regular basis. The evidence before such courts has been almost entirely oral evidence presented by witnesses. Frequently the outcome was a compromise indicated by the court. 'But this need not surprise us, for the essence of customary law, at least in south-west Arabia, is reconciliation and the acceptance of compromise.'[89] Let us consider the procedure in one case[90] which raised a claim of priority to obtain a share of water established by custom.

It was alleged that the defendant had blocked a *fuqr* (a cut in the bund of a plot of land) on his land. As a result the plaintiff's plot of land did not receive its customary share of water when the flood waters came down. The water flowed on to the plot of the defendant. In the court the defendant denied the existence of a customary right which gave priority to the plaintiff's land. The defendant admitted blocking the *fuqr* but claimed that he had done so in exercise of his own right. The defendant produced no evidence in support of his claim to this right. The plaintiff produced evidence in the form of two witnesses who supported his claim. The court confirmed that the plaintiff's land had a prior right established by custom to obtain a share of the water. The court ordered the defendant to pay a sum of money equivalent to the rent of the land as compensation. It stipulated that the amount should be estimated according to whether the land was dry or watered.

Interestingly, although the right in question was a customary right, the issues were analysed in much the same way as they would be analysed in accordance with Roman law, civil law, the common law or the codified arrangements presently under discussion. In other words, the existence of a right, the means for protecting the right, any correlative duty, any infringement of the duty, and any consequential infringement of the right. In essence, therefore, a relatively uniform structural arrangement of norms.

(d) The Constitution and Islamic Law in Iraq

In Iraq the constitution 'enshrines Islamic principles as a main source of law that no law can contradict – this is certainly in keeping with basic principles of Islamic law'. These principles of Islamic law require 'governments to protect the environment and conserve natural resources'.[91] In addition Islamic law regards water resources as community resources. The application of these rules may indicate that Islamic law has the capacity to protect the Mesopotamian marshlands that are located at the confluence of the rivers Tigris and Euphrates. Major infrastructural works associated with these marshlands have caused substantial damage to their ecosystems and 'the loss of the marshes has caused the upheaval of the Marsh Arabs, the human community that traditionally inhabited the area'.[92] The application of Islamic law would, it has been suggested, assist in their conservation. The waters of the marshlands would be protected as a community resource of the Marsh Arabs. In this sense the marshlands, together with the rivers Tigris and Euphrates, would represent common-pool or shared resources capable of being protected under Islamic law. In particular two norms would be relevant:

- the implied duty not to overuse any resource;
- the implied duty not to use water so that it takes away or interferes with the right of other members of the community to do so.[93]

While this is speculation, it is a significant example of the potential interrelationship between the norms of a traditional system such as Islamic law and the norms incorporated in a contemporary constitution. In this case it is the two sets of norms acting together that arguably provide a legal solution to a water resources governance problem.

CODIFIED NORMATIVE ARRANGEMENTS IN THE CIVIL LAW

(a) Introduction

The formal structure of Roman law recognised essentially three categories of water: water as *res publica*, as *res communis* and as *res privata*. Towards the end of the Roman empire increasingly waters were classified as *res publicae*. Accordingly waters classified in this way were managed according to a regime of public law rather than of private or civil law. The civil law nevertheless had an important part to play in setting the normative standards that governed the relationship between individuals within the community. It comes as no surprise, therefore, that several hundred years later the doctrine of the public domain formed the basis of the codified systems for water resources governance that emerged in Europe.

(b) Codified Public Domain Regimes

According to Article 538 of the French Civil Code of 1804, 'rivers or streams which will carry floats . . . which are not susceptible of private proprietorship are considered as dependences on public domain'.[94] Public waters were thus subject to public law. The owners of land, nevertheless, had rights in relation to water that flowed through or across their land or to water located on or below the surface of the land. These rights were governed by the Civil Code. For example, rainwater was required to be collected by the owner of the land whereon it fell to enable it to be controlled through rights of ownership.[95] The owner of land on which there was a spring had a right to use the water, but it was limited 'by the respect due to the rights of others and the general welfare'.[96] However, once the water had become 'a waterway of public and running water', the owner of the land was under a duty not to interfere with the flow of the water 'to the detriment of lower lying estates'.[97] In other words, at that point the water had become *res publica* or *res communis*. Running water – *aqua profluens* – was *res communis*. The use of such waters was 'common to everybody'. But how such water was used was governed by 'police regulations'.[98] Subject to this condition, these waters could be used for domestic, industrial or agricultural – including irrigation – purposes.

Water that was *res communis* was available to be used by members of the group with the common interest. The rights of riparian owners were limited rather than absolute. This is entirely consistent with the doctrinal foundations of the notion of common property. Thus:

> The law does not go so far as to give riparian owners the right of disposing of the water of the river. None of them is permitted to absorb it entirely or even to take from it a quantity so large that its withdrawal would be detrimental to the rights of the other riparian owners. Consequently, whenever a riparian owner draws a great quantity of water from the river, as he would do for irrigating his property, he must return the water to its natural stream after having used it. He cannot alter the water, or put deleterious substances in it that would make it unfit for other uses. In other words, riparian owners have solely the right, as article 644 puts it, to use the water during its passage.[99]

In other words, the riparian owner is under a duty not to interfere with the flow of the water in terms of either quantity or quality. In addition the riparian owner must not build constructions in or over the water if these constructions interfere with the flow of the water or cause damage to other riparians.

The arrangements in Italy are similar. The doctrine of the public domain is perhaps even stronger. According to the Italian Civil Code of 1865 – as in the French code – 'riparian land owners were entitled by law to use non-navigable streams for irrigation and industry provided waters so used were returned to their original course before leaving riparian land'.[100] However the Italian Civil Code went further than the French Civil Code by declaring both navigable and non-navigable streams to be public waters. Accordingly this was 'fundamental in implicitly establishing the principle that all waters fit to meet the public interest constituted public waters and that a concession was to be obtained for their use'.[101] The concept of private waters had essentially disappeared. The rights of riparian land owners had in effect been limited to a right to use freely 'non-public waters flowing in natural channels provided these cannot be used for public interest purposes'.[102] Italy has specialised water tribunals and courts. It is one of their functions to settle disputes between riparian and non-riparian land owners. The basis upon which the courts settle these disputes is equity.[103] In Italy, therefore, rights in relation to public waters are protected by public law in accordance with the doctrine of the public domain, and rights in relation to waters other than public waters are protected by the remedies available under the civil law: for example, 'the various possessory actions and suits for damages'.[104]

The doctrine of the public domain is just as dominant in Spain. All water resources are publicly owned and controlled by the state. The only exceptions are:

- groundwater under private land;
- surface waters other than rivers and streams originating on the land of the land owner.[105]

Riparian rights are generally not recognised by Spanish law. A right to use water arises, therefore, as a result of an administrative grant or in the very limited circumstances recognised by the civil law. Accordingly much depends upon how the public domain regime is administered. In Spain the law recognises two classes of water use: common uses and special uses. Common uses are given priority over special uses.[106] Common uses are the use of water for domestic, household, fishing, small navigation and floating purposes. There are six special uses of water and the law specifies an order of priorities. The uses are namely:

- city water supply;
- water supply for railroads;
- irrigation;
- navigational canals;
- mining and other industrial uses;
- reservoirs for animal watering and fish breeding.[107]

(c) The Function of the Civil Code

One of the functions of a civil code is to state the circumstances in which liability arises for wrongful acts or omissions. This applies as much to the use of water resources as it does to any other activities. The practical application of such rules of liability emerges from the way the Civil Code of the Netherlands has been interpreted in a number of cases. 'Under Dutch tort law', it has been stated, 'conduct causing environmental damage may be deemed unlawful on three grounds: breach of statutory duty, infringement of a right or breach of due diligence'.[108] Of particular relevance are the infringement of a right and the breach of due diligence. Significantly, Article 5:37 of the Civil Code states:

> An owner of land may not cause nuisance to owners of other properties to a degree or in a fashion which is unlawful according to article 162 of book 6.[109]

The concept of nuisance in this context accords with the concept of nuisance in the common law. In other words liability in nuisance protects how the mutual interest of land owners may be affected by each other's activities in relation to the land. Breach of due diligence, on the other hand, involves essentially acts or omissions which amount to a breach of the duty of care owed to those within the scope of this duty.

First there is the *Jochems case* decided in 1944.[110] Land owned by the municipality of The Hague was subject to 'intensive drainage'. The plaintiff owned land nearby. As a result of the activities undertaken by the municipality the groundwater level on the plaintiff's land was lowered considerably. Was the municipality liable to the plaintiff for the loss of the groundwater in question? According to the court the activities of the municipality amounted to the infringement of a property right of the plaintiff. The activities of the municipality caused the loss or damage sustained by the plaintiff, and this loss or damage was foreseeable.

The municipality argued that the reason for the drainage of the land was relevant. The land in question comprised dunes which were used to drain water for the production of drinking water for the benefit of the community. The municipality thus engaged in these activities for the benefit of the population of the municipality. Did this justify the infringement of the plaintiff's property right?

> The court held that the course followed by the municipality in the production of drinking water leading to drainage of adjacent land could well be justified or even be required from the point of view of municipal interests, but the municipality nonetheless should have taken the detrimental consequences as regards owners of property in jeopardy into account.[111]

By failing to consider the interests of the adjacent land owners, the municipality had breached the duty owed to them and infringed their rights. Whether or not a duty has been breached is determined to some extent by a juridical analysis of why and how the activities in question were undertaken and the process according to which the decision was made to undertake them. There is, in other words, an obligation of behaviour or conduct and an obligation of methodology related to decisions leading to the behaviour in question.

The regime for water resources governance in the Netherlands is based upon the doctrine of the public domain. The state on the face of it is, therefore, responsible for the quality of public waters and for the way they are managed. In the *Huizen case*[112] the state as the plaintiff applied for an injunction to restrain the municipality of Huizen from discharging untreated sewage into a lake in the middle of the Netherlands called the IJsselmeer. The lake in question was part of the public domain. So, arguably, the rights of the state as owner of the lake had been infringed. In a sense the issue was whether the rights of the state as an incident to its governance of the public domain were capable of being protected by the provisions of the Civil Code which protected the rights of individuals within the community. One of the issues was whether such a matter properly fell within the civil, as distinct from the administrative, jurisdiction of the courts in the Netherlands. However, the remedy sought in this case was not a judicial review – a matter for the administrative jurisdiction of the courts in the Netherlands – but rather the civil remedy of an injunction. According to the court:

> Article 577 [of the civil code then in force], by conferring upon the state – subject to rights of others acquired by title or possession – the ownership fo the goods mentioned in that article, which includes the IJsselmeer, ascribes to the state, regarding the IJsselmeer, the same rights as article 625 et seq ascribe to any owner of an immovable good.

And so it was concluded:

> It is incompatible with the above that the state should tolerate the use of that lake as a recipient of – also untreated – sewage, a use not emanating from the state.[113]

Accordingly the municipality by discharging this sewage into the lake had infringed the right of the state by 'trespassing' upon its right of property.

Then there is the French *Potassium Mines case* decided finally in 1988.[114] A group of Dutch horticulturalists used water from the river Rhine to irrigate their land for the purposes of their horticulture. A French mining company discharged saline waste into the river Rhine upstream in France before the river crossed the boundary into the Netherlands. The plaintiffs brought an action in the Dutch courts for a declaration of unlawfulness of the activities of the French mining company in France and for an award of damages for the losses and costs incurred by them as a result of the discharge of the waste into the river. The river Rhine is an international watercourse. The problems of jurisdiction and applicable law were successfully circumvented by the plaintiffs in this case. In the event the plaintiffs succeeded.

Essentially the claim by the plaintiffs was based upon a breach of due diligence or, in other words, a failure to discharge the defendant's duty of care owed to the plaintiffs. For present purposes the critical issue was whether the discharge of saline waste into the river Rhine by the French mining company was unlawful. The Dutch court approached the matter in this way:

> The criterion should be that whether a party who discharges substances into a river fails to observe its duty of due care towards those using the river-water downstream depends on the nature, seriousness and duration of the damage caused to the latter and on the other circumstances of the case; among the factors to be considered here are, on the one hand, the nature and importance of the interests served by the discharges and, on the other hand, the interests served by the use of water downstream and the extent to which this use is liable to be affected by the substances discharged. It should be born [*sic*] in mind that in weighing up the respective interests, special importance must be attributed to the interests of the user downstream in that such a user may in principle expect the river not to be polluted excessively by large discharges.[115]

The court noted that the action was brought not in the interests of the environment but in the interests of the downstream users of the water, in this case clearly an issue concerning a common-pool or shared water resource. In other words, the interests of the Dutch horticulturalists and of the French mining company were 'of a financial and commercial nature'.[116] The action was thus not directed at protecting rights in the context of the public domain but protecting rights of individual citizens. That is the specific function of a civil code.

Although the river in question was an international watercourse and although the wrongful acts took place in France with injurious consequences in the Netherlands, the issue ultimately turned not upon the interests of either state or of the public domain in either state but upon the private interests of corporate and individual citizens. Again, there has emerged a right, an infringement of a right, a breach of a duty and a relevant remedy. The obligations in this case were obligations of behaviour or conduct – the actual activities of the French mining company – and obligations of methodology – how the French mining company reached its decision to undertake these activities.

The concept of the public domain produces a set of rights vested in the state or an agency of the state as a part of the formal structure according to which water resources are governed. It is assumed that the exercise of these rights is dictated by the public interest and the public welfare. Whether the way in which these rights are exercised does so or not is determined in the first instance by the rules of the public law. But the rules of the civil law have a particular part to play as part of the normative structure for the governance of water resources. There is, therefore, no clear distinction between public law regimes and private law regimes. This is certainly true of states which have adopted arrangements based upon codes. While civil and administrative jurisdictions in such states may be in theory distinct, they are to some extent linked in the sense that the choice of an appropriate remedy to protect rights and ensure compliance with duties in relation to the governance of water resources is made in accordance not only with the circumstances of the case but also consistently with the compatibly of the various remedies that are available.

CONCLUSION

Traditionally the approach in many jurisdictions has been the management of water resources by local communities in accordance with custom and practice, but subject always to centrally administered directions to ensure the ongoing and effective management of the water resource. In other words formal central direction but informal local practice. The polarity of these arrangements was overtaken in some jurisdictions by the emergence of more formal sets of rights and obligations. These were directly enforceable through the legal system. This did not inhibit central administration or local custom, but in a sense it intervened between the two. There emerged, therefore, on the one hand the exercise of a set of public rights by a central administration acting in their perception of the public interest – a public domain regime. And, on the other hand, there emerged a set of arrangements that governed the relationship between those exercising their rights of access to the water resource in question: in other words a system of individual rights and obligations recognised and enforced through the rules of the common law and those of the civil law jurisdictions. The emphasis was thus upon duties owed by individuals to other individuals, but not to the community at large. It was the body politic which had responsibilities towards the public at large, but responsibilities that did not emerge as enforceable legal duties. It is against this background of relatively fragmented systems of public rights and duties and private rights and duties that sustainable development has begun to emerge.

NOTES

1. Caponera (1978, p. 97).
2. Caponera (1978, p. 98).
3. Caponera (1978, p. 100).

4. Economic Commission for Asia and the Far East (1968, p. 210) and Caponera (1960, p. 248).
5. Caponera (1960, p. 256).
6. Caponera (1960, p. 263).
7. Caponera (1960, p. 248). See also Caponera (1960, pp. 247 and 253).
8. Caponera (1960, pp. 250 and 252) and Economic Commission for Asia and the Far East (1968, p. 210).
9. Caponera (1960, pp. 250 and 255).
10. Caponera (1960, p. 249).
11. Caponera (1960, p. 254).
12. Caponera (1960, pp. 250 and 251).
13. Caponera (1960, p. 255).
14. Caponera (1960, p. 252).
15. Caponera (1992, p. 37).
16. Food and Agriculture Organisation (1975, p. 118).
17. Borkowski and du Plessis (2005, p. 161).
18. Food and Agriculture Organisation (1975, p. 118).
19. Borkowski and du Plessis (2005, p. 161).
20. Borkowski and du Plessis (2005, p. 161).
21. Food and Agriculture Organisation (1975, p. 118).
22. Food and Agriculture Organisation (1975, p. 118).
23. Mallatt, Chibli (1995, p. 128).
24. Mallatt, Chibli (1995, p. 128).
25. Economic Commission for Asia and the Far East (1968, p. 211), Food and Agriculture Organisation (1973, p. 15), Mallatt, Chibli (1995, p. 129), Ahmad (2000, p. 171) and Roughton, (2007, p. 6).
26. Economic Commission for Asia and the Far East (1968, p. 211).
27. Roughton (2007, p. 2).
28. Roughton (2007, p. 3).
29. Ahmad (2000, p. 172).
30. Ahmad (2000, p. 173).
31. Ahmad (2000, pp. 174–178) and Roughton, (2007, pp. 6–8).
32. Food and Agriculture Organisation (1973, p. 19).
33. Maktari (1971, p. 21).
34. Maktari (1971, p. 21).
35. Maktari (1971, pp. 22 and 23).
36. Food and Agriculture Organisation (1973, p. 17).
37. Food and Agriculture Organisation (1973, p. 17).
38. Maktari (1971, p. 23).
39. Maktari (1971, p. 27).
40. Maktari (1971, p. 27).
41. Maktari (1971, p. 27).
42. Maktari (1971, p. 28).
43. Mallat, Chibli (1995, pp. 129 and 130).
44. Mallat, Chibli (1995, p. 128).
45. Getzler (2006, p. 45).
46. Getzler (2006, p. 54).
47. Getzler (2006, p. 52).
48. Getzler (2006, p. 55).
49. Getzler (2006, pp. 113 and 114).
50. Getzler (2006, p. 154).
51. *Embrey's case* (1851, p. 586).
52. *Embrey's case* (1851, p. 587).
53. *Miner's case* (1858, p. 870).
54. *Miner's case* (1858, p. 870).
55. *Miner's case* (1858, p. 870).
56. *Young's case* (1893, p. 698).
57. Howarth (1992, pp. 121–126 and 128–129) and Fisher (2000, pp. 64–74).
58. *Birmingham case* (1858, p. 226).
59. *Birmingham case* (1858, p. 225).
60. *Pennington's case* (1877, pp. 769–774).
61. *Pennington's case* (1877, p. 772).
62. *Pennington's case* (1877, p. 774).
63. Fisher (2000, pp. 74–89 and pp. 122–131).

64. Mallat, Chibli (1995, p. 130).
65. Mallat, Chibli (1995, p. 130).
66. Roughton (2007, p. 121).
67. Caponera (1992, p. 72).
68. Caponera (1992, p. 72).
69. Devif (1995, p. 141).
70. Devif (1995, p. 141).
71. Devif (1995, p. 141).
72. Devif (1995, pp. 142 and 143).
73. Devif (1995, p. 143).
74. Devif (1995, p. 145).
75. Devif (1995, p. 146).
76. Devif (1995, p. 149).
77. Devif (1995, p. 148).
78. Mallat, Hyam (1995, p. 155).
79. Mallat, Hyam (1995, pp. 152 and 153).
80. Mallat, Hyam (1995, p. 161).
81. Mallat, Hyam (1995, p. 161).
82. Mallat, Hyam (1995, p. 161).
83. Food and Agriculture Organisation (1973, pp. 171 and 172).
84. Maktari (1971, p. 70).
85. Maktari (1971, p. 71).
86. Maktari (1971, pp. 75 and 76).
87. Maktari (1971, pp. 79–81).
88. Maktari (1971, p. 94).
89. Maktari (1971, p. 98).
90. Maktari (1971, pp. 125–129).
91. Roughton (2007, p. 13).
92. Roughton (2007, p. 14).
93. Roughton (2007, p. 14).
94. Hu (2006, p. 24) and Planiol (1939, p. 437).
95. Planiol (1939, p. 428).
96. Planiol (1939, p. 428).
97. Planiol (1939, p. 429).
98. Planiol (1939, pp. 430 and 435).
99. Planiol (1939, pp. 431 and 432).
100. Food and Agriculture Organisation (1975, p. 119).
101. Food and Agriculture Organisation (1975, p. 119).
102. Food and Agriculture Organisation (1975, p. 119).
103. Food and Agriculture Organisation (1975, pp. 119, 159–161).
104. Planiol (1939, p. 431).
105. Dzurik (1990, p. 41).
106. Dzurik (1990, pp. 41 and 42).
107. Dzurik (1990, p. 42).
108. Betlem (1993, p. 440).
109. Betlem (1993, p. 377).
110. Van Dunné (1997, pp. 198 and 199).
111. Van Dunné (1997, p. 199).
112. Betlem (1993, pp. 386 and 387).
113. Betlem (1993, p. 386).
114. Betlem (1993, pp. 393–399) and Van Dunné (1997, pp. 204 and 205).
115. Betlem (1993, p. 395).
116. Betlem (1993, p. 396).

GENERAL REFERENCES

Ahmad, Ali (2000), 'Islamic water law as a comparative model for maintaining water quality', *The Journal of Islamic Law and Culture* 5, 159–185.

Betlem, G. (1993), *Civil Liability for Transfrontier Pollution: Dutch Environmental Tort Law in International Cases in the Light of Community Law* (London; Boston: Graham & Trotman/M. Nijhoff; Norwell, MA, USA: Kluwer Academic Publishers).

Borkowski, Andrew and Paul du Plessis (2005), *Textbook on Roman Law* (3rd edn, Oxford: Oxford University Press).

Caponera, D.A. (1960), 'Water law principles in the Chinese legal system', *Indian Journal of International Law* 239-275.

Caponera, D.A. (1978), 'Water laws in hydraulic civilizations', in G.L. Ullmen (ed.), *Society and History* (The Hague: Mouton Publishers), 91–106.

Caponera, D.A. (1992), *Principles of Water Law and Administration* (Rotterdam: A.A. Balkema Publishers).

Devif, Michèle Zirari (1995), 'The law of small and large scale hydraulics in Morocco', in J.A. Allan and Chibli Mallat (eds), *Water in the Middle East: Legal, Political and Commercial Implications* (London: Tauris Academic Studies), pp. 139–150.

Dzurik, Andrew A. (1990), *Water Resources Planning* (Savage, Maryland: Rowman and Littlefield Publishers Inc).

Economic Commission for Asia and the Far East (1968), *Water Legislation in Asia and the Far East*: Water Resources Series No. 35 (New York: United Nations).

Fisher, D.E. (2000), *Water Law* (Sydney: LBC Information Services).

Food and Agriculture Organisation of the United Nations (1973), *Water Laws in Moslem Countries* (Rome: Food and Agriculture Organisation of the United Nations).

Food and Agriculture Organisation of the United Nations (1975), *Water Laws in Selected European Countries* (Rome: Food and Agriculture Organisation of the United Nations).

Getzler, Joshua (2006), *A History of Water Rights at Common Law* (Oxford: Oxford University Press).

Howarth, William (1992), *Wisdom's Law of Watercourses* (5th edn, Crayford: Shaw).

Hu, Desheng (2006), *Water Rights: An International and Comparative Study* (London: IWA Publishing).

Maktari, A.M.A. (1971), *Water Rights and Irrigation Practices in Lahj* (Cambridge: Cambridge University Press).

Mallat, Chibli (1995), 'The quest for water use principles: reflections on Shari'a and custom in the Middle East', in J.A. Allan and Chibli Mallat (eds), *Water in the Middle East: Legal, Political and Commercial Implications* (London: Tauris Academic Studies), 127–137.

Mallat, Hyam (1995), 'Water laws in Lebanon', in J.A. Allan and Chibli Mallat (eds), *Water in the Middle East: Legal, Political and Commercial Implications* (London: Tauris Academic Studies), pp. 151–176.

Planiol, Marcel (1939), *Treatise on the Civil Law: Volume 1, Part 2* (12th edn), trans. by the Louisiana State Law Institute with the authority of Librairie Générale de Droit et de Jurisprudence Paris (Louisiana State Law Institute).

Roughton, Geoffrey E. (2007), 'The ancient and the modern: environmental law and governance in Islam', *Columbia Journal of Environmental Law* 32, 99–139.

Van Dunné, Jan M. (1997), 'Liability in torts for the detrimental use of fresh water resources under Dutch law in domestic and international cases', in Edward H. P. Brans et al. (eds), *The Scarcity of Water: Emerging Legal and Policy Responses* (London: Kluwer Law International Ltd), pp. 196–211.

CASES

Embrey's case: *Embrey v Owen* (1851) *English Reports*, 155, 579–587.

Miner's case: *Miner v Gilmour* (1858) *English Reports*, 14, 861–871.

Young's case: *Young v Bankier Distillery Company* (1893) *Appeal Cases*, 691–703.

Birmingham case: *Attorney-General v Birmingham Borough Council* (1858) *English Reports* 70, 220–229.

Pennington's case: *Pennington v Brinsop Hall Coal Company* (1877) *Chancery Division* 5, 769–774.

PART V

Evolving international arrangements for water resources governance

So far we have considered in some detail the language, the structure and then the normative content of a number of international and national instruments which inform and direct how water resources should be managed. In addition there has been a similar consideration of the range of practices and customs according to which water resources are used. On the face of it each jurisdiction has approached the management of water resources in its own way, for example by following or perhaps adapting the fundamental principles of Roman law, of Islamic law or of English law. The evolution of these principles preceded the emergence of international legal arrangements. Yet international legal arrangements have been influenced by the existence of these sets of legal principles: for example the test of what is reasonable or what is equitable. The function of international law is quite different from the function of national law. However the emerging ideology of sustainable development is influencing legal arrangements for the governance of water resources across the globe. Do the arrangements that are emerging disclose any patterns or approaches which may lead to the sustainable use and development of water resources? We turn now from the formal and normative arrangements for the management of water resources to the evolving dynamic capacity of the law to respond to the challenges of sustainable development. Let us begin with the international legal system.

In Chapter 12 we consider the dynamics of the international legal system. There is first a discussion of the function of international law in providing a strategic framework for the management of a shared resource such as shared water resources. First there is the principle of equity and second the principle of prevention of harm. What has emerged as a particularly significant, and in this sense dynamic, issue is the relationship between these two principles. Then there is the relationship between these two principles on the one hand and sustainable development on the other hand. Are they compatible or is a new approach required to ensure the achievement of sustainable use and development of water resources?

Chapter 13 examines the dynamics of interstate agreements about the use and development of shared water resources. This involves an examination of the functions of interstate agreements. The principle of cooperation is fundamental to the achievement of a negotiated solution in the form of an agreement. A synoptic review of some examples of negotiated arrangements points to some of the difficulties encountered in reaching agreement, and at the same time suggests more generally the conditions necessary for a successful interstate agreement about water resources. A particularly interesting dynamic perspective of interstate agreements both internationally and within federations is the set of arrangements according to which water is physically moved or transferred or authorised to be physically moved or transferred across the boundary from one state into the territory of another state. The nature of these arrangements has become increasingly important. Are they controlled by the rules about the use of shared water resources, the rules about the protection of ecosystems and the environment or the rules about international trade? The question, again, is whether these rules impede the achievement of sustainable development.

12. The dynamics of the international legal system

INTRODUCTION

So far we have considered in some detail the language, the structure and the normative content of a number of international and national instruments that inform and direct how water resources should be managed. This has been complemented by a similar consideration of the range of practices and customs according to which water resources are used. Now is the time to consider whether these arrangements disclose any patterns or approaches that may lead to the sustainable use and development of water resources. Let us begin with the international legal system.

THE CONTEXT

The doctrinal foundations of the international legal system in relation to water resources include the concepts of sovereignty and *pacta sunt servanda*. It has been suggested that sovereignty over natural resources is the only principle that 'can be considered as binding international law'.[1] One aspect of sovereignty is entering into agreements. In this sense these are norms of competence. It has simultaneously been suggested that the principle of sovereignty 'places states under an obligation to respect the environment of other states and areas beyond a state's national jurisdiction'.[2] Thus sovereignty as a norm of competence carries with it two correlative obligations. The first is to comply with any obligations undertaken by agreement – *pacta sunt servanda*. The second is to respect the environment of other states and areas beyond the state's national jurisdiction. Doctrinally, therefore, sovereignty as a norm of competence is not absolute, for it carries with it by reason of its own nature certain obligations. Much of the detail of the international legal system is concerned with the existence of these obligations, their status and their capacity to bind states as a matter of law.

But what are these obligations? Essentially, as we have seen, these are the norms of procedure and the norms of substance that find their origins in a variety of sources and with varying degrees of normativity: for example, binding rules of law, expectations of performance or outcome, methodologies of decision-making and expressions of value. These have been described as sets of legal and paralegal rules. Each impacts upon the others; one illuminates the others; and each provides substance and direction for the others. It is the nature of international law that makes the interplay between these various types of norms so significant. In particular, in the absence of any international institution with the power to enact laws of general application, the legal arrangements for the management of water resources are sourced either in the rules of customary international law or the rules prescribed by multi-lateral, plurilateral or bilateral agreements to which the state is a party. There is often controversy about the precise substance of the rules of customary international law and about the proper interpretation and application of agreements. With the principal exception of customary rules, states are bound only by the obligations that they have agreed to undertake.

It is significant, therefore, whether or not there are any requirements to be fulfilled prior to entering into agreements – particularly agreements relating to specific uses and developments of water resources rather than more strategic arrangements: for example the range of functions and responsibilities that form part of the 'obligation to cooperate' on the part of states interested in and concerned about water resources. It has been suggested that 'while cooperation is the underpinning of other obligations, failure to cooperate could constitute an independent internationally wrongful act, engaging the responsibility of the non-cooperating state'.[3] It would seem, therefore, that rules of procedure and rules of substance impact upon each other just as the doctrinal foundations of international law impact upon each other. This is so whatever the normative status of these rules and doctrines: in other words their interstitial normativity.

The use of water resources and their protection as ecosystems are increasingly the focus of much of the international environmental legal system as it relates to water resources. As we have seen, the approach of the international legal system is inspired by consensual arrangements among states in the form of multi-lateral or bilateral agreements, the rules of customary international law, the declarations of the United Nations General Assembly, declarations resulting from United Nations sponsored conferences, statements by international courts and tribunals, and the opinions of influential jurists. Whatever their normative status, they are influential individually and collectively.

The emerging rules have been stated to include:

- the obligation to prevent transboundary pollution;
- the rules relating to responsibility and liability for such pollution;
- the obligation to cooperate;
- the requirement for environmental impact assessment for projects having transboundary effects.[4]

The emerging customary principles include:

- the precautionary principle;
- sustainable development;
- intergenerational equity;
- common but differentiated responsibility.[5]

It is their mutual interaction that is important. Thus:

> The key significance of such rules and principles lies in the fact that, as the accumulated legal expression of environmental protection concerns by the international community, they indicate the issues which are likely to be identified and articulated as central in the environmental protection of international rivers and the means by which such issues are likely to be considered. The normative content of the rules and principles of customary and general international law on the environment is likely to inform the interpretation and application of the rules and principles which are set out in outline in the environmental provisions of the 1997 Convention [the United Nations Convention on

> the Law of the Non-navigational Uses of International Watercourses 1997] and other relevant instruments.[6]

Their relationship is therefore organic and symbiotic.

THE FUNCTIONS OF INTERNATIONAL LAW IN RELATION TO WATER RESOURCES

It is against this doctrinal background that it is appropriate to identify the role of international law in water resources governance. One response has been this:

- to ensure the stability and facilitate the implementation of effective strategies for dealing with water scarcity;
- to prevent and resolve disputes, both within and between countries, over increasingly scarce water resources.[7]

The first clearly is strategic. But what is contemplated on the one hand is stability of strategy and on the other hand implementation of strategy. Stability is expected to be a phenomenon of the legal system together with implementation not just of strategy but of the whole system of which strategy is a part. This highlights the second of these two functions: the prevention and resolution of disputes. This requires by implication a set of standards and criteria according to which agreements are negotiated and disputes resolved. This, too, is a particular function of the legal system. While these particular functions do not necessarily require enforceable rules of law, their effective performance must be based upon some normative structure. Such a structure has been evolving over the last few decades.

A different but consistent function has been identified for international law. Thus:

> International law has a role not only in managing and preventing conflicts between states relating to water but also in managing water for use in a sustainable manner. Until now, international law has sought to prevent conflicts relating to water by providing rules for water sharing between the riparian states. It is resource management rather than demand management that can address the competing demands for water between states and balance water rights with environmental needs for the long- term welfare of the people residing in the watersheds.[8]

Clearly dispute prevention and resolution are functions of international law. But its function includes, or perhaps even focuses upon, sustainable water use management. Significantly the focus has been on water sharing rules – demand management. But the focus now is on resource management directed towards the sustainable use and development of water resources.

The language that describes these functions of international law is instructive. There is talk of strategy, management and dispute resolution. Rules – specifically legally enforceable rules – have a part to play. But these rules are subsidiary to strategy and management, at least in the sense of vertical relationships. But these relationships may be more than vertical. Perhaps also horizontal. Horizontal in the sense that one is just as important as the other, and vertical in the sense that one is at a higher level of generality than the other. Why is this? One response is the need for a certain degree of flexibility in the international arrangements for water resources governance. Accordingly it has been stated that 'flexibility is certainly preferable to a body of fixed rules calling for mechanical application'.[9] And flexibility is achieved by 'the equitable utilisation doctrine [that] is now widely accepted and is central to the management of international water resources'. Consistently with this, 'a major strength of the rule of equitable and reasonable use is its flexibility; or its ability to take into account all relevant factors and circumstances when balancing competing interest'.[10]

Interestingly, equitable use is described on the one hand as a doctrine and on the other hand as a rule. Whatever the correct descriptor, it is not a fixed or rigid rule. It has consequently been asked whether 'the lack of a rigid rule regulating the law of international watercourses means that states are left with little guidance as to how their competing interest should be reconciled'.[11] Clearly the answer is no. It is the relative flexibility of the principle – to change again the terminology – of equitable use that enables it to be applied meaningfully in particular sets of circumstances in accordance with the legal and paralegal rules which comprise this matrix of norms. Again, it should be emphasised, this is a reflection of the nature of international law on the one hand and of sustainable development on the other hand.

WATER AS A SHARED RESOURCE IN INTERNATIONAL LAW

The functions of international law in relation to water resources governance reflect the idea that water is a common-pool resource, and hence in more juridical terms a shared resource in which the relevant community has an interest. It is indeed the need for flexibility rather than rigidity in the management of shared resources that has led to the recognition of the doctrine of equitable use for the purpose of the quantitative allocation of rights in relation to these shared water resources. The approach of international law in this respect is not new. It has evolved essentially from communal practices and customs over hundreds of years, and more recently it has been formalised in the decisions of national and subsequently international adjudicative institutions. Thus:

> The common import of all these decisions is that where a river flows between two or more states or similar entities, each of those states or entities has a vital interest in the waters of the river and none has an absolute right to them; consequently, every state must, in the exercise of its sovereignty, consider the needs of the neighbouring state.[12]

What these needs and interests may be depends upon particular circumstances: for example, in the context of sustainability, these may be economic, social or ecological. Consequently, as we have already noted, it is the principle of sovereignty itself that 'places states under an obligation to respect the environment of other states'.[13] It is thus the doctrine of sovereignty itself in the context of water resources as shared resources within the international community that leads inevitably to limitations upon the exercise of these rights of sovereignty. In other words – despite an apparent contradiction – the exercise of rights of sovereignty over water resources as shared resources is intrinsically limited or qualified. One of the qualifications that has emerged is, as we have seen, the principle of equitable use. Equitable use is merely one limitation. It is for the most part quantitative. Other limitations are essentially qualitative.

It has been argued that the community of interest in shared water resources effectively expands the doctrine of limited territorial sovereignty. Thus:

> Whereas the doctrine of limited territorial sovereignty merely connotes unilateral restraint, the concept of a community of interest evokes shared governance, joint action. Since all riparian states have interest in the watercourse, it follows that they should act jointly in managing it.[14]

Significantly the emerging notion is shared governance. It is not shared ownership or a regime of common property. While sovereignty and property share some characteristics – particularly the competence to control – they are by no means synonymous. Sovereignty connotes the wider, political interests of the public, while property connotes the narrower, civil interests of the individual owner. Notions of ownership or property do not align themselves comfortably with the doctrine of sovereignty. Indeed 'shared ownership is simply not a part of the law of international water rights – in the absence of an agreement making it so'.[15] But this does not invalidate a notion of shared governance. Whatever form is appropriate to the function of shared governance, its justification is the recognition of a community of interest in water resources, but not of common property in water resources. It has consistently been commented that:

> It would be going too far in the current state of international law to suggest that all fresh water is *res communis*. But it is critical that states begin to conceive of the *hydrologic cycle* in this way.[16]

One of the functions of international law is to incorporate this way of thinking into the arrangements for the governance of water resources. This is likely to be achieved not so much by the creation and enforcement of rigid rules of law but by the creation of objectives and strategies that are clear enough to be comprehensible, stable enough to provide continuity, flexible enough to enable adaptation to particular circumstances and certain enough to form part of a matrix of relevant legal and paralegal rules. It is, therefore, a recognition of this community of interest in shared water resources – certainly at the international level – that is the justification for the arrangements emerging within the international legal system. Indeed it has been concluded that the United Nations Convention on the Law of the Non-navigational Uses of International Watercourses of 1997 is 'the first global instrument that has been adopted to deal with the management of water as a shared common pool resource among states'.[17] This system of management is 'based on equity, the needs of countries affected, and a list of other nonexclusive criteria':[18] in other words, a degree of stability tempered by flexibility. The duty – if it is a duty – to exercise rights of sovereignty over shared water resources in an equitable manner has evolved from the earlier doctrine of the equitable and reasonable use of water resources.

EQUITY AS AN OPERATIONAL PRINCIPLE

If equity is the principal substantive limitation upon the exercise of rights of sovereignty, it becomes import to identify what it means as an operational principle. According to Article 38(2) of the Statute of the International Court of Justice the Court may decide a case *ex aequo et bono* if the parties agree. The power to make a decision *ex aequo et bono* effectively invites the exercise of an open discretion unrestricted by the law. Clearly this is quite different from making a decision based upon equity. Article 38(1)(c) of its statute requires the International Court of Justice to apply the general principles of law recognised by civilised nations. The principles of equity are probably an example of such general principles. Broadly, equity has emerged as a standard or a criterion within national legal systems to moderate the strict application of rigid rules of law.

It is its flexibility as well as its capacity to look to the future as well as the past which justifies its function in the context of water resources governance. It has been suggested, for example, that equity:

> Represents a fruitful approach to questions concerning access to shared resources, such as international watercourses, where it is more important to secure a solution which can be sustained in the future than one based upon a strict vindication of legal rights and duties which arose in the past.[19]

Such an approach may indeed be even more relevant in relation to the achievement of sustainable development, simply because sustainable development is all about securing 'a solution which can be sustained in the future'. Equity does not lack substance. It incorporates within it norms of procedure and norms of substance, but these are essentially flexible in nature. In other words, it is capable of adapting to particular existing sets of circumstances and capable of adapting to changes in these circumstances in the future. The functions of equity have been accordingly summarised in these three ways:

- equity as a means of achieving a desired equitable result;
- equity as a process of taking account of all the relevant circumstances;
- equity as a means for rendering specific laws of general application.[20]

So much for equity in general: let us turn more specifically to the principle of equitable utilisation of water resources. Essentially it presents a methodology for determining the most appropriate uses of water resources in particular sets of circumstances. The methodology may be procedural in the sense of the processes according to which a decision is made, and substantive in the sense of what the decision on its merits should achieve. In the first sense equitable utilisation has been described as 'a dynamic process':

> Which depends heavily upon active cooperation between states sharing freshwater resources. A state's regime of utilization that is equitable vis-à-vis its coriparians today may not be next year. A new use in one state may change the equitable utilization calculus as among all riparians and therefore should be the subject of prior notification, consultation, and, if necessary, negotiation.[21]

It is therefore an ongoing process that responds to changing circumstances, and consequently a range of procedural norms involving notification, consultation and negotiation. Such a process is a methodology of procedure.

Just as important, in the second sense, is the methodology of substance. It has been suggested that, while there is 'no hierarchy between the various uses that may be considered in the establishment of an equitable regime for the utilization of international rivers', if there is a conflict between such uses, this 'must be resolved on the basis of the principle of equitable utilization, with special regard being given to the requirements of vital human needs because they are directly linked to human survival'.[22] While there is, according to this analysis, no legal duty to give priority to vital human needs, there is an implied duty to give these needs 'special regard' in the methodology of either decision-making or conflict resolution.

Consistently with this approach it has been suggested that the principle of equitable utilisation 'reflects three fundamental concerns', namely:

> Firstly, it takes into account the socio-economic needs of the basin states through an objective consideration of various factors and conflicting elements relevant to the use of the waters; secondly, it aims at distributing the waters among the basin states in such a manner as to satisfy their needs to the greatest extent possible; and thirdly, it seeks to distribute the waters so as to achieve the maximum benefit for each co-basin state with the minimum of detriment to each.[23]

The reference to satisfying 'vital human needs' may be described as a matter of social justice, and this indirectly invokes the principle of intra-generational equity as well as the principle of inter-generational equity. The reference to the satisfaction of other needs incorporates notions of economic sustainability as well as social sustainability. What is missing from each of these analyses is ecological sustainability. Equity in general and the principle of equitable utilisation of water resources in particular thus emerge as essentially and solely anthropocentric values.

THE PRINCIPLE OF HARM AVOIDANCE

The international legal arrangements for the governance of water resources, however, are not limited to the principle of the equitable utilisation of water resources. As early as 1995 it was observed that customary international law recognised four 'important rules for the use of shared waters', namely:

- the duty to cooperate and to negotiate in good faith with a genuine intention of reaching an agreement;
- the prohibition of management practices likely to cause substantial and lasting injury to other states;
- the duty of prior consultation;
- the principle of equitable utilisation of shared water resources.[24]

A more recent response is that modern international water law discloses three main principles: 'the principle of equitable and reasonable utilization; the no-harm principle; and the principle of cooperation'.[25]

In any event it is the principle of equitable utilisation that has been 'universally accepted as the prevailing rule of international law relating to the non-navigational uses of international watercourses'.[26] If the right to use a watercourse is a right to do so equitably, there are inevitably limitations upon its exercise. Hence the no injury or the no harm rule. An obligation 'not to cause significant harm to other watercourses states' has been 'clearly established in customary international law'.[27] Is the injury or harm restricted to physical harm? Does it include harm to the environment or ecosystems? Does it extend to interference with the interests of other states? What has been emerging is a no injury rule, a no harm rule and a no significant harm rule. One of the unresolved problems is the relationship between them and the principle of equitable utilisation.

While these three rules reflect the doctrinal foundations of the civil law of delict and of the common law of tort, it is generally accepted that principle 21 of the Stockholm Declaration of 1972 and principle 2 of the Rio Declaration of 1992 declare their incorporation in contemporary international law. Accordingly, states have the responsibility to ensure that activities within their jurisdiction or control do not cause damage to the environment of other states or of areas beyond the limits of national jurisdiction. While the interest of a state in the quality of its environment is protected in this way, it does not extend protection to other interests that the state may have in relation to the use and development of water resources. For example, while an activity of an upper riparian state which restricts the flow of water through the watercourse may well damage the environment of the watercourse, and hence the environment of a downstream riparian state, such an interference may also detrimentally affect the economic benefits accruing to the downstream riparian state as a consequence of the interrupted flow.

While the protection of the environment is an objective of the relevant international legal arrangements, are other interests protected? For example, does the principle of equitable utilisation of water resources protect the social and economic interests of states in their water resources? What is the relationship between the interests protected by the no harm rule and the interests protected by the equitable utilisation rule? If this is transposed to the strategic level, what is the relationship between the objective of equitable use and the objective of environmental protection?

This has more or less been the conundrum facing legal systems – both national and international – since the 1960s. As we have seen, during the earlier part of the 20th century the function of the legal system was to facilitate the development of natural resources to ensure economic growth. During the 1960s this was counterbalanced by the need to protect the environment. There were simultaneously, therefore, two separate independent legal regimes seeking to achieve different outcomes. The function of sustainable development as an outcome is to bring together these two sets of values so that they are both doctrinally and operationally part of one system. If this is transposed to water resources, the social and economic uses of water and the ecologically uses of water must be managed in accordance with one system.

The answer to these questions is not uncontroversial. It may be helpful to crystallise the issues by expressing the relevant normative arrangements as duties. Thus:

- a duty imposed upon a state not to cause significant harm to another state;
- a duty imposed upon a state not to interfere with the right of another state to the equitable use of shared natural resources.

Thus, in the context of international watercourses as shared water resources, the issues have been formulated in this way:

> There now appears to be a norm of customary international law that protects a riparian state against the deprivation of its equitable share of the uses and benefits of an international watercourse. The question that must now be considered is whether in the field of international watercourses the no-harm principle is an independent norm or is part of the principle of equitable utilization. That is, do riparian states have an independent legally protected interest in being free from significant, *factual* harm that is occasioned by another state's use of an international watercourse? Or is factual harm merely a factor to be considered in determining whether a state's utilisation is equitable and reasonable?[28]

Structurally, is it a duty not to harm or is it a duty to take into consideration the risk of harm? The response has been this:

> It is the not the causing of significant harm *per se*, but the unreasonable causing of such harm that is prohibited. In other words, conduct of a state that results in some harm – even significant harm – may be reasonable under the circumstances, and thus the harm may have to be tolerated by the

> harmed state. Analytically, this is equivalent to treating harm as but one factor in an equitable utilization determination: significant harm may have to be tolerated in order to achieve an overall regime of equitable and reasonable utilization.[29]

In other words, the outcome to be achieved is the equitable use of shared water resources. This may be achieved either by causing significant harm or by avoiding significant harm. If the harm is caused unreasonably then it is prohibited, but not otherwise. It becomes relevant, therefore, to consider what is reasonable in a context such as this. This is a matter of circumstance, and the conclusion is that the duty in relation to significant harm is not absolute but one of due diligence. The conclusion reached is significant:

> First, there is no doubt that such an obligation [the no-harm obligation], as reflected in the *sic utere tuo* maxim, exists in general international law and, in particular, in the law of the environment. Second, the obligation, whether applied to international watercourses or more generally, is not an absolute one. It is mitigated by several factors, all of which depend upon the circumstances of the particular case: the harm must be 'significant'; the obligation is one of due diligence; and the harm must be unreasonable.[30]

Ultimately the relationship between the equitable utilisation rule and the no-harm rule depends upon the nature of the duties or obligations that comprise the rules in question.

THE RELATIONSHIP BETWEEN EQUITY AND HARM AVOIDANCE

(a) The Rules in the 1997 Convention

The most authoritative statement of these rules in codified form is the set of relevant provisions in the United Nations Convention on the Law of the Non-navigational Uses of International Watercourses of 1997. Let us remind ourselves of these provisions. While Article 20 imposes an obligation upon watercourse states to protect and preserve the ecosystems of international watercourses and Article 21 an obligation to prevent, reduce and control the pollution of an international watercourse which may cause significant harm to other watercourse states or to their environment, it is Articles 5, 6 and 7 in particular that address the relationship between the duty of equitable of utilization and the duty not to cause significant harm.

The duties imposed upon watercourse states by the convention in relation to an international watercourse in their territory are these:

- to utilise it in an equitable and reasonable manner according to Article 5(1);
- to use and develop it with a view to attaining optimal and sustainable utilisation thereof and benefits therefrom according to Article 5(1);
- to do so by taking into account the interests of the watercourse states concerned consistent with adequate protection of the watercourse in accordance with Article 5(1);
- to take into account all relevant factors and circumstances, including those specified, in utilising it in an equitable and reasonable manner in accordance with Article 6;
- to take all appropriate measures to prevent the causing of significant harm to other watercourse states in utilising it in their territory in accordance with Article 7(1);

- to take all appropriate measures, having due regard to the foregoing duties, to eliminate or mitigate significant harm caused to another watercourse state in accordance with Article 7(2).

(b) The Relationship between the Rules

The first duty in Article 5(1) prescribes the manner in which the water resource must be utilised. This adverbial phrase may be reduced to two adverbs. Thus it is a duty to utilise the water resource equitably and reasonably. This is consistent with and no doubt will be applied in accordance with the rules of customary international law that have been evolving over recent years. The adverbial condition attached to this duty is expanded by Article 6. This requires the relevant factors and circumstances to be taken into account. There is no specification of factorial priority. A balancing approach is mandated. It is, therefore, a duty that prescribes a methodology of procedure and a methodology of substance in the senses already discussed in relation to the utilisation of the water resource and in relation to making decisions about the utilisation of the water resource. The focus of Articles 5 and 6 is upon how water resources are to be utilised and how decisions are to be made about their utilisation. The subject matter of the duty is use – a verb. The normative restrictions upon use are – in the truncated form – the adverbs equitably and reasonably. The application of these adverbial limitations is regulated by the duty of methodology in Article 6.

Article 5(1), as we have seen, mandates three duties in addition to the duty to utilise the water resource in an equitable and reasonable manner. These relate to the objective to be achieved in using and developing water resources. Significantly, it incorporates future use as well as current use in consequence of the use of the word develop. The overall objective is the optimal and sustainable utilisation of the water resource and the benefits accruing from its utilisation. The juxtaposition of the adjectives optimal and sustainable is significant. Optimal suggests that the outcome will be the best use of the water resource. But it must simultaneously be a use that will be sustainable.

How this is done is qualified in two ways. First, the interests of concerned watercourse states must be taken into account: in other words, a duty of methodology in the sense discussed. Second, the optimal and sustainable utilisation of the water resource must be consistent with adequate protection of the water resource – in this case the watercourse itself. The expression consistent – apparently an adjective – is not unambiguous. On the face of it, the achievement of optimal and sustainable utilisation of the water resource must be consistent with the achievement of the adequate protection of the water resource. Presumably the test of consistency means that the two outcomes can be accommodated within the final decision rather than in conflict with each other. In this context, it is the protection of the water resource itself that must be consistent with its optimal and sustainable utilisation. If this is translated into an implicit duty, it means that there is a duty not to harm the water resource beyond the requirement for its adequate protection.

This takes the analysis to Article 7. There is the duty, first, to take all appropriate measures to prevent significant harm to other watercourse states and the duty, second, in the event of such significant harm, to eliminate or mitigate it. The duty arises in utilising the water resources of the watercourse. The beneficiaries of the duty in Article 7 are other watercourse states, whereas the beneficiary of the duty in the second part of Article 5(1) is the water resource of the watercourse itself. Harm to the water resource itself may well fall within the notion of harm to the state in which the water resource is located. But harm to the state is a wider concept in the sense that the interests of the state so protected are more extensive than the interest in protecting the water resource itself. Hence the significance of the context in which the duty arises: namely the use of the water resource of the watercourse.

This takes us closer to the relationship between these normative arrangements, in particular the norms in Article 7 and those in Article 5. Thus, so the argument runs, while the duty in Article 7 is significant, it is subsidiary to the duties in Article 5(1). In this sense the duty in Article 5(1) imposed upon a watercourse states to utilise the international watercourse in an equitable and reasonable manner correlates with a right of each and every watercourse state to utilise the international watercourse in an equitable and reasonable manner. This encapsulates the essential nature of a community of interest among all watercourse states in the international watercourses in which they have an interest. By an analogy of reasoning, the duty associated with the attainment of the optimal and sustainable utilisation of the water resource correlates with a right to do so. But the discharge of this duty and the exercise of this right are subject themselves to the two duties already noted: to take into account the interests of the other watercourse states and to do so consistently with adequate protection of the watercourse. This analysis of Articles 5, 6 and 7 suggests a complicated arrangement of duties – and implied correlative rights – each of which performs a different function in accordance with the structure of these Articles and the language used to create them.

The meaning of these provisions is by no means uncontroversial. Jurists have subjected the relevant customary rules of international law and the provisions of the 1997 convention to detailed and careful scrutiny over the years. One of the most recent of these assessments has described the relationship between the equitable utilisation rule and the no significant harm rule in this way:

> The rule against causing significant harm enjoys a privileged position within the corpus of rules of international law relating to the non-navigational uses of international watercourses. This rule is not an absolute one as to result, but a relative rule, based on standards of due diligence, applying to the conduct of the riparian states. However, it is equally clear that the principle of equitable utilization, based on an equitable balancing of all factors relevant in the particular circumstances of each case, remains the prevailing rule, providing the normative framework within which all others apply. In other words, while the rule against causing significant harm, and a number of other substantive rules primary concerned with aspects of environmental protection, enjoy some priority, this could only ever amount to the interests which these substantive rules represent being given a privileged weighting within the balancing process or some other form of advantage, such as that of giving rise to a rebuttable presumption of inequity.[31]

Incorporated within this assessment are – at least by implication – allusions to obligations of different kinds, including obligations of result and obligations of conduct. The nature of these obligations warrants some further analysis.

(c) The Nature of the Obligation

Duties or obligations contain standards often identified as those of conduct, of performance or of result. These may be described very simply in the vernacular as:

- what you should be doing – conduct;
- how you should be doing it – performance;
- what you should be trying to achieve in doing it – result.

More particularly:

- a duty to dig a ditch one kilometre long, two metres deep and one metre wide between points A and B in irrigation area X – conduct;
- a duty to dig a ditch in such a way that it carries water effectively and efficiently between points A and B in irrigation area X – performance;
- a duty to dig a ditch so as to achieve the effective and efficient irrigation of irrigation area X – result.

Classifications such as these are controversial. This approach, however, highlights the degree of rigidity or flexibility about how the duty is discharged: in other words, the degree of discretion available in the discharge of the duty. The rules of customary international law and those in multilateral agreements for the most part are so structured as to give considerable flexibility in the application of the relevant rules. In other words, the standards tend to be those of performance or result rather than conduct. Accordingly, the duties tend to be those of due diligence rather than absolute duties. Much depends upon the origins of the rule – custom or agreement – that incorporates the duty and the way it has accordingly been structured. Duties, therefore, may overlap in the context of these classifications or indeed not fit comfortably within any of them. This is particularly true of duties finding their source in rules of customary international law simply because of its nature.

An interesting example is the rule of customary international law to prevent transboundary harm. It has been pointed out that customary international law does not prohibit all transboundary harm. The reasons are significant:

> It is widely understood that this rule applies subject to two considerable limitations. First, the harm or potential harm involved must exceed the threshold of 'significant' or 'substantial' harm in order to come within the ambit of the prohibition. Second, the prohibition is normally understood as reflecting an obligation as to performance, based on standards of 'due diligence', rather than an absolute obligation as to result.[32]

If this is translated into a formal rather than a customary rule, perhaps it may read like this:

> A state must take reasonable measures to ensure that activities within its territory do not cause harm beyond the boundaries of its territory.

Compare a different form of words:

> A state must ensure that activities within its territory do not cause harm beyond the boundaries of its territory.

The interposition of the verb ensure in both cases connotes that the subject of the duty is a result or outcome. That is, no harm. If the duty is to ensure, then it is probably absolute. But if it is to take reasonable measures to ensure, then it is not absolute. Although the duty continues to contain an outcome or a result, the interposition of the words take reasonable measures qualifies the duty. In other words, it explains how the duty is to be discharged – performance. Written rules can be as ambiguous as customary rules. A degree of caution is justified in either context.

THE EMERGING RELEVANCE OF SUSTAINABILITY

(a) The Relevance of Sustainability for Use and Development

The relationship between the equitable utilisation rule and the no harm rule is simply one example of the tensions that have been emerging in this branch of the law – both international and national – since the last decades of the 20th century. On the one hand, there is a culture that promotes the development of natural resources for the purposes of economic growth, and on the other hand a culture that seeks to protect and conserve the ecological values of the environment. Indeed, as Judge Weeramantry has noted in the *River Danube case*, 'the problem of steering a course between the needs of development and the necessity to protect the environment is a problem alike of the law of development and of the law of the environment. Both these vital and developing areas of law require, and indeed assume, the existence of a principle which harmonises both needs'.[33] The principle turns out to be sustainable development. The harmonisation of resource development and environmental protection is particularly significant in the context of water resources. This is as true for international arrangements as for national arrangements. Thus:

> The delicate balance that the international community has tried to achieve between economic development and environmental protection through the introduction of the principle of sustainable development is relevant to the issues relating to the utilization of international watercourses.[34]

While the direction of water resources governance during most of the 20th century was historically their use and development – indeed their 'optimum development' – this was tempered by the perceived needs to sustain the quality of water – particularly for drinking – and the ecosystems of which the water resources were a part. And so sustainable development has emerged:

> As the proper approach to reconciling economic development with protection of the environment. It aims at ensuring that economic development will not exhaust the very resources on which it, and human welfare itself, depend. Clean water and healthy aquatic ecosystems are cornerstones of this effort. International law has now progressed to the point that it protects those values.[35]

The implication, therefore, is that sustainable development plays some part – formal or not – in protecting these values.

We have noted that Article 5(2) of the United Nations Convention on the Law of the Non-navigational Uses of International Watercourses of 1997 refers to the attainment of the optimal and sustainable utilisation of the water resources of international watercourses. It has been suggested accordingly that the concept of sustainability 'may enjoy real priority as a factor and might even to some extent limit the freedom to balance interests'.[36] In other words, it would be difficult to justify an unsustainable use of water. Perhaps even more significantly, the introduction of the concept of sustainability in this way and in similar contexts may have introduced 'important environmental constraints into this part of international law relating to natural resources'.[37] Hitherto protection of the environment of water resources was a discrete part of international law and was largely irrelevant to making decisions about the use and development of water resources in the context of planning for their future use: in other words, the vital distinction between protection and planning as functions given legal form.

The significance of this approach – the introduction of sustainability into the legal arrangements for the management of water resources in the context of planning – is its impact upon 'existing legal concepts of equitable utilization of shared resources'.[38] In other words, sustainable development has become an integral part of the principle of equitable utilisation of water resources. Consistently with this 'sustainable utilization is merely a goal which states ought to attempt to, but are not legally obliged to, achieve'.[39] Even as a legally unenforceable goal, it has become nevertheless a factor of relevance and perhaps even of priority in the methodology of decision-making about the future use and development of water resources in international watercourses. Sustainability thus becomes a factor of relevance in making decisions about use and development. Do use and development become relevant factors in making decisions about sustainable development?

(b) The Function of Sustainable Development

This takes us back to the fundamental question of the function of sustainable development in managing water resources. Sustainable development, as Judge Weeramantry indicated, harmonises the need for development and the need for environmental protection. The need for development seeks the enhancement of the welfare of society as a whole and also of the individuals who comprise the society in question. It has economic as well as social perspectives. The need for environmental protection has economic perspectives because the environment is the source of natural resources; social perspectives because society has both economic and cultural interests in the environment; and ecological perspectives because of the intrinsic qualities of the environment as well as those attributed to it by humankind. Hence the emergence of the three pillars of sustainable development: economic, social and ecological. In this context water has been described for analytical purposes as water as a social good, as an economic good and as an ecological good. From the legal perspective water as a social good focuses upon who has access to water; as an economic good upon who controls water; and as an ecological good how water is protected.[40]

Of these three it is perhaps the norms of competence relating to the control of water that represent the point of doctrinal departure for an analysis of the whole system. States in exercise of their rights of sovereignty control water resources, but the exercise of these rights has become increasingly constrained by a complex set of legal and paralegal rules that seek to recognise and protect the range of economic, social and ecological interests in water resources. While water resources may be seen generally as shared resources, water resources in international watercourses are regarded in this way. So, while states share a community of interest in water resources, this apparently coherent community of interest has collapsed into a set of disconnected economic, social and ecological values. Sustainable development seeks to reconnect these values.

(c) Conclusion

So far – at least – the international legal system does not appear to have brought these values together in a coherent system of water resources governance. The use and development of water resources – essentially the economic and social perspectives of sustainable development – are governed in the first instance by the principle of equitable utilisation, the application of which is itself subject to a subsidiary principle in the form of the duty to take reasonable measures to avoid significant harm to other states. The duty not to harm the environment of other states has both anthropocentric and ecocentric perspectives, but it is a duty which arises independently of the exercise of the right to use water resources equitably. The emerging duty to protect the ecosystems of watercourses focuses more on the protection of their intrinsic values. It similarly is a duty which arises independently of the right to use water resources equitably. Then there is the emerging human right to water. This is particularly significant because a human right potentially transcends the rights generally exercisable by states.

THE HUMAN RIGHT TO WATER

Let us take the right to water in this context a little further. The right to water, it has been suggested, 'would provide a means for people to claim access to water for their basic needs'.[41] While the substance of such a right is probably no more than a correlative duty to make available a source of supply for the satisfaction of basic human needs, it is the function of such a right within the legal system that is important. It has been suggested that 'if people have a right to clean water this right is valid *erga omnes*'.[42] The corresponding duty is thus owed *erga omnes*. If this is a human right, it enjoys a privileged position within the legal system, and in a sense other rights – including those of states – are subsidiary to it. If it is a right valid *erga omnes*, then any correlative duty similarly occupies a position of privilege.

According to this analysis, the implications of a human right to water are these:

- the right is available within and across national boundaries;
- the state of the territory must recognise and not violate this right;
- the state of the territory must realise this right;
- states other than the state of the territory have the same responsibilities;
- these responsibilities apply to the water resources of international watercourses.

While the primary responsibility of the state is to realise this right for its citizens, it has been suggested that this responsibility extends to the citizens of other states. Thus:

> If the right to clean water is a right valid *erga omnes* a watercourse state would not be in a position to commit any action that would deny this right to the people residing in other watercourse states. This obligation invokes the principle of equitable and reasonable utilization even if the state in question is not a party to the 1997 UN Convention. This combined obligation on the part of both of these states gives rise to the duty jointly to develop the water resources of the basin of a watercourse by the watercourse states, to cooperate to protect and preserve the ecosystems and marine environment of the international watercourse, and to prevent, reduce and control pollution of the watercourse.[43]

This highlights the relationship between a human right, the right of a state to use water resources equitably and the duty of a state to protect ecosystems and the environment. The implications of the existence of a human right to water are thus significant.

CONCLUSION

The international arrangements for the management of water resources have traditionally been concerned with the reciprocal rights and duties of nation states. First, there is the principle of sovereignty and the concomitant power to enter into agreements with other states. Secondly, there has been increasing recognition of water resources as shared resources with an accompanying set of common responsibilities in managing these resources. Hence the principle of equitable utilisation and the principle of no harm. While each of these principles reflects aspects of sustainable development, they do not amount to sustainable development itself. In this sense, as has been suggested, the international legal system does not currently acknowledge the fundamental need for coherence and consistency between and among the economic, social and ecological values that underpin sustainable development.

The emergence of a human right to water adds a potentially significant dimension to these arrangements. International law has been concerned traditionally – with some exceptions – with the relationships between and among states. The emergence of human rights in general has in some cases effectively brought individuals within the protective arrangements of the international legal system: for example, the interpretation of some human rights in a way that incorporates a human right to an environment of a particular quality. The emergence of a human right in relation to water may have a similar application. Much depends upon how such a right is realised within a national legal system. But the existence of a human right to water would constitute a constraint upon how water resources are managed within a state and would confer a consequential benefit – but perhaps not necessarily an enforceable legal right – accruing to members of the community. The significance of a human right to water must therefore not be underestimated.

NOTES

1. Rieu-Clarke (2005, p. 96).
2. Rieu-Clarke (2005, p. 96).
3. McCaffrey (2007, p. 470).
4. McIntyre (2007, p. 191).
5. McIntyre (2007, p. 191).
6. McIntyre (2007, pp. 191 and 192).
7. McCaffrey (1997a, p. 52).
8. Subedi (2005, p. 18).
9. Godana (1985, p. 60).
10. Rieu-Clarke (2005, p. 131).
11. Rieu-Clarke (2005, p. 131).
12. Godana (1985, pp. 47 and 48).
13. Rieu-Clarke (2005, p. 96).
14. McCaffrey (2007, p. 165).
15. McCaffrey (2007, p. 165). See also Godana (1985, pp. 48 and 49).
16. McCaffrey (2007, p. 54).
17. Louka (2006, p. 88).
18. Louka (2006, p. 88).
19. Lowe (1989, p. 73).
20. McIntyre (2007, p. 132).
21. McCaffrey (2007, pp. 404 and 405).
22. Kaya (2003, p. 119).
23. Godana (1985, p. 55).
24. Caponera (1995, p. 123).
25. Hildering (2004, p. 44).
26. McIntyre (2007, p. 85).
27. McIntyre (2007, pp. 85 and 86).
28. McCaffrey (2007, p. 408).

29. McCaffrey (2007, p. 436).
30. McCaffrey (2007, pp. 445).
31. McIntyre (2007, p. 119).
32. McIntyre (2007, p. 205).
33. *River Danube case* (1988, p. 205).
34. Subedi (2005, p. 15).
35. McCaffrey (2007, p. 462).
36. McIntyre (2007, p. 111).
37. Birnie et al. (2009, p. 124).
38. Birnie et al. (2009, p. 124).
39. McIntyre (2007, p. 243).
40. Hildering (2004, p. 16).
41. Hildering (2004, p. 75).
42. Subedi (2005, p. 15).
43. Subedi (2005, p. 15).

GENERAL REFERENCES

Birnie, Patricia, Alan Boyle and Catherine Redgwell (2002), *International Law and the Environment* (3rd edn, Oxford: Oxford University Press).

Caponera, D.A. (1995) , 'Shared waters and international law', in G.H. Blake et al. (eds), *The Peaceful Management of Transboundary Resources* (London: Graham and Trotman/Martinus Nijhoff), pp. 121–126.

Godana, Bonaya Adhi (1985), *Africa's Shared Water Resources* (London: Francis Pinter Publishers).

Hildering, Antoinette (2004), *International Law, Sustainable Development and Water Management* (Delft: Eburon).

Kaya, Ibrahim (2003), *Equitable Utilization: the Law of Non-navigatinal Uses of International Watercourses* (Aldershot: Ashgate Publishing Limited).

Louka, Elli (2006), *International Environmental Law* (Cambridge: Cambridge University Press).

Lowe, Vaughan (1989), 'The role of equity in international law', *Australian Yearbook of International Law* 12, 54–81.

McCaffrey, Stephen C. (1997a), 'Water scarcity: institutional and legal responses', in Edward H.P. Brans et al. (eds), *The Scarcity of Water: Emerging Legal and Policy Responses* (London: Kluwer Law International), pp. 43–58.

McCaffrey, Stephen C. (2007), *The Law of International Watercourses: Non-Navigational Uses* (2nd edn, Oxford: Oxford University Press).

McIntyre, Owen (2007), *Environmental Protection of International Watercourses under International Law* (Aldershot: Ashgate Publishing Ltd).

Rieu-Clarke, Alistair (2005), *International Law and Sustainable Development: Lessons from the Law of International Watercourses*, London: IWA Publishing.

Subedi, Surya P. (2005), 'Regulation of shared water resources in international law: the challenge of balancing competing demands', in Surya P. Subedi (ed.), *International Watercourses Law for the 21st Century* (Aldershot: Ashgate Publishing Ltd), pp. 7–18.

LEGAL INSTRUMENTS

Rio Declaration (1992): Declaration of the United Nations Conference on Environment and Development (1992), *International Legal Materials* 31, 874–880.

Stockholm Declaration (1972): Declaration of the United Nations Conference on the Human Environment (1972), *International Legal Materials* 11, 1416–1421.

United Nations Convention on the Law of the Non-navigational Uses of International Watercourses 1997 (1997), *International Legal Materials* 36, 700–718.

CASES

River Danube case: Judgment in the case concerning the Gabcikovo – Nagymaros Project (1988) *International Legal Materials* 37, 162–242.

13. The dynamics of interstate agreements

INTRODUCTION

(a) The Power to Enter into Agreements

The rights and duties of states in relation to water resources derive not only from the general principles of international law but also – and in many cases almost totally – from the arrangements in interstate agreements that relate to the use and development of specific water resources in particular locations: in other words, the exercise by states of their right to enter into agreements with other states and in accordance with the fundamental doctrine that requires compliance with consensual obligations – *pacta sunt servanda*. States are free to enter into agreements with other states about the use and development of their water resources. But is this right constrained in any way? The question is:

> Whether transboundary water allocation is an unprincipled, *ad hoc* development based on political, economic, and authoritative configurations in a region, or whether equity considerations and the 'no harm' principle, embodied in international instruments, have played a role in the allocation of water resources.[1]

Whatever may be the answer to this question, it has been observed that the tendency in practice is to describe the 'benefits and allocations as reasonable and equitable'.[2]

Assuming that the principle of equitable and reasonable utilisation is a guiding principle of international law in this context, it may reasonably be expected that it would be the principle to be observed by states when entering into negotiations and consequently agreements. It may theoretically be assumed that the states in question regard the detailed arrangements in the agreement as equitable and reasonable in the circumstances. Is there any guidance on this issue? The United Nations Convention on the Law of the Non-navigational Uses of International Watercourses of 1997 is regarded by many as a codified version of the relevant rules. The convention includes the principle of equitable and reasonable utilisation as one of its fundamental rules. Article 3(2) of the convention encourages states to harmonise existing agreements with the basic principles of the convention and – no doubt by implication – to give effect to these principles in any subsequent agreements even if one or more of the states is not bound by the convention. Accordingly, whether or not a state is bound by the convention, the way in which the rules, including those relating to harmonisation, are structured suggests that states can include within the agreement whatever rights and duties seem to them to be appropriate, and hence equitable and reasonable.

This approach has led to three conclusions. First, in relation to existing agreements:

> The text of article 3 paragraph (1) is unequivocal and categorical: the rights and obligations under such agreements remain unaffected.[3]

Second, in relation to future agreements:

> It is clear from the text of article 3 paragraph (3) that states have a considerable degree of freedom; the need to depart from the provisions of the Convention is anticipated and legitimised.[4]

And third, more generally:

> The Convention does not contain any formula to resolve disputes or provide any definitive answers to them. Indeed, it is categorised by deliberate normative ambiguity, leaving space for the disputing parties to arrive at an agreement through political compromise.[5]

The right to enter into agreements and to be bound by them in this way leaves intact the fundamental concept of sovereignty.

(b) The Relevance of Interests External to the Agreement

But what about the interests of states that are not parties to the agreement? The community of interest which lies at the foundation of the principle of equitable and reasonable utilisation may well extend to these other states. Are these interests protected? For example the implementation of an agreement between states A and B may cause substantial harm to the environment in state C. Or the implementation of the agreement may interfere so significantly with the flow of water through the watercourse that state C effectively receives no benefit whatever from it. The answer depends upon how the general principles of international law apply to the circumstances of the case.

How may these rules and principles become relevant? The rules supporting the community of interest doctrine may extend to protect the interests of a third party to the agreement. Similarly the rules for the protection of the environment and of ecosystems may extend to protect the interests of a third party state. Significantly for the future, the way in which the rules in the interstate agreement are applied may contravene the provisions for free trade in international trade law and the provisions protecting human rights in international human rights law. This may be particularly an issue if the use of economic instruments such as trading in water or in water entitlements is contemplated by the agreement or if a human right to water emerges and is protected.

The existence of these interests, rights and duties independently of the principle of equitable and reasonable utilisation illustrates how these legal arrangements operate in a fragmented rather than in an integrated way. The protection of the environment and of ecosystems is undoubtedly the expression of an ecological value. The recognition of a human right to water is the expression of a social value. And the encouragement of free trade and the use of economic instruments is an expression of an economic value. Even if they operate within the principle of equitable and reasonable utilisation, they also operate more positively outside it. If sustainable development is concerned to bring together economic, social and ecological values in an integrated way, then more needs to be done. The existing arrangements do not require this approach. This takes us back to the fundamental issue in relation to water resources governance. The form of legal arrangements must reflect the function that they are designed to perform. What are these functions?

THE FUNCTION AND STRUCTURE OF INTERSTATE AGREEMENTS

(a) Introduction

The normative structure of interstate agreements reflects the functions intended to be performed by the parties to the agreement. These functions are as diverse as the uses to which water resources are put. Several stages, nevertheless, have been identified in the evolution of these functions. These have been described in summary form as:

- stopping short of allocating water;
- allocating water;
- jointly managing water.[6]

Let us consider each in more detail. The first involved 'a commitment to share information'[7] about the uses to which the shared water resources could be put. This might involve consultations before a project was authorised. The next step was to require the consent of the other state to any proposed project 'if the proposed project would sensibly impair the water-body, or at least sensibly interfere with the waters uses in the other state'.[8] Such an approach impliedly acknowledged aspects of the community of interest doctrine, but probably not much more. The second category of agreement comprised those that 'divide the transboundary waters by volume'.[9] This would amount to a set of substantive rules about water entitlements based upon the mutual provision of adequate information and the effects that the use of these water entitlements would have upon the shared water resources. Such arrangements were static rather than dynamic, and did not in themselves have the capacity to respond to changing circumstances. The third step, therefore, was some form of institutional arrangements for joint management on an ongoing – and in this sense a dynamic – basis.[10]

(b) Water Resource Development as the Principal Function

The evolving arrangements in Africa are particularly interesting. The essence of the arrangements negotiated for the river Senegal in the early 1960s was 'to recognize the right of each riparian state to initiate projects within the portion of the river lying within its territory but with the obligation to exchange information and to coordinate the initiatives'.[11] The arrangements negotiated in 1972 facilitated control of the flow regime and the transfer of water for irrigation and hydroelectric power generation purposes. How was this done? In effect by establishing:

- institutional arrangements;
- conditions for financing common construction works;
- a system for the settlement of disputes.[12]

The agreements concluded in the early 1960s about the River Niger Basin focused essentially on the use of the river for the purposes of navigation and transport.[13] However, in consequence of the creation of the Niger Basin Authority in 1980 institutional arrangements were accordingly put in place to achieve a set of much more wide ranging objectives. These included 'harmonisation and coordination of national development policies and activities in order to ensure equitable sharing of the surface and groundwaters of the Niger Basin'.[14] This involved the use and control of water; the regulation of flow; the regulation of drainage; flood control; control of land degradation; and environmental protection. The availability of resources was critical to the performance of such functions. A development fund was established to assist with the provision of financial resources to support these functions.

A number of general conclusions have been drawn from this survey of the consensual arrangements that have evolved in Africa. These are:

- most of the treaties required the parties to get together to do things jointly;
- the agreements were multi-purpose in character;
- African countries seemed to have accepted the principle that cooperation in drainage basin management should be a basis for development in non-water sectors;
- there was a growing trend towards shared financial responsibility in the construction of common works.[15]

The doctrine of community of interest was thus made operational by the acceptance of common responsibility for management, planning and activities designed to achieve a range of purposes in the use of the shared natural resources including the water resources of the basin. For the most part, the interstate agreements in Africa focussed upon the use and development of water resources for economic and social purposes.

Similarly in Asia the purposes have been principally economic and social. In relation to the sub-continent of South Asia, for example, this has been suggested:

> The parties attempt, through the treaties, to develop water resources for common economic growth, to fix and limit rights of use by the parties, and to confirm amounts of water to be used by the parties on the shared rivers. Essentially they intend to address the paucity of water availability in the region, both in qualitative and quantitative terms.[16]

The focus was thus upon ensuring enough water for economic and social purposes. Water quality became relevant in relation to its use for economic and social purposes rather than for its ecological values. Interstate agreements were in this sense statements of public policy cast in the form of strategies, entitlements and duties, but essentially based upon stated or unstated political priorities.

(c) Water Resource Protection as a Subsidiary Function

Values have changed over time. Accordingly, it would be misleading to suggest that interstate agreements do not seek to protect the environment. This is an important aspect of sustainable development – namely the use of water for ecological purposes. As we have seen, the arrangements included in interstate agreements suggest that the form in which the several duties are expressed indicates the type of function that the particular duty is intended to perform. In relation to environmental protection, some prohibit pollution of the water resource generally; others prohibit pollution so as to protect a particular interest in the water resource; others regulate pollution; some prohibit environmental damage so as to protect particular interests; and others impose procedural duties.

The specific functions performed by these rules have been described in these ways:

- the avoidance of potential disputes in frontier waters;
- the protection of specific uses or interests of the parties;
- the mitigation of the polluted condition of certain watercourses;
- the enforcement of an agreed water apportionment.[17]

These are rules of specific and not general application partly because of their inclusion in an interstate agreement and partly because of the language which has created them. In addition, according to this analysis, they are subsidiary to the principle of equitable and reasonable utilisation of international watercourses. Accordingly their inclusion in interstate agreements does not point to a 'principle of general international law prohibiting significant damage to the environment of international watercourses'.[18]

(d) An Emerging Pattern of Functions

Is there any evolving pattern of functions disclosed by these sets of arrangements in interstate agreements? Perhaps there are four:

- procedural;
- institutional;
- strategic;
- legislative.

The first two are concerned principally with putting in place arrangements according to which the states can get together to discuss the issues; to collect and present the information and analysis upon which their claims are made; to provide the basis for the involvement of third parties either to assist in the negotiations or provide financial support for any outcomes; and to establish arrangements to solve any disagreements which may arise in the future: in short, the functions essential to support cooperation. Strategic and legislative functions focus more upon the achievement of substantive outcomes. The formulation of strategy points out the general but critical direction that decision-making at a more specific level is intended to take, for example, equitable use or sustainable use.

The legislative function is more specific. It involves the creation of a set of enforceable rules that it is hoped will achieve the direction of decision-making set out at the strategic level: for example, specific quantities of water able to be abstracted or required to be returned with reference to particular locations in watercourses or the purposes for which water in a watercourse may be used. The calculation of specific quantities and the identification of particular purposes of use are no doubt difficult in practice. Hence the need for procedural and institutional functions in support of strategic and legislative functions and similarly the other way round. It is the effective performance of all of these functions simultaneously in relation to the shared water resource in question that can be the particular strength of interstate agreements. This is especially so if the interstate agreements include all, or at least most, of the states with an interest in the shared water resource. Many, however, are bilateral and lack the cooperation of all the basin states or all those with an interest in the shared watercourse.

For example, it has been suggested that the 'tension' between Egypt and Ethiopia in the context of the Blue Nile 'can be managed only if the water is cooperatively managed by the several neighbouring states in such a way as to assure equitable participation in the benefits derived from the water from all communities sharing the basin'.[19] The references to cooperation, management, assurance and equitable participation describe in their own ways functions which are procedural, institutional, strategic and legislative. Customary international law, it was suggested, 'in its somewhat primitive state of development, simply cannot solve the management problems of the region'.[20] What was required was 'a new treaty, one that includes all basin communities, creates appropriate representative basin-wide institutions, and has the clout to enforce its mandates'.[21] Again, a set of procedural, institutional, strategic and legislative functions. Interstate agreements can be and have been structured to do so. What is required is simply the determination of the nation states to cooperate and ultimately to agree.

THE CHALLENGES FACING COOPERATION

(a) The Need for Political Will

To enter into an interstate agreement is an exercise of the right of sovereignty. It cannot be compelled – even if there is a general duty to cooperate. In other words, the principles and rules of international law apply and form the basis for the resolution of any dispute between states. Although these rules and principles – as they evolve from time to time – represent the juridical context in which interstate agreements are negotiated and have effect, it is clear that states are free – in the sense of freedom from binding restrictions – to put in place consensual arrangements inconsistent with and contrary to the rules and principles of international law. For example, in relation to the interstate agreements among the states of the subcontinent of South Asia, it has been commented:

> The treaties have all been entered into during the past 50 years, starting in the 1950s and ending in the 1990s. However, despite significant developments in the field of international law with regard to water or the environment, even the recent treaties have not fully incorporated, in a uniform fashion, the more current thinking about equitable and sustainable use of international watercourses. The lack of political will as well as the unpredictability of hydro-politics in the sub-continent are perhaps responsible for such outcome. Such unpredictability has also significantly influenced the overall negotiating process of these agreements.[22]

On occasions the necessary political will has been present rather than absent: for example, the arrangements put together in 1995 by the four lower riparian states of the river Mekong for the sustainable development of the Mekong River Basin located within their territory. The arrangements were directed at cooperation in accordance with a complex range of objectives and principles. According to one analysis:

> Among those legal concepts, the most significant is the concept of sustainable development to the effect that the contracting parties agreed to cooperate and make every effort to use their natural resources in a manner in which the environment and ecological balance of the river is also conserved and maintained.[23]

In this sense the concept of sustainable development – the outcome of the cooperation – comprised the range of objectives, principles and rules in the interstate agreement itself. If these rules in particular were to be implemented in accordance with the agreement, then the concept of sustainable development would emerge as an achievement rather than a concept. In effect, sustainable development would have been achieved simply because that was the will of the body politic.

But how can the legal arrangements prescribed by an interstate agreement be structured so as to ensure the achievement of sustainable development? Possibly there are two ways according to which sustainable development may be achieved. The first is for the international legal system to recognise a rule of international law to this effect and for states to comply with it in designing the agreement, and the second is for states to include within their agreements such a rule. Either way, sustainable development of water resources may be achieved.

The importance of cooperation cannot be overemphasised. Thus, in relation to south Asia:

> The realization of the importance of cooperation is obvious in the treaties concluded thus far, and in the ongoing discussions for agreements on other shared rivers as well. And although treaties and other legal instruments over international rivers will not, by themselves, solve all the problems in such rivers, it is equally true that there can be no resolution of water disputes and conflicts without them. Such instruments are indeed the only means for translating cooperative political will into applicable and enforceable action.[24]

Significantly, the emphasis is again upon political will and cooperation in the exercise of the right to enter into agreements and upon the capacity for such agreements to incorporate at an operational rather than a strategic level enforceable rules of law that will achieve sustainable development. Once in force, interstate agreements will be more than political compacts by including within them protectable normative arrangements such as duties capable of enforcement and rights capable of protection.

(b) Conflicting Perceptions of the Relevant Legal Rules

Achieving consensus is not simply a matter of political will. At stake are issues of economic and legal complexity. States may quite reasonably have different views on such matters. The inherent community of interest in the shared water resources may accordingly become fragmented. The issues which interstate agreements are trying to solve are as diverse and challenging as the hydrological cycles affecting the states in question. Some have more than enough water during the periods of the monsoon and and not enough water at times of scarcity, during other periods. Some suffer from permanent scarcity while others enjoy a reasonable balance between availability and need. Some states may find it easier to manage their interstate water resources than others. In any event, whatever the issues, there are a number of practical difficulties in the way of securing a set of consensual arrangements likely to be successful in solving the problems.

One such difficulty is understanding and applying the relevant principles and rules that comprise the international legal context. For example, 'even if each interested state agrees that the principles of equitable utilization and due diligence are the controlling law, states would still dispute how to apply these standards properly'.[25] Let us, therefore, not underestimate the practical difficulties associated with the relative generality and uncertainty about the relevant rules of international law. Identifying the status and substance of potentially relevant rules of customary international law is no easy task. Even if these rules have been codified in one form or another, their interpretation and application are no easy task either.

It has been suggested that 'states perceive rigid and detailed rules as inhibiting rather than encouraging international cooperation'.[26] What about flexible rules? Thus:

> There are few rules that may be deemed to be part of customary international law. And even if certain core rules are characterised as such (for example, the principle of reasonable and equitable utilization) it is of no great help for these are marked by a high degree interpretive ambiguity.[27]

Interpretive ambiguity, however, is an almost necessary consequence of flexible rules. It is no surprise, therefore, that states perceive differently – and often substantially differently – the substantive legal parameters within which they negotiate.

An instance is how the relationship between the equitable utilisation rule and the no harm rule operates in practice. The perception of an upper riparian state may be markedly different from that of a lower riparian state. A hypothetical example has been the possible impact on a lower riparian state of the construction of a large dam by an upper riparian state on the Blue Nile. One consequence would be the reduction of the flow through the watercourse to the lower riparian state. The arguments based upon the equitable utilization rule and the no 'appreciable' harm rule available to the lower riparian state – Egypt – and to the upper riparian state – Ethiopia – have been hypothetically presented in this way:

> [The Egyptians] would certainly invoke the principle of no appreciable harm, narrowly construed, and reject Ethiopian arguments based on rights conferred by equitable utilization and upper riparian status. If the principle of appreciable harm prevailed equitable utilization would be negated, but at the same time, Ethiopia would be denied the legitimate right of economic development, thus causing it appreciable harm.[28]

In this way it is probably easier to point to 'appreciable' harm as the determining factor rather than to justify an acceptable balance in accordance with the equitable utilisation rule.

The application of the equitable utilisation rule depends first upon the flexibility of equity and second upon the range of factors included – to some extent on a discretionary basis – in the prescribed methodology of decision-making. Consider as an example the waters of international watercourses that flow through Nepal and across the boundary into India. 'Perceptional differences'[29] have made it more difficult to solve these problems. On the one hand, it was a Nepalese perception 'that India is draining Nepal's watershed for its own benefit'.[30] On the other hand:

> The Indian stance has been to defend, as a lower riparian country, its equitable use of these international rivers according to international law and practice. India has contended that it has the right to use the water in accordance with its needs, with the term 'need' embracing its unlimited socioeconomic requirements dependent on the waters of the rivers flowing from Nepal into India.[31]

Perceptional differences have similarly not been uncommon in the Middle East. For instance, in relation to the Jordan–Yarmouk Basin it has been suggested that rules have been 'applied selectively according to the interests of the analyst'.[32] Accordingly, some other method is required 'to embrace all the relevant factors and at the same time attribute weights to the respective principles and factors included in the analysis'.[33]

So the application of the equitable utilisation rule has been and is controversial. However, it is sometimes a matter of dispute whether it applies at all. This has emerged very clearly from this observation about the perceived status of the waters of the river Euphrates from the perspective of the relevant states:

> The Arab–Turkish dispute over the legal status of the waters of the Euphrates can be summed up as follows. Syria and Iraq consider the Tigris and the Euphrates to be international rivers and, consequently, claim a share of their waters. Turkey, on the other hand, does not recognise the international character of these two rivers and only speaks of rational and optimal utilization of the single and unique cross-frontier river basin, the transboundary waters or cross-border waters. Moreover, Turkey sees an unlimited use of these waters according to her needs as her most natural right.[34]

In effect, each nation state has 'embarked on separate projects on the section of the rivers flowing through their respective territories'.[35] An example, ultimately, of no cooperation.

Nevertheless, cooperation – at least to some extent – is less the exception than the norm. In relation to Africa, for example, the 'experience related to the three drainage basins analysed here [the Niger, the Nile and the Senegal basins] has yielded remarkable achievements as regards the will of the states concerned to harness basin resources in a cooperative spirit and within appropriate legal and institutional frameworks'.[36] This success has come about despite cooperation being 'still in its initial phase' and without any experience of implementation. These joint arrangements should be seen, it was suggested, as part of regional economic cooperation and a growing respect for relevant general principles such as the need 'to treat the whole water system of a drainage basin as one integrated unit' and the recognition of a duty 'to respect the rights of every other such state':[37] in more general terms, the recognition of a community of interest in these water resources.

(c) Confronting Hydrological Issues

Entering into interstate agreements often presents a range of other but less esoteric problems to confront. Many of the issues are controversial not only politically but also technically. The adoption of a hydrological approach rather than one based on political or administrative convenience is commonly suggested. In relation to the river Jordan, for example, 'the fact that after decades of strife Israel and Jordan were able to come to agreement on sharing a resource that is vital to their very existence, yet is insufficient to meet the needs of both of them, is truly remarkable'.[38] How a scarce resource should be shared for the future on an ongoing basis is perhaps an even greater challenge. It has been suggested, therefore, that:

> The parties to the Peace Treaty seem to have taken the approach of trying to hide existing problems under the mirage of additional water, rather than coming to grips with such difficult issues as present misallocation and unsustainability of existing use patterns. It can only be hoped that the framework of principles and the joint mechanism established by the Peace Treaty will prove up to the task of adjusting to changing needs and conditions.[39]

Accordingly, a contextual and institutional framework needs to exist for successful negotiations to proceed. Such a framework should perhaps align with the hydrological perspectives to be effective.

Sometimes, however, the contextual framework selected for an interstate agreement is not the one most likely to be effective in the longer term, perhaps because it fails to align with hydrological reality. There is much support for the integrated management of water resources – and indeed other resources – within a basin or watershed context. Such arrangements do not always eventuate. It has been argued, for example, that 'the Euphrates should be looked on as a river and its tributaries, rather than a sole river'.[40] The argument then proceeds:

> Curiously, no mention of the river basin or fluvial basin concept is ever made in discussions about the common utilization of the Euphrates. Using this notion [the river basin concept] would negate claims that no treaty regulating the sharing of the waters between the different riparian states exists.[41]

But no such notion or concept was used. There was no negotiated regime that 'regulates the sharing or common exploitation of the Euphrates'.[42] There were merely declarations of rights in relation to water entering the territory of the relevant states.

The waters of the river Nile have for many years been managed in accordance with the arrangements between Egypt and the Sudan – the two lower riparian states – and these arrangements distributed the benefits equally between the two states. Consequently, it has been noted:

> No participation of any upstream state was sought either in that, or any other, basin-wide scheme. Furthermore, the needs, immediate as well as long term, of these states were not taken into consideration. However, in appraising this state of affairs, account must be taken of the fact that none of the upstream states had, until recently, made any tangible claims to the waters of the Nile.[43]

Again, a failure to align with the hydrological realities.

The consensual arrangements for the management of the waters of the river Indus – with its source in Tibet – and its tributaries have been negotiated by Pakistan and India. In effect, the exclusive use of the waters of the river system so far as it is located within these two states was granted either to Pakistan or to India in accordance with the actual location of the waters of the river, irrespective of whether the river flowed within the territory of one or other of the states. The arrangements have been described in this way:

> The fact that there were six rivers in the system offered the simple solution of the three western rivers (the Indus, Jhelum and Chenab) being reserved for consumptive use by Pakistan, and the three eastern rivers (the Ravi, Beas and Sutlej), being reserved for consumptive use by India.[44]

The usufructuary entitlements were thus based upon physical location and not volumetric quantity. Apparently, '79 percent of the total volume of waters (the statistical average of the three rivers) was made available to Pakistan, whereas the eastern rivers earmarked for India equalled only the balance of 21 percent'.[45] Clearly there was no contemplation that the river Indus and its tributaries were part of a river basin or formed a single hydrological unit.

In any event a number of reasons have been suggested for 'the successful conclusion' of these arrangements. These included:

- the help of an international organisation during the negotiations;
- avoiding the difficulties associated with joint basin management;
- the uniqueness of the solution;
- the provision of financial resources in support of the projects involved.[46]

The involvement of outside agencies has been seen as a reward for cooperation. Thus:

> By providing Pakistan with the resources needed to control its own water supply, the World Bank in effect bought one party its objective, while relieving the other, India, of the burden of the dispute.[47]

In juridical terms, it is the nature of the entitlement to the waters of the river system that is significant: namely, an entitlement to the exclusive use of waters in a specified location.

(d) The Need for Comprehensive Information

The quality of decisions, it is often observed, is a reflection of the quality of the information upon which they are based. This applies as much to strategic planning, watershed or water basin planning, operational activities and even day- to-day activities. The issues are particularly complex at the levels of strategic and basin planning. In relation to the river Mekong, for example, it has been suggested that there is one overriding need, namely:

> This is for the basic data which are indispensable for rational planning and the operation of the water resources projects. The exploitation of water resources could hardly be contemplated unless all the relevant facts are known about rainfall and run-off, not just for one or two years but for decades, and not just within national boundaries but for the whole basin; otherwise resources would be wasted on reservoirs that might never fill up or dams that might be just washed away.[48]

Further, the difficulty in applying relatively general and potentially ambiguous legal principles 'is compounded by insufficient knowledge about the hydrological behaviour of river basins'.[49]

The need for information, knowledge and understanding goes beyond planning and implementing water resource projects. It applies with ever greater force to the interconnections between rainfall, runoff, river flows, aquifer absorption and aquifer recharge as well as to the implications of human abstraction and discharge. The problem is even more complex if this is translated to water resources, watersheds or water catchments and underground aquifers that interconnect within nation states and across the boundaries of nation states. If cooperation and eventually common management arrangements are to be effective, then at least this much is required.

For example, for a 'successful and healthy partnership to develop between India and Nepal', it has been suggested that:

> An environment of trust that facilitates unfettered exchange of information is a prerequisite that can only lead to purposeful dialogue between the two countries.[50]

What is required is not only dialogue but also good faith and an appropriate 'culture of behaviour and thinking'.[51] All of this hopefully will lead ultimately to 'well-balanced and well-drafted comprehensive legal instruments'.[52] The form of these legal instruments must reflect the function that they are intended to perform. These arrangements must themselves be based upon a complete and comprehensive understanding of the issues, which is itself based upon comprehensive knowledge and information appropriately and effectively analysed so that these legal instruments perform their function effectively.

(e) Conclusion

The obligation to cooperate is a general responsibility recognised by international law. The challenges in responding to it relate not only to potential interstate agreements between two states with a clearly identified community of interest but also to all states with an interest in and affected by the way water resources are managed – even if the community of interest goes beyond water basins and watercourses in two or more states. Many of the rivers and river systems in southern and eastern Asia rise in the Tibetan plateau. One of these leads ultimately to the river Ganges, which itself is part of a wide river basin system. There are interstate agreements affecting this river system between India and Nepal and between India and Bangladesh. But there is no set of consensual arrangements for the management of the Ganges Basin as a basin.

The need for such an arrangement which addresses comprehensively many of these challenges has been strongly advocated in this way:

> The Ganges Basin is one of the most populated in the world, with a population that exceeds that of North America. The bulk of the population is the poorest of the world. The challenges of the basins are tremendous: floods, drought and environmental degradation, including bank erosion. The challenges of the other shared water resources between India and Bangladesh are equally staggering. Those challenges can only be met through cooperation, and not just between India and Bangladesh. Nepal, China and Bhutan should be included in a wider Ganges–Brahmaputra–Meghna collaborative basin management arrangement. Such arrangement should go beyond the issue of quantitative sharing of the waters of those rivers to include other areas such as lean season augmentation, hydropower generation, flood control and environmental protection of those rivers. Any basin management arrangement will need to address immediately the issue of expansion and modernisation of the irrigation system and efficient use of the waters of those rivers so as to pull the more than half a billion inhabitants of this basin from their misery and poverty.[53]

A scheme of governance, in other words, directed towards the sustainable development of the Ganges Basin and one that includes the three pillars of sustainable development – economic, social and ecological.

THE RESPONSE TO THE CHALLENGE OF COOPERATION IN EUROPE

(a) Scandinavia

While the managers of international watercourses in Europe have been confronted in some respects with different issues, the solution has on many occasions been an interstate agreement. The evolution of arrangements in Scandinavia illustrates a range of interesting developments. The arrangements entered into in 1929 between Sweden and Norway constituted 'a regime based upon cooperation and consultation, without any permanent or supranational organisational framework'.[54] It was based upon three principles: consultation, cooperation and national treatment. These, it has been suggested, 'constitute the most common application of *voisinage*'.[55]

The distinguishing feature of these arrangements is national treatment.[56] The arrangements are administered by the competent regulatory authorities in each state, and deliberately not by officials operating at the strategic or public policy level. Notification is given of any proposed project to the other state. Appropriate information is made available so that the impact of the proposed project in each of the states can be taken into account. The decision on the application for authorisation is made in accordance with the law of the state concerned. Accordingly, the interests of both states are taken into account in the formal decision-making process undertaken in the state concerned. This 'localisation' of regulatory procedures has been described as 'characteristic of the regime of *voisinage*'.[57] These arrangements clearly recognised a community of interest procedurally and substantively. But equally clearly they did not amount to a system of joint management.

The arrangements negotiated between Finland and Sweden in 1971 were quite different. The interstate agreements specifically incorporated public interest criteria in making decisions on particular projects. For example, in determining an order of priority, 'preference shall be given to the project which may be assumed to be of the greatest public and private benefit'.[58] Similarly, 'hydraulic works which permanently diminish the comfort of local residents or which cause significant nature conservation losses' may be authorised 'only if the works are of particular importance from some public standpoint'.[59] Clearly this prescribes how the function of substantive decision-making in relation to particular projects is to be performed. The interstate agreement places this obligation upon the competent authorities in the state in a way that presumably overrides, to the extent of any inconsistency, the decision-making process required to be undertaken within the state.

The nature of this regime has been described in this way:

> No longer are frontier waters treated almost like a geological fault line, where chaffing could send serious tremors deep into the territory of the neighbouring state, and where friction had to be eliminated by the delicate adjusting of disputes between individuals. In the 1971 agreement, the riparian states marched boldly in to the frontier zone, regulating it, and imposing and preserving a balancing of interests in the zone for what is perceived to be the greater public good.[60]

The performance of these functions is supported by a sophisticated set of institutional arrangements. What emerges, therefore, is a self-imposed regulatory regime that determines whether and how a proposed development can go ahead. An appeal lies to 'the supreme water right courts of the state where the interests concerned are located'.[61] These arrangements, nevertheless, go some way towards devolving responsibility for managing water resources in which there is a community of interest to a range of institutions, some of which sit above the state in the hierarchy of institutional relationships.

(b) The Rhine Basin

The evolving arrangements for the protection of the river Rhine are perhaps even more ambitious. By the 1970s, as a result of industrial, agricultural and urban developments, 'the ecosystem of the river was virtually dead'.[62] Since then attitudes have changed and 'all states accepted equal levels of commitment to reduce pollution loads and to undertake projects to rehabilitate the ecosystem'.[63] In 1999 these new approaches were incorporated formally into the regime for managing the Rhine Basin established by the plurilateral agreement for its protection. Most importantly, the basis of the arrangements had been dramatically changed. 'Its underlying principle of equal apportionment has been exchanged for the principle of sustainable development.'[64] Protection of the ecosystems of the river basin and of its environment had become just as important as – perhaps even more important than – the economic benefits accruing from the use of the river.

Perhaps even more important in legal terms, 'the regime no longer consists of a set of obligations to which riparian states must adhere in different ways depending on whether they are upstream and downstream states. It contains similar obligations for the protection of the ecosystem for all states – largely irrespective of their being upstream or downstream states.'[65] In other words, a set of specific obligations has perhaps modified – or perhaps even superseded – the principle of equity and its application to international watercourses through the relationship between the equitable utilisation rule and the no harm rule. The emerging obligations are thus less negative and more positive and less reactive and more proactive. The implementation of these arrangements remains, however, a major challenge for the states in the basin of the river Rhine and its tributaries.

CONDITIONS NECESSARY FOR SUCCESSFUL INTERSTATE AGREEMENTS

(a) A Mutuality of Interest

If international water resources, their ecosystems and their environments are to be used and developed sustainably, then the artificial barriers created by the territorial boundaries of nation states need to be removed, or at least their impact modified. One way is for the states in question to enter into agreements along the lines under discussion. Is it possible to identify the conditions to be satisfied before such agreements can be successfully negotiated and then implemented?

Perhaps a few can be identified. Most important of all is for each state to recognise that other states have a legitimate interest in the water resources in question. The mere fact that a watercourse flows across two or more states or is the boundary between two or more states should in itself be sufficient; similarly, where it is known that an underground aquifer underlies the superficial territory of two or more states. Perhaps even the flow of water across the surface of land from the territory of one state to that of another may be sufficient, for example, where a range of mountains is located fairly close to the boundary with another state lying in the floodplain.

The interest of each state in these shared water resources need not be the same. The interest of one state may be the quantity of water for irrigation; that of another the quality of water for the processing of food products; that of another for urban water supply; or that of another for hydroelectric generation. In this sense the recognition of a diversity of interests is the first step towards recognising a community of interest. Similarly, a community of interest does not necessarily mean an equality of interest. But circumstances change. This, too, needs to be recognised as part of the community of interest. Unless there is a recognition of a community of interest in this way, the likelihood of successful negotiations is reduced.

(b) Good Faith

Once this test has been satisfied, then there is at least a reason to cooperate. This is based upon a recognition that what one state does is likely to impact upon another state and the interests of that other state. Hence it is important to know, to be fully informed, to participate and hence to cooperate. And it is cooperation in good faith. To cooperate in good faith implies an understanding that any perception of exclusivity in the water resource must be abandoned, and this includes exclusive control of the water resource while it is located within the territory of the state. This necessarily involves acceptance of a limitation upon the exercise of rights otherwise regarded as rights of sovereignty.

(c) Observable Principles

To facilitate negotiations within such a framework, one or more principles could be adopted to guide the negotiators in their deliberations. More particularly, a set of principles according to which the arrangements for managing the water resources in question would be determined. The states themselves might declare these principles: for example, equity, equality, priority of interest, priority of need, perhaps even sustainability. It would be even more helpful if the general principles of international law as they relate to water resources governance were accepted. While these might be difficult to identify, the negotiations could be facilitated by an appropriately qualified international jurist. In any event, it would not be appropriate for any such principles to be seen to be binding. In this way they would represent, in the first instance at least, the parameters within which discussions would take place.

(d) Detailed Knowledge

Just as important as a set of guiding principles would be a detailed and comprehensive knowledge and understanding of all the facts, information and circumstances relevant to the issues. Acquiring such information, analysing it and understanding it from all perspectives would be no mean task. But without it, the arrangements concluded in any subsequent agreement would, more likely than not, be flawed. Such information would relate to what has happened in the past, what was happening then, and what was likely to happen in the future. This would involve strategic analysis, policy analysis, economic analysis, social impact analysis and environmental impact analysis – particularly if a sustainable outcome were contemplated. All of this would require substantial intellectual as well as financial resources. Hence the existence of such resources or the capacity to have them available would be critical not only for any successful negotiations but also for the successful implementation of any subsequent agreement. This set of conditions is somewhat ambitious. But unless most of them are satisfied, interstate agreements are unlikely to be effective.

TRADE, TRANSFERS AND INTERSTATE AGREEMENTS

(a) Introduction

While states are free to manage their water resources as they see fit, including by way of interstate agreements, they are limited by whatever obligations are owed to the other party or parties to the agreement, to any other state with whom they share a community of interest or to any other member of the international community, but always in accordance with international law. The range of obligations increasingly imposed upon states relate to the protection of ecosystems and the protection of the environment. However, obligations are owed to other members of the international community in the context of trade. Although these obligations are designed to facilitate free trade, almost inevitably and somewhat paradoxically there are restrictions on activities that would thwart free trade.

(b) The Issues

Where do water resources and water fit within this scenario? This is an issue for water resources as well as for natural resources generally, because the use of economic instruments or market mechanisms is increasingly seen as one way of achieving sustainable development. From this perspective trade is a means to an end and not an end in itself. But what is trade? One aspect is the buying and selling of a commodity or of a service. We have already discussed if and when water becomes a commodity. The question then becomes whether trading in water or in interests in water is subject to the rules of international trade law. A critical question is what is being traded. Water, a water entitlement, an interest in water or something else.

A distinction perhaps needs to be drawn between a trade and a transfer. A trade is an arrangement to buy and sell something or to do something else with it. A transfer is the completion of the arrangement to buy and sell it. In relation to water resources and water, it is more complicated. Water has been physically conveyed from one location to another by means of one hydraulic system or another for thousands of years.[66] A hydraulic system simply makes the water in this way available in one place rather than another. There is no transaction in a commercial sense. Although their size and function may have changed, such arrangements continue: for example, the storage of water in a reservoir, its physical conveyance through pipes or channels to another location and its availability for use through a reticulation system. These arrangements may be by way of public supply, commercial transaction or some other form. Arrangements such as these take place within states and may also take place between and among states. One way of formalising these arrangements is by interstate agreement.

Traditionally, interstate agreements have been put in place to facilitate the use and development of shared water resources in support of economic growth. The focus has now changed to include the protection of ecosystems and the environment. An ecosystem may need an increase in the available water. To increase the available water may similarly involve interference with the physical flow of the water and the actual movement of water from one location to another. The water may remain under the control of the same institution throughout the process from abstraction to delivery and use. Or it may not. The water may cross state boundaries. It may be transferred physically from the location of storage to the location of distribution, to the location of supply and to the location of use or consumption.

The legal character of water may change from water as a resource to water as a commodity. These physical movements accordingly may affect the legal rights and duties of the range of institutions and persons involved. For example, the institution or person in lawful control of the water may have the physical capacity to move the water, but not through the pipes or channels by means of which it is proposed to move. If these pipes and channels already exist, then arrangements will need to be put in place to obtain the consent of those in control of them. Alternatively the necessary reticulation facilities will need to be constructed, and this may well involve a range of consents and approvals from the owners of the land and the relevant public agencies responsible for planning and environmental protection. Moreover, the nature of the transaction may change. Is the actual movement of the water simply a part of a public supply service? In this case, the legal character of the arrangements has not changed. Is the actual movement of the water a transfer of water in the sense that a contract to buy and sell water is completed through a transfer of ownership by delivery or *traditio*? If it is a trade followed by a transfer in this sense, are these transactions subject to any international legal constraints?

The transfer of water across international boundaries has been analysed in three ways:

- water flowing in its natural state across national borders;
- export of water products or products containing water;
- export of bulk water.[67]

The first and the third are relevant for present purposes. The use and development of the water resources of international watercourses are governed in the first instance by the rules of international law – particularly the equitable utilisation rule in association with the no harm rule – or by the rules included in interstate agreements. It has been suggested that 'if water flowing in its natural state in an international watercourse were to be subject to international trade rules, it would insert yet another criterion in determining whether such use was reasonable and equitable, and that criterion might override the others'.[68] Further:

> If the trade rules were to apply, decisions and disputes about international watercourses could be decided by trade dispute bodies. This would contradict countries' intentions in negotiating the [United Nations Convention on the Law of the Non-navigational Uses of International Watercourses of 1997] and the hundreds of regional and bilateral international water agreements.[69]

This raises directly the relationship between the rules of international trade law, the rules of international water resources law and the rules in interstate water agreements. For the reasons already discussed, it seems reasonably clear that the rules forming part of the arrangements for the use and development of international water resources in interstate agreements govern the legal rights and duties of the states that are parties to the agreement. Accordingly it has been suggested that:

> While international water agreements might prevail against trade agreements to the extent of the inconsistency, it is not clear how rules of customary international law regarding the allocation, access and use of fresh water would fare if they were deemed inconsistent with provisions in trade agreements.[70]

(c) Arrangements for the Transfer of Water

An interesting example is the set of arrangements entered into between Lesotho and South Africa. The interstate agreement authorised the construction and implementation of the Lesotho Highlands Water Project. This involved the construction of the infrastructure necessary to generate electricity in Lesotho and ensure the delivery of water across the border into South Africa. The relevant institutions in Lesotho and South Africa were set up for this purpose and the Joint Permanent Technical Commission was established to oversee the activities of the institution in each state and to ensure ongoing implementation of the project in accordance with the interstate agreement. The agreement required each state to enact appropriate legislation to enable the treaty to be put into effect within each of the jurisdictions.

Then there were the obligations relating to the use of the water. The agreement placed an obligation on Lesotho to ensure the delivery of the specified quantities of water to South Africa. An obligation was also placed upon Lesotho to ensure that minimum rates of flow would be maintained in the natural river channel downstream of each of the dams. The agreement included detailed arrangements for the payment of the costs associated with the project, for the financing of the project and for the payment of a royalty to be shared 56% to Lesotho and 44% to South Africa. The royalty payment comprised the net benefit accruing from the project to each of the two states.

The treaty contains five provisions that are particularly noteworthy from a wider perspective. Article 6(18) is something in the nature of a free trade provision. Thus:

> Neither party shall apply quantitative restrictions on goods, materials, plant, equipment or services, whether produced or manufactured in the territory of any of the parties or imported from outside such territories, required for the implementation, operation or maintenance of the project or any phase thereof.

Accordingly the procurement of all such goods and services was to be without any restriction on the source of supply.

Article 15 incorporates provisions not directly related to the economic and commercial arrangements put in place by the treaty. Thus:

> The parties agree to take all reasonable measures to ensure that the implementation, operation and maintenance of the project are compatible with the protection of the existing quality of the environment, and in particular, shall pay due regard to the maintenance of the welfare of persons and communities immediately affected by the project.

The relationship between social and environmental considerations on the one hand and economic and commercial considerations on the other hand is regulated by the criterion of compatibility. This is similar to but perhaps not exactly the same as the criterion of consistency. In any event there is a duty – not an absolute but a qualified duty – to take all reasonable measures to ensure the compatibility of these outcomes.

Article 16 provides in some detail for the prevention and settlement of disputes. The overriding consideration to be taken into account in resolving any dispute is directed at two matters. The first is to do so in a spirit of conciliation and the second is to avoid any impairment of the implementation, operation and maintenance of the project. Any disputes are determined in the first instance by the relevant institution in each of the states. If that fails, it becomes the responsibility of the Joint Permanent Technical Commission. If that fails, it is a matter of negotiation between the parties. And if that fails there is a reference to an Arbitral Tribunal. The award of the tribunal is definitive and binding. And each of the states is under a duty 'duly and expeditiously' to give effect to the award.

The law to be applied by the tribunal is stated in Article 16(17). Subject to any agreement otherwise by the states, the law to be applied is the treaty. If it is necessary to go beyond the treaty, the rules of law to be applied are, in order of priority, these:

- international agreements entered into by both states;
- customary international law universally recognised or having received the assent of both states;
- Roman Dutch customary law;
- all such other rules of law in force in both Lesotho and South Africa.

The significance of the order is this. The laws to be applied first are those that contain obligations that the states have voluntarily undertaken. The next two are customary laws – either international or Roman Dutch. The last is a rule of law in force in both states – a recognition that each state effectively recognises the existence of such a rule of law.

The fifth point of note is Article 17(2). It states:

> The provisions of this treaty are without prejudice to the rights under public international law of riparians of the Senqu/Orange River other than the Kingdom of Lesotho and the Republic of South Africa.

This acknowledges that the river is an international watercourse in which states other than Lesotho and South Africa have an interest. It is consequently a recognition of the doctrine of a community of interest in an international watercourse and that this community of interest has itself been recognised and capable of being protected under the rules of international law. Accordingly, the structure and form of the treaty reflect the functions that the international legal system in relation to shared water resources seeks to perform.

How should these arrangements be characterised? As water flowing in its natural state across national boundaries or as exported water? Subject to international water resources law or subject to international trade law? The treaty clearly involves the movement of water from one state to another for economic, social and ecological purposes. There is no suggestion – at least directly – that water is being 'traded' in the commercial sense. But there are commercial aspects to the arrangements. While the treaty provides essentially for the use and development of their shared water resources, the arrangements that comprise the treaty include restrictions designed at least to some extent to protect the environment and to ensure free trade. Accordingly the perspectives reflected in the arrangements include those that are economic, commercial, social, environmental and ecological. The treaty in particular authorises the transfer of water from Lesotho to South Africa to be delivered in specific quantities and in specific locations. This may be seen to be to some extent an 'export' of water. But export in a different sense from water transfers that are increasingly taking place between other states – particularly those with water scarcity problems, for example, the export of water from Turkey to states in the Middle East. In these instances the water seems to be traded directly as a commodity rather than as a transfer of water or of the benefits of water as one of the resources in a watercourse.

Arrangements to export or to transfer water such as these point significantly to the growing relationship between the rules of international water resources law, the rules of international environmental law and the rules of international trade law. What is becoming increasingly relevant, therefore, is a trilogy of international legal regimes:

- the regime that facilitates the use and development by the state of its natural resources;
- the regime that focuses upon the protection of the environment and increasingly upon the protection of ecosystems;
- the regime that facilitates free trade.

The relationship between these three regimes is important. But perhaps even more important is their relationship with sustainable development.

(d) The Relationship between Use Rules and Trade Rules

Let us turn now to how the relationship between the rules of international water resources law and those of international trade law has been analysed. In this context the rules of international trade law are for the most part those set out in the General Agreement on Tariffs and Trade of 1994. Essentially, there must be no prohibition or restriction on the importation or on the exportation of any product destined for the territory of any other state which is a party to these arrangements. This does not apply to duties, taxes or other charges. And a state may take measures necessary to protect human, animal or plant life or health, as well as measures relating to the conservation of exhaustible natural resources provided such measures are applied in a manner which – to paraphrase the provision – would be neither discriminatory nor restrictive.

How do these rules apply to water? First of all, water must be a product. Even if water is perceived as a product or a commodity, the water may be used for a range of different purposes at different times. Each of these purposes may be perfectly justifiable. In certain circumstances, it may be sensible to permit the export of water. But a change in circumstances may suggest a different response. This is particularly so if the water becomes valued not so much for economic purposes as for ecological purposes.

Assuming, therefore, that water is a commodity capable of being traded across international boundaries, what are the implications of the application of the rules of the General Agreement on Tariffs and Trade to water? Three have been identified from the perspective of facilitating trade in water and four from the perspective of facilitating the multiple purposes for which water may be used.[71] To focus first upon facilitating trade, the rules:

- recognise the status of water as a commodity;
- level the playing field for trade in water resources;
- protect against disguised barriers to trade.

Focusing next upon facilitating the diverse uses of water, the rules:

- constrain a state's ability to protect ecological systems;
- make long-term management of water resources more difficult;
- give trade considerations a significant role in reconciling conflicting uses of water;
- give trade dispute settlement mechanisms authority to resolve water claims.

The promotion of free trade and of multiple purpose water use may both be seen as objectives in themselves. But they are conceptually quite different. The function of trading is a means to an end. And the end is economic growth with a view to improving the welfare of the individual and the economic health of the community. It is to some extent linked to the achievement of economic and social justice, and perhaps even ecological justice, but in an indirect way. The need to achieve multiple purpose water use recognises that water may be used for economic, social and ecological purposes. But each of these purposes for which water may be used is itself an objective.

Let us extend the analysis to include sustainability. If sustainable development is to be achieved, then each of these subsidiary objectives must be so related to each other that the ultimate objective of sustainable development is achieved. Although in this sense they are individually a means to an end, they are more than that, for they are part of the overall objective. Although free trade can no doubt be seen as end in itself in some contexts, it is a means to an end rather than an end in itself when it is linked to the way natural resources and their environment should be used and developed. In other words, the application to water of the rules in the General Agreement on Tariffs and Trade mandates a methodology of decision-making which is fundamentally inconsistent with the methodology of decision-making appropriate to the achievement of sustainable development. Certainly, the methodology of decision-making under the General Agreement on Tariffs and Trade does not exclude the social and ecological pillars of sustainable development. It simply renders their function unequal and ambiguous.

It is not only the nature of sustainable development which is critical to determining the relationship between rules about trade and rules about use and development. The nature of water itself is critical. It is trite to note that it is both unique and essential. It is accordingly beginning to emerge as a human right recognised by international law. These characteristics suggest that an approach of 'anticipatory caution' should be adopted. What does this mean?

> Anticipatory caution means that in the face of uncertainty about the future, a country should be able to exercise its sovereign authority to maintain its fresh water resources without having to convince the trade community of the legitimacy of its actions. This is especially important because of the difficulty in predicting water demand and water supply, and of anticipating adequately the natural variations that will affect water availability and the stability and integrity of the Earth's ecosystems.[72]

The rules of international trade law, those of international environmental law and those of international water resources law often pull and push in different directions. There are perfectly understandable historical and political reasons for this. If sustainable development is to be achieved, including the sustainable use and development of water resources, the rules of the international legal system – and of a national legal system for that matter – ought to be at least consistent and coherent, if not fully harmonised.

CONCLUSION

The concept of sovereignty enables a state to enter into agreements with other states about how water resources are managed. This has proved to be a very significant aspect of how internationally shared water resources are managed. These agreements between states have disclosed a number of features. Some operate at the level of strategy or planning. Others provide for specific projects in particular sets of circumstances. A small number are legislative in the sense that they create a framework of rights and duties in relation to how shared water resources are managed. Agreements are clearly the result of negotiation and diplomacy. But these negotiations take place within the context of the general principles of international law. For example the principle of equitable utilisation and the principle of no harm. While the expectation is that these principles are observed in designing these agreements, this does not always happen in practice.

These agreements almost necessarily involve cooperation among the states and the joint implementation of the arrangements. What one state does impacts upon another state. In this sense water flowing through an international watercourse and crossing territorial boundaries into the watercourse located within a neighbouring state necessarily affects the quantity and the quality of the water received by the neighbouring state. In this sense the natural flow of the water amounts to a transfer or a conveyance of the water from one state to another. The classification of transboundary movements of water is becoming a significant issue. Are these circumstances governed by the rules of international law about the use of international watercourses, about the protection of the environment or about trade between states. Significantly the relationship between social, economic and ecological values in such arrangements reflects the different values underpinning each of these sets of legal arrangements. These questions are assuming greater importance. There is however no clear answer to them.

NOTES

1. Louka (2006, p. 177).
2. Salman and Uprety (2002, p. 199).
3. Chimni (2005, p. 93).
4. Chimni (2005, p. 94).
5. Chimni (2005, p. 94).
6. Dellapenna (1994, pp.42–47).
7. Dellapenna (1994, p. 42).
8. Dellapenna (1994, p. 43).
9. Dellapenna (1994, p. 44).
10. Dellapenna (1994, p. 46).
11. Okidi (1997, p. 171).
12. Okidi (1997, p. 173).
13. Okidi (1997, p. 171).
14. Okidi (1997, p. 174).
15. Okidi (1997, p. 178).
16. Salman and Uprety (2002, p. 199).
17. Fuentes (1998, p. 102).
18. Fuentes (1998, p. 103).
19. Dellapenna (1997, p. 133).
20. Dellapenna (1997, pp. 133 and 134).
21. Dellapenna (1997, p. 134).
22. Salman and Uprety (2002, p. 196).
23. Pichyakorn (2005, p. 175).
24. Salman and Uprety (2002, p. 202).
25. Dellapenna (1997, p. 123).
26. Chimni (2005, p. 63).
27. Chimni (2005, p. 64)
28. Goldberg (1995, p. 163, quoting Thomas Naff).
29. Salman and Uprety (2002, p. 119).
30. Salman and Uprety (2002, p. 119).
31. Salman and Uprety (2002, pp. 119 and 120).
32. Al-Kloub and Al-Shemmeri (1996, p. 196).
33. Al-Kloub and Al-Shemmeri (1996, p. 196).
34. Chalabi and Majzoub (1995, pp. 209 and 210).
35. Chalabi and Majzoub (1995, p. 198).
36. Godana (1985, p. 333).
37. Godana (1985, pp. 335 and 336).
38. McCaffrey (1997b, p. 165).
39. McCaffrey (1997b, pp. 164 and 165).
40. Chalabi and Majzoub (1995, p. 192).
41. Chalabi and Majzoub (1995, p. 192).
42. Chalabi and Majzoub (1995, p. 192).
43. Tamrat (1995, p. 181).
44. Salman and Uprety (2002, p. 57).
45. Salman and Uprety (2002, p. 58).

46. Salman and Uprety (2002, p. 59).
47. Salman and Uprety (2002, p. 59).
48. Chomchai (1995, p. 253).
49. Chimni (2005, p. 102).
50. Salman and Uprety (2002, p. 122).
51. Salman and Uprety (2002, p. 122).
52. Salman and Uprety (2002, p. 122).
53. Salman and Uprety (2002, p. 191).
54. Fitzmaurice and Elias (2004, p. 95).
55. Fitzmaurice and Elias (2004, p. 96).
56. Fitzmaurice and Elias (2004, pp. 96 and 97).
57. Fitzmaurice and Elias (2004, p. 97).
58. Fitzmaurice and Elias (2004, p. 115).
59. Fitzmaurice and Elias (2004, p. 115).
60. Fitzmaurice and Elias (2004, p. 116).
61. Fitzmaurice and Elias (2004, p. 116).
62. Nollkaemper (2005, p. 153).
63. Nollkaemper (2005, p. 162).
64. Nollkaemper (2005, p. 165).
65. Nollkaemper (2005, p. 165).
66. McCaffrey (2007, pp. 8 and 9).
67. Brown Weiss (2005, pp. 63–67).
68. Brown Weiss (2005, pp. 64).
69. Brown Weiss (2005, pp. 64 and 65).
70. Brown Weiss (2005, pp. 66).
71. Brown Weiss (2005, pp. 80–83).
72. Brown Weiss (2005, pp. 83).

GENERAL REFERENCES

Al-Kloub, B. and T.T. Al-Shemmeri (1996), 'Application of multi-criteria decision aid to rank the Jordan-Yarmouk basin co-riparians according to the Helsinki and ILC rules', in J.A. Allan (ed.), *Water, Peace and the Middle East* (London: Tauris Academic Studies), pp. 185–207.

Brown Weiss, Edith (2005), 'Water transfers and international trade law', in Edith Brown Weiss et al. (eds), *Fresh Water and International Economic Law* (Oxford: Oxford University Press), pp. 61–89.

Chalabi, Hasan and Tarek Majzoub (1995), 'Turkey, the waters of the Euphrates and public international law', in J.A. Allan and Chibli Mallat (eds), *Water in the Middle East: Legal, Political and Commercial Implications* (London: Tauris Academic Studies), pp. 189–235.

Chimni, B.S. (2005), 'A tale of two treaties: the Ganga and Mahakali agreements and the watercourses convention', in Surya P. Subedi (ed.), *International Watercourses Law for the 21st Century* (Aldershot: Ashgate Publishing Limited), pp. 63–103.

Chomchai, Prachoom (1995), 'Management of transboundary water resources: a case study of the Mekong', in Gerald H. Blake et al. (eds), *The Peaceful Management of Transboundary Resources* (London: Graham and Trotman Limited), pp. 245–259.

Dellapenna, Joseph W. (1994), 'Treaties as instruments for managing internationally shared water resources: restricted sovereignty vs community of property', *Case Western Reserve Journal of International Law* 26, 27–56.

Dellapenna, Joseph W. (1997), 'The Nile as a legal and political structure', in Edward H.P. Brans et al. (eds) (1997), *The Scarcity of Water: Emerging Legal and Policy Responses* (London; Boston: Kluwer Law International), pp. 121–134.

Fitzmaurice, Malgosia and Olufemi Elias (2004), *Watercourse Co-operation in Northern Europe: a Model for the Future* (The Hague: TMC Asser Press).

Fuentes, Ximena (1998) 'Sustainable development in the equitable utilization of watercourses', *British Yearbook of International Law* 69, 119–200.

Godana, Bonaya Adhi (1985), *Africa's Shared Water Resources* (London: Francis Pinter Publishers).

Goldberg, David (1995), 'World Bank policy on projects on international waterways in the context of emerging international law and the work of the International Law Commission', in Gerald H. Blake et al. (eds), *The Peaceful Management of Transboundary Resources* (London: Graham and Trotman Limited), pp. 153–166.

Louka, Elli (2006), *International Environmental Law* (Cambridge: Cambridge University Press).

McCaffrey, Stephen C. (1997b), 'Middle East water problems: the Jordan River', in Edward H.P. Brans et al. (eds), *The Scarcity of Water: Emerging Legal and Policy Responses* (London: Kluwer Law International), pp. 158–165.

McCaffrey, Stephen C. (2007), *The Law of International Watercourses: Non-Navigational Uses* (2nd edn, Oxford: Oxford University Press).

Nollkaemper, André (2005), 'The evolution of the regime for the River Rhine', in Surya P. Subedi (ed.), *International Watercourses Law for the 21st Century* (Aldershot: Ashgate Publishing Limited), pp. 151–166.

Okidi, C.O. (1997), 'International law and water scarcity in Africa', in Edward H.P. Brans et al. (eds), *The Scarcity of Water: Emerging Legal and Policy Responses* (London: Kluwer Law International), pp. 166–180.

Pichyakorn, Bantita (2005), 'International watercourses law: the experience of the Mekong River Basin', in Surya P. Subedi (ed.), *International Watercourses Law for the 21st Century* (Aldershot: Ashgate Publishing Limited), pp. 167–192.

Salman, Salman M. A. and Kishor Uprety (2002), *Conflict and Cooperation on South Asia's International Rivers: a Legal Perspective* (The Hague: Kluwer Law International).

Tamrat, Imeru (1995), 'Constraints and opportunities for basin-wide cooperation in the Nile: a legal perspective', in J.A. Allan and Chibli Mallat (eds), *Water in the Middle East: Legal, Political and Commercial Implications* (London: Tauris Academic Studies), pp. 177–188.

LEGAL INSTRUMENTS

General Agreement on Tariffs and Trade 1994, (1994) *International Legal Materials* 33, 9–12.

Treaty on the Lesotho Highlands Water Project between the Government of the Kingdom of Lesotho and the Government of the Republic of South Africa 1986, in Food and Agriculture Organisation of the United Nations (1997), *Treaties Concerning the Non-navigational Uses of International Watercourses Africa* (Rome: Food and Agriculture Organisation), pp. 171–212.

United Nations Convention on the Law of Non-navigational Uses of International Watercourses (1997) *International Legal Materials* 36, 700–718.

PART VI

Evolving national arrangements for water resources governance

The doctrinal foundations of contemporary international law emerged only some 350 years ago. The principles and rules of international law governing the use and development of natural resources – including water resources – have increasingly been a phenomenon of the second half of the 20th century. But the ways in which local communities manage their water resources have been evolving over thousands of years – probably longer – in response to changing philosophical, religious and cultural perceptions. Accordingly, the arrangements for the governance of water resources in particular communities and their associated practices, customs and rules have responded to the diverse and constantly changing circumstances which confronted the community at any particular time.

Roman law evolved from custom, through rules to a more dominant involvement by the state in a regulatory as well as an operational capacity. Islamic law has always paid particular attention to water resources. Their use has been influenced by the customs acknowledged by local communities, which have themselves been guided by the religious underpinnings of their culture. English law similarly evolved from customary practices confirmed by sets of legal rules and found to be unsatisfactory only towards the end of the 20th century. Communities in other parts of the world have been influenced by regimes such as these. But they similarly have responded in various ways through their legal systems to their own particular needs and circumstances. Concepts of property have moulded the approaches of many of these regimes. What has emerged is a matrix of rights, duties and liabilities associated with the public domain and with individuals within the community.

While nation states are subject to the rules of international law, the way in which they exercise their rights and discharge their obligations lies to a large extent within their discretion. However the internal arrangements within nation states often constrain how this apparent discretion is exercised. In this sense these restrictions are self-imposed in accordance with the constitutional, institutional and legal arrangements within the state in question. Nevertheless they are critical. The norms of international law are relatively flexible except in the case of those aspects of interstate agreements which deal specifically with particular sets of circumstances. In this sense the circumstantial arrangements are operational rather than strategic. While the same is to some extent true of legal arrangements within nation states, the effectiveness of a water resources governance system is likely to depend upon the capacity of their operational rules to be enforced. In the context of sustainability, the effectiveness of the system will depend upon the answer to the question whether an existing use of water or a proposed use of water is sustainable. While this question can be answered in terms of its own particular set of circumstances, such an assessment does not have effect in a legal vacuum. A decision in particular circumstances must respond to the legal framework which applies to it. It is against this background that we turn to a review of the legal arrangements that are evolving in nation states.

The traditional functions of a state are legislative, executive and judicial. These involve norms of competence to create rules, to implement rules and to enforce rules. The constitutional and institutional structures for doing so vary enormously. These involve not just norms of competence but also norms of obligation – rights and duties of various kinds. One of the challenges is to ensure that the formal structures comprising these arrangements reflect the function they are designed to perform. In relation to water resources this essentially requires a recognition that everyone within the relevant community has an interest in them: in other words, the doctrine of the community of interest in water resources. It is thus a function of the legal system to recognise or to create a set of normative relationships which includes such a range of public and private interests.

In this Part we review and assess how a number of national legal systems are responding to the challenge of water resources governance. Despite the diversity of these responses – in keeping with the historical development of their discrete legal systems – there are several themes around which these responses can be organised. Ultimately it is a question of control: in particular, control of what, when, where and how. The arrangements for responding to this challenge may be constitutional, institutional, strategic, principled, planned or strictly legal, and the instruments may be rights, duties and liabilities operating in the context of either public or private management and responsibility. The focus of any system of water resources governance is presumably to satisfy the needs for water of the community and of the environment. The system is multifunctional. Each and every person has a part to play. The legal arrangements may emerge from local practice and custom, from centrally dictated solutions or from a combination of both. In terms of legal doctrine, a system of appropriation or one of allocation.

In Chapter 14 there is a review of the fundamental approaches to water resources management; either an allocation or an appropriation of interests in water or a combination of both. A system based on the notion of the public domain or one based on individual rights or a combination of both. The way in which interests in water are allocated or appropriated is governed by the law. In either case the state performs a number of functions either legislatively, executively or judicially. Interests in water may exist simply as a matter of legal right without further regulation or determination. Interests in water may be acquired by appropriation simply in exercise of this right which exists as a rule of law. Or interests in water may be granted by way of allocation by the relevant public agency. The way in which interests in water are granted is increasingly controlled by the law, sometimes in accordance with principles or directions that guide the decision-making process but in an unenforceable way; sometimes in accordance with formal requirements in the context of a regulatory planning regime. The question ultimately is whether and to what extent the exercise of their rights by the states on the one hand or by individuals on the other hand are controlled by the law.

Chapter 14 considers how the law guides the management by the state of the way in which interests in relation to water are granted or created by the state. The guidelines or the directions may appear in a constitution – if there is one – or in the legislation that provides for the strategic management of water resources. Constitutional constraints upon the state in the exercise of its powers may manifest themselves as substantive or procedural rights conferred by the constitution upon individuals within the community. Increasingly in a number of jurisdictions the constitutional rights of individuals within the state are being not only recognised but also protected by the courts. In some instances there is or may be a tension between constitutional rights of this kind and the rights that exist under the common law, particularly – as in the case of the United States of America – rights in the nature of private property rights. Chapter 14 then considers how the law guides the exercise by the state of its rights in a public domain regime. In this case the question is whether and to what extent the administration of a public domain regime is directed at the sustainable development of water resources.

In Chapter 15 attention turns to planning for water resources development. Planning may be undertaken informally and voluntarily in a way which is incidental to performing other functions or undertaking activities linked to the use and development of water resources. Planning may take place in this way prior to the making of strategic policy decisions or prior to the undertaking of specific projects. An example discussed in Chapter 15 is the performance of a planning function in the context of water transfers. Chapter 15 then analyses planning as a formal and mandatory function which creates as a consequence a series of enforceable rules of law. Such a system incorporates a duty to prepare a plan in a particular context and to include in the plan rules about the use and development of water resources. These duties are essentially enforceable rules of law. These rules effectively regulate how water resources are used and developed.

Water rights or, more accurately, water entitlements held by persons or juridical entities exist as a matter of law, by appropriation or by allocation. In Chapter 16 we consider how water rights or water entitlements perform as regulatory instruments themselves. If water rights are created by way of allocation, the way in which rights are allocated is increasingly governed by sets of principles and rules in the statutory arrangements in the jurisdiction in question. This enables not only the creation of these rights but also their exercise to be controlled and regulated in accordance with the legislation. Water rights – whatever their source whether by appropriation or by allocation – are intrinsically capable of being traded in a commercial context. Trading in water rights, or at least trading in certain circumstances and in certain conditions, may be seen as an indirect method of regulation of how the water resource in question is used or developed. Increasingly water rights are created or used for ecological purposes. Water rights are in addition protected by the legal system, in some cases as property rights and in other cases as rights of their own kind. Chapter 16 considers whether and to what extent all of these evolving approaches acknowledge or incorporate sustainability or the sustainable use and development of water resources within their framework.

14. *Grundnorms* underlying water resources development

THE RANGE OF APPROACHES

(a) Introduction

The foundations of any legal system comprise the constitutional, doctrinal, strategic and institutional arrangements within each state. While it is useful to deal with them separately, each impacts upon the others in similar fashion to the interrelation between the legislative, executive and judicial functions of a state. Let us begin by considering the fundamental principles which guide the use and development of water resources within a state. Some assume the form of constitutional instruments while others may be in statutory or doctrinal forms. What we are looking for, in essence, are the *Grundnorms* of the water resources governance system in question. It may be a public domain regime or one based on individual water rights.

At which end of the doctrinal spectrum do we begin? The notion of the public domain and its associated rights and duties or a focus upon the rights and duties of individuals. For reasons that are both theoretical and practical, each is a counterpoint to the other. According to one view:

> One of the basic issues in water law is that of rights: what are the rights of the people and what are the rights of the state. The whole edifice of water law rests upon this basic issue of rights. The extent of the state's obligations to the people and the people's obligations to each other cannot be determined unless the legal framework of the related rights in water is clear.[1]

In other words, rights and obligations are correlative whether they attach to the community that comprises the state or to individual – or indeed corporate – members of this community.

(b) The Emergence of the Public Domain

Historically, individual rights in relation to water have always been recognised – mainly in earlier civilisations as customary rights. But the exercise of these rights has been either protected or regulated by complementary sets of legal arrangements. More recently, however, for reasons as much political as legal, the focus switched to public rights: in other words, the doctrine that has emerged in some civil law jurisdictions as the doctrine of the public domain. For example, in relation to the European states in and around the Mediterranean sea:

> Common features of water resources management laws are characterised by considering water as a public domain, protecting the resource, for example through ensuring a minimum flow, controlling use and pollution, increasing users' participation and dealing with water rights and markets.[2]

Such arrangements incorporate the public interest as well as the interest of individuals. While water rights and markets have a function to perform, it is the notion of the public domain that dominates. Control of water resources in the public interest is one aspect of the public domain, and a second aspect is the resurrection of the concept of water as a *res communis* or common property. Consistently with this approach:

> There is a trend towards a single legal regime, towards the establishment of the river basin as the water management unit, and towards water becoming a common asset (concerning both surface waters and ground waters).[3]

The notion of the public domain in one form or another underpins the legal arrangements for managing water resources in many – probably most – states across the five continents.

The justification for the doctrine of the public domain is the community of interest in water resources. Such an idea has no doubt underpinned customary arrangements in place over many years. While customary arrangements have an important function to perform in many jurisdictions, they have in some respects been complemented by statutory arrangements based upon the interest of the community in how water resources are used. In Nepal, for example, the use of water resources was originally organised in accordance with customary arrangements. Codified arrangements introduced more recently tended to connect rights in relation to water with the ownership of land. 'However, in 1967 for the first time a system of licensing in the use of water resources was introduced.'[4] A licensing system permits the control of water resources in the public interest. In this respect it is an aspect of the doctrine of the public domain. In Nepal, 'the reason for introducing such a law seemed to have emanated from the increasing pressure on the demand for water, which led the state to intervene and regulate it properly'.[5]

The approach introduced in the Republic of Yemen by the Water Law of 2002 similarly reflects a community of interest in water resources. It is extensive in its application:

> The Water Law provides in article 6 that all potential beneficiaries of any water resources shall enjoy the right to benefit from the relevant water resource, provided they do so in a way that does not harm the interests of other beneficiaries and they carry out all duties relating to the conservation and safeguarding of the water resources. The government regulates the rights and responsibilities of those benefitting from these resources.[6]

There is, therefore, a relevant water resource; a right to benefit given to all potential beneficiaries; a duty not to harm the interests of other beneficiaries; and duties for the conservation and safeguarding of the resource. This reciprocal set of rights and duties either creates or acknowledges the existence of a community of interest. While these rights exist as a matter of law, they are 'regulated' by the public sector. Again, even in the absence of a formal recognition of the doctrine of the public domain, these arrangements mean that the legislature and the executive acting together in this way give effect to this doctrine.

(c) The Source of Individual Rights

At the other end of the spectrum, as it were, are a number of states in the United States of America. The courts adapted the rules of the common law to fit the circumstances in the relevant states: in effect the adaptation of the common law of riparian rights and the introduction of the acquisition of rights by prior appropriation. In either case, the law evolved from an acknowledgement of individual rights in relation to water which either existed in accordance with a rule of law or were recognised by the operation of a rule of law. While 'water rights are a form of property rights', they are 'different from land and other forms of property; they reflect a web of connected interests that often makes it hard to move water away from the watershed of origin'.[7] While the reference to connected interests connotes the relevance of different interests in the water, it probably does not go as far as the recognition of a community of interest in the water resource. Significantly, however, 'water rights are correlative rather than exclusive property rights; a user's right to use the resource is a function of the impact of the use on other uses'.[8] This more directly acknowledges a community of interest in the use of the water and in the impact of the use of the water. This doctrine protects the interests of each individual person who shares the community of interest in the water resource.

But what about the public interest? While the rules that protect the community of interest do not protect the public interest as such, it is in the public interest that these rights or interests should be protected. The public interest in how water resources are used and developed has been promoted and protected in accordance with a range of additional legal mechanisms. There emerged, therefore, what have been described as the 'two primary functions' of water law:

- the creation of correlative private property rights in scarce resources;
- the imposition of public interest limitations on the exercise of these rights.[9]

In other words, the exercise of rights in relation to water is controlled – either by mechanisms of prohibition or restriction – so as to ensure that the public interest in protecting water, its ecosystems and its environment is ensured. These arrangements thus accommodate both private interests and the public interest. In this way the public interest operates as a burden upon the exercise of private rights, whereas in public domain regimes private rights are the outcome of the exercise of public rights. This doctrinal perspective is critical. It represents the analytical point of departure for the interpretation and application of the norms and rules in question. Attention is directed, first of all, towards how the law guides the management by the state of its interests.

CONSTITUTIONAL *GRUNDNORMS*

(a) Introduction

In theory – and in most cases in practice – an institution of the state or the state itself is as much subject to the law as anyone else. The state may nevertheless be governed by a specific set of rules that apply to it. Often it is the constitution or a similar basic law that creates these parameters: for example, the nature of the right conferred upon the state or the nature of the right conferred upon an individual which correlates to a burden upon the exercise of the state's right. These constitutional arrangements are manifested in various ways. These include:

- statements of fundamental values;
- statements of civil rights inhering in the individual;
- statements of human rights inhering in the individual;
- statements of procedural rights inhering in the individual;
- statements of rights inhering in the state;
- statements of duties correlative to rights inhering in the state.

The formal status of these statements is a matter either for the constitution itself or for the way the provision is interpreted and applied by the judiciary. Clearly there is neither a uniform nor a consistent approach to this.

(b) The Rights of a State

It is not uncommon for constitutional arrangements to focus upon a form of ownership or control which itself is the medium through which obligations may be imposed. The ownership may be variously vested in the people, the public, the nation or the state. What is important is that it is not an individual. In this respect it is a formal manifestation of the public domain doctrine.

Consider, first, the Constitution of the Socialist Republic of Vietnam. According to Article 17 rivers, lakes, water sources and underground natural resources are under the ownership of the entire people.[10] Article 29 imposes a duty upon the institutions of the state and upon individuals to implement state regulations on the rational use of natural resources and protection of the environment. Further, all acts depleting natural resources and destroying the environment are strictly prohibited. These environmental values are certainly included, but they will be difficult to enforce because of the limited availability of procedures to seek a remedy for compliance. In any event, elements of sustainability are implicit in these duties. The Constitution of the Republic of Yemen[11] provides for ownership by the state rather than ownership by the people. According to Article 8, 'all types of natural resources and sources of energy, whether above ground, underground are owned by the state, which assures their exploitation for the common good of the people'. The constitution thus contemplates anthropocentric exploitation rather than sustainable use.

In federations the constitutional arrangements tend to be more complicated. The Constitution of Brazil[12] relies upon the concept of property for the governance of water resources in Brazil. According to Article 26(I) the property of the states includes 'surface or underground waters, whether flowing, emerging or in reservoirs'. But the property of the states excludes waters in reservoirs constructed by the Union. The property of the Union thus includes these waters but also, in accordance with Article 20(III), additional waters, namely:

> Lakes, rivers, and any watercourses on lands that it owns; interstate waters; waters that serve as borders with other countries; waters that extend into or come from a foreign country, as well as bordering lands and river beaches.

Despite the division of property in this way, Article 21(XIX) requires the Union to 'establish a national system for management of water resources and define criteria for granting rights of their use'. The national system so created declares, among others, that water is public property. According to Article 27(5) of the Constitution of the Mexico[13] waters owned by the nation are national waters. A concession enables individuals and corporations to use national waters, while a grant enables municipal, state and federal agencies to use these waters. In the case of these two federations, therefore, state or national ownership is the basis for a public domain regime.

The effect is much the same according to the Constitution of Ecuador. It provides that 'water is a national good of public use; its domain is inalienable and imprescriptible and its use or exploitation is in charge of the state or whom gets the rights in accordance with the corresponding law'.[14] This is clearly a statement of public domain for public use. There are no rights in relation to water except in exercise of this norm of competence. The constitutional arrangements in Ghana point in a similar direction but in accordance with slightly different mechanisms. Section 269 of the Constitution of Ghana calls for the establishment of the Water Resource Commission. It was established by the relevant legislation. The Water Resource Commission Act of 1996 'puts all the water resources of Ghana under the control of the WRC [Water Resource Commission]. The WRC acts on behalf of the President, in whom all water rights are vested in trust for the people of Ghana'.[15] The creation of this statutory trust emphasises that these constitutional and statutory arrangements are designed to secure the public interest in the management of Ghana's water resources. In addition the consent of the Commission is required to do anything in relation to water including the construction of infrastructure. In effect, therefore, this is a public domain regime without, as in the case of Ecuador, any exceptions.

(c) The Substantive Rights of Individuals

These constitutional arrangements deal essentially with the rights and responsibilities of the states. While the citizens of the states are no doubt intended to be the beneficiaries of the exercise of these rights, there is on the face of it no way of assuring that the rights are exercised in this way. The position is different if rights are conferred upon individuals by the relevant constitutional arrangements. We have already noticed that a human right to water may be emerging in international law. There is no reason why a nation state cannot in exercise of its rights of sovereignty create a right to water as an element of its legal system.

As we have noted, section 27(1) of the Constitution of South Africa provides that everyone has the right to have access to sufficient water. And according to subsection (2) the state is under a duty to take reasonable legislative and other measures within its available resources to achieve the progressive realisation of this right. Clearly there is no absolute right to water and no absolute duty to provide it. More common – but still very much the exception – is the heavily circumscribed responsibility of a state in relation to the provision of water. For example, according to the Constitution of Zambia 'the state shall endeavour to provide clean and safe water for all persons, and take measures to constantly improve such facilities and amenities'.[16] According to the Constitution of Ethiopia, 'to the extent the country's resources permit, policy shall aim to provide all Ethiopians access to clean water'.[17] Although the Zambian provision is more positive than the Ethiopian provision, neither is likely to correlate to a protectable and enforceable legal right.

(d) The Procedural Rights of Individuals

In some cases, a constitution goes beyond the formulation of a human right supported by a correlative duty upon the state, and includes the recognition of the capacity of a court to provide a remedy for a violation of a right or a breach of a duty. Or alternatively a court may determine that its jurisdiction includes the power to grant a remedy for the violation of a constitutional right or a breach of a constitutional duty. There have been some interesting examples from India. Article 21 of the constitution provides that no person shall be deprived of his life or personal liberty except according to procedure established by law. And Article 32(1) guarantees the right to move the Supreme Court by appropriate proceedings for the enforcement of the rights conferred by the part of the constitution that includes Article 21.

Our attention in this respect has been drawn to a decision of the High Court of Kerala in 1990:

> This was a public interest litigation from Lakshadweep islands, in which the residents claimed that the excessive pumping of groundwater by the rich farmers was threatening the very availability of groundwater for all. They claimed, under article 21 of the constitution, that their life opportunities were being threatened since the depleting groundwater resource was likely to become saline. The court upheld their claim. Such a decision once again makes the right to water a natural or fundamental right under article 21 – right to life.[18]

The water resources in this case could be considered to be common-pool or shared resources. Under the common law, it seems clear that the owner of the surface of the land could lawfully extract the groundwater to which the owner of the surface of the land had access. In the circumstances in this case, the exercise by the owner of the land of this right constituted a violation of the substantive right in the constitution. This right was itself protected by the procedural right in the constitution to seek a remedy. It is the constitutional juxtaposition of substantive rights and procedural rights that enables a remedy of this kind.[19]

The significance of the relationship between substantive rights and procedural rights has become apparent through the emerging jurisprudence in Latin America. The constitutional arrangements in some jurisdictions contemplate rights and duties in relation to the environment. These may emerge either as an individual or a collective right in relation to the environment. Traditionally individual rights and collective rights have been protected according to different procedures. It appears that these procedural distinctions have not impeded the development by the Constitutional Court of Colombia of its jurisprudence to assure the environment – and by implication water resources and their ecosystems – of constitutional and judicial protection. Thus:

> The constitutional court initiated its own doctrinaire development in order to recognize the right to a healthy environment as a fundamental right susceptible to individual titularity. Although the court did not specify the difference established in the Colombian Constitution between fundamental human rights and environmental protection, it nevertheless recognised their interrelationship, and particularly, the relationship between the right to a healthy environment and other fundamental rights.[20]

Traditionally a collective right had been protected by the *actio popularis* and individual rights by the *accion de tutela*. In the event it mattered little whether rights in relation to the environment were collective rights or individual rights. Thus:

> Due to the characteristics of indivisibility, connexity and interdependency inherent to fundamental human rights, the right to a healthy environment is protected in Colombian constitutional practice, not only through *actio popularis* established for 'collective rights' but also through actions of tutelage exclusive of fundamental individual rights.[21]

The Supreme Court of Peru went even further. In Peru rights and duties in relation to the environment were incorporated in the Code on Environment and Natural Resources. This 'confers on all persons the right to interpose actions before the nation's courts demanding rapid and effective protection of the environment and natural resources'.[22] There is in effect no need for a person seeking a remedy to stipulate a specific – for example economic – interest in the matter. In this sense the rights protected were collective rather than individual rights. The court was invited in one case to apply the *accion de amparo* to protect collective as well as individual rights. This action provides a preventive rather than a reactive remedy. The court accepted that the *accion de tutela* had evolved sufficiently to protect diffuse interests – including the environment – but went further to determine that the *accion de amparo* was available for the same purpose. This was based upon 'the right of every citizen to the environment, to the landscape, and to health' as 'consecrated constitutional rights'. Hence:

> The *accion de amparo* assumes the function of a preventive jurisdictional tutelage since more than repairing a damage already committed, it tends to prevent future damages, or in any case to avoid the damage being repeated or indefinitely prolonged.[23]

In another case an aquaculture enterprise authorised by a concession was located adjacent to a mangrove forest. Areas adjacent to the mangrove forest were 'dried up' as a result of these aquaculture enterprises and these areas of land were allocated to the aquaculture enterprises. In an *accion de amparo* the court issued orders relating to the protection of the area adjacent to the mangrove forest. This was subsequently regarded as reasonable and the relevant authorities were ordered 'to comply strictly with the laws regarding the protection of the mangrove forest'. In addition:

> It also ordered them to initiate the procedures corresponding to the declaration of expiration of the concessions granted in the jurisdiction of the salt marshes and mangroves, so that the affected areas might be regenerated and returned to their natural state by the violators.[24]

It was thus the relationship between substantive rights and procedural rights in the context of the constitutional arrangements in Peru that enabled the court to protect the quality of the environment – in this case salt marshes and mangroves – irrespective of whether the substantive right was a collective or an individual right.

(e) The Relationships between Constitutional and Common Law Rights

The doctrinal foundations of the legal arrangements for water resources governance in the United States of America are different. These arrangements evolve from rights – for the most part individual rights – arising from the application of the riparian doctrine or the prior appropriation doctrine. Nevertheless, in some jurisdictions, the regime for water resources governance finds support in relevant constitutional arrangements.

Let us consider first the constitution of the state of Colorado. Colorado was one of the first jurisdictions to adopt the doctrine of prior appropriation. This effectively separated interests in water from interests in land. Article XV(5) of the Constitution of Colorado states:

> The water of every natural stream, not heretofore appropriated, within the state of Colorado, is hereby declared to be the property of the public, and the same is dedicated to the use of the people of the state, subject to appropriation as hereinafter provided.

Does this give to the people of the state a right to use the water – for example for recreational purposes – or does it confirm the rights of individuals to use the water according to the doctrine of prior appropriation? Three words are particularly significant: property, use, and appropriation. There is, on the one hand, water that is the property of the public and, on the other hand, water that is the property of an individual after it has been appropriated. And then there is use by the people. Which comes first? Or, in more juridical language, which right has priority? The right associated with the property of the public; the right of the individual after appropriation; or the right of use of the people?

The answer to this question depends upon not only the language of Article XV(5) but also the context of the constitution as well as the legal system in Colorado as a whole. The Article refers twice to appropriation. In each case it appears as a limiting factor. The first is the reference to water not appropriated and the second is use subject to appropriation. The structure suggests that the property of the public is subject to prior appropriation, and similarly the use of the water by the people is subject to prior appropriation. If this is so, then access to and use of water that has been lawfully appropriated takes precedence over everything else. The Supreme Court of Colorado has so decided.

First, in *Hartman's case* as early as 1906, this constitutional provision 'declares the unappropriated waters of our natural streams to be the property of the public, and dedicates the same to the use of the people of the state, subject to appropriation'.[25] And, second, in *Emmert's case* in 1979, it 'simply and firmly establishes the right of appropriation in this state'.[26] The provision in the constitution, in other words, simply confirmed the common law. The Supreme Court found support in another provision in the constitution – namely Article XV(6) – to the effect that 'the right to divert the unappropriated waters of any natural stream to beneficial uses shall never be denied'. Consequently, both textually and contextually, the Constitution of Colorado confirmed the priority of the prior appropriation doctrine in the legal arrangements for the governance of water resources in this state.

In this respect it is useful to compare the Constitution of Wyoming. According to Article 8(1), 'the water of all natural streams, springs, lakes or other collections of still water, within the boundaries of the state, are hereby declared to be the property of the state'. This is a simple declaration without any qualification. In particular there is no reference whatsoever to appropriation. In addition the reference is to the property of the state as distinct from the property of the public. This may be of little consequence simply because the reference to the state or to the public is in one sense symbolic. But in another sense it is more than that because it clearly indicates that no person – individual, corporate or institutional – has property in the water. Use, perhaps, but not property.

Indeed the Supreme Court of Wyoming in *Day's case* has determined that the public has a right to the recreational use of the surface waters of non-navigable streams bounded by private land.[27] In this sense it is a right inhering in a member of the public as a member of a particular community. It is a collective right rather than an individual right. But it is nevertheless a right which is exercisable by individual members of the public or of the collectivity in question. In any event an analysis of these provisions emphasises the importance from the perspective of the law of the structure and language of the provision in question. Not only that – it draws attention to the ambiguities inherently associated with the use of the concept of property not only in constitutional instruments but in any legal instrument.

Let us turn now to the constitutional arrangements in Florida. The doctrinal foundations of the arrangements for water resources governance have been traditionally the riparian doctrine of reasonable use. The expectations of society changed during the 1960s – as in many other places – so as to acknowledge the harm that was being done to the environment. More specifically, a recognition of the need to conserve natural resources and to protect the environment. The Constitution of Florida was accordingly amended in 1968 'to express a state policy of conservation and protection of Florida's natural resources and scenic beauty'.[28] No rights; no duties; an expression of policy representing a value. The incorporation of a provision such as this in the constitution gave formal legitimacy to these values. It was consequently the function of the legislature to put in place the legal mechanisms for doing so. The outcome, as we shall see, was a set of arrangements for planning and controlling the use of water resources and for protecting their quality.

While arrangements such as these do not necessarily abrogate the rules that comprise the riparian doctrine, their application is in practice substantially modified. To some extent, therefore, while the focus of these legal arrangements has changed, the doctrinal foundations have remained constant. In other words, there would be little or no reason to suggest that Florida – or Colorado or Wyoming for that matter – has evolved into a public domain regime. The arrangements in the United States of America comprise essentially sets of 'correlative private property rights' the exercise of which is controlled by 'the imposition of public interest limitations'.[29]

THE *GRUNDNORM* OF THE PUBLIC DOMAIN

(a) A Public Domain by Public Ownership

We turn now to the way the law directs rather than merely guides how a state manages its water resources. This is essentially a controlled public domain regime. How is such a regime created? A constitution may entrench a public domain regime, a regime based upon individual rights or an accommodation of both. A constitution – if there is one – may equally be silent on any of these matters. A public domain regime may be the *Grundnorm* of the system simply because it is recognised in that way or indirectly incorporated in a set of statutory arrangements.

But what is a public domain regime? It is certainly akin to the concept of *dominium publicum*. But it is a regime neither of ownership nor of administration per se; rather a regime *sui generis* that shares characteristics of each. Its function quintessentially is the management of natural resources in the public interest. A decision of the Constitutional Tribunal in Spain has defined the public domain 'as *res extra-comertium*, i.e. a good outside the market mechanisms'.[30] That is, things not able to be traded or – in the context of Roman law – *res extra commercium*. More often than not there is no specific reference in the relevant legal instruments to the public domain as such. It manifests itself as:

- state ownership;
- public ownership;
- public trusteeship;
- exclusive public control.

Let us consider some examples.

The characteristics of the public domain approach in China are emphasised because of their incorporation not only in the constitution but also because they are incorporated in the Water Law of 2002. While the constitution deals with natural resources generally, Article 3 of the Water Law directly declares that water resources are owned by the state. This right of ownership is exercised by the state council and the taking of water is authorised by a system of licensing administered under Article 7 by the state council except in the case of water that belongs to rural economic collectives. In other words, there are no individual rights in relation to water except in accordance with this system.[31]

According to Article 1 of the Water Resource Act of 1992 of Nepal the ownership of the water resources in Nepal is vested in the state. According to Article 4 the use of water resources is prohibited without a licence.[32] Similarly in Senegal water is owned by the state and any use of water requires an authorisation granted by the Minister.[33] According to Article 1 of the Water Law of 1995 of Morocco water is a public asset. The description of water as a public asset is extensive. It includes:

- all water tables whether surface or underground;
- lakes, ponds and other surface water resources;
- wells built for public use;
- canals built for navigation, irrigation and drainage;
- waterworks assigned for public use;
- permanent or non-permanent watercourses and their sources;
- banks up to the levels attained by flood waters.[34]

Exceptions to this rule in the laws of Morocco are rights of ownership or usufruct properly acquired under specified earlier legal arrangements. Otherwise the use of water in the public domain requires an authorisation or a concession.[35]

In France – perhaps the ancestral home of the public domain – water resources are an element of *le patrimoine commun de la nation* – the national or common heritage.[36] The state manages water resources on behalf of the nation while the use of water is controlled by a complex set of rules contained either in the relevant code or in the relevant regulatory scheme. This applies to groundwater as it does to other water resources. Consequently landowners do not own the water under the surface of their land. Their right is a right to use in accordance with the relevant rules.[37]

While there is no reference in the arrangements for water resources governance in Germany to the public domain or to ownership of water resources, the regime in Germany – a federation – is fundamentally a regime of the public domain. The *Grundgesetz* – the basic law – provides that water resources are managed in the first place in accordance with the regulatory framework established at the federal level 'while the *Länder* [the states] are responsible for implementing and supplementing the federal regulations and enforcing all statutory provisions in the field of water pollution control'.[38] Accordingly any use of water requires a permit or licence unless this is not required by a law of the *Länder*.

In Canada – a federation – the provinces are vested with the ownership of most of the water resources in Canada and with the ownership of the land, the minerals and the forest resources which impact upon water resources. The Dominion of Canada – the federal entity – has no direct involvement in the governance of water resources located in or flowing through the provinces. Nevertheless the water resources in Canada are for the most part managed on the basis of joint arrangements and cooperation on the part of both the Dominion and the provinces.[39]

(b) A Public Domain by Public Control

The constitutional arrangements in Australia – like Canada – leave the responsibility for water resources governance to the states unless the Commonwealth is able to become involved through the indirect exercise of its powers. During most of the 19th century the water resources of Australia were managed in accordance with the rules of the common law. However, towards the end of the 19th century it became increasingly recognised that the climatic and hydrological environment in Australia was sufficiently different from that of Great Britain to justify a different approach. Each of the states responded by enacting a statutory regime that invested the Crown – the executive branch of government – with the right to the use, flow and control of most of the water resources in the jurisdiction. Effectively this reduced the range of circumstances to which the rules of the common law applied.[40] The legislation for the most part did not specify the purposes for which this right was to be exercised. In practice the water resources in question were managed so as to facilitate the development of Australia's natural resources and its economic growth. This meant in particular the use and development of water resources for irrigation in support of primary production.

As in other jurisdictions during the second half of the 20th century, legal arrangements were introduced in Australia to control and prevent water pollution and to protect the environment generally. And towards the end of the 20th century the concept of sustainable development – or ecologically sustainable development as it is known in Australia – began to play a part in these legal arrangements.[41] In Australia, therefore, there has emerged a regime similar to that of the public domain.

The doctrinal basis for this emerges from this judicial analysis in *Harper's case*:

> When a natural resource is limited so that it is liable to damage, exhaustion or destruction by uncontrolled exploitation by the public, a statute which prohibits the public from exercising a common law right to exploit the resource and confers statutory rights on licensees to exploit the resource to a limited extent confers on those licensees a privilege analogous to a profit à prendre in or over the property of another. A limited natural resource which is otherwise available for exploitation by the public can be said truly to be public property whether or not the Crown has the radical or freehold title to the resource.[42]

In relation to water resources this approach has brought about the use and control of most water resources by the executive branch of government through the construction and operation of infrastructure on the one hand and through the allocation of water rights on the other hand.

Apart from the constitutional right of everyone in South Africa to have access to sufficient water, the *Grundnorm* of the arrangements is the doctrine of public trusteeship. According to section 3(1) of the National Water Act of 1998:

> As the public trustee of the nation's water resources the national government, acting through the Minister, must ensure that water is protected, used, developed, conserved, managed and controlled in a sustainable and equitable manner, for the benefit of all persons and in accordance with its constitutional mandate.

The constitutional mandate includes not only the right of access to sufficient water but also the right to a healthy environment. Accordingly, section 3(2) of the Act of 1998 places the ultimate responsibility upon the Minister 'to ensure that water is allocated equitably and used beneficially in the public interest while promoting environmental values'. While the Act itself states the overall purpose to be achieved in its implementation, it additionally contemplates that the allocation, and hence the use, of water in particular sets of circumstances are themselves to be determined by what is in the public interest. Thus any use of water must be authorised directly by the Act or in accordance with the Act.

The doctrine of public trusteeship in South Africa is further emphasised by the provisions of section 3(3). It provides that the national government, acting through the Minister, has the power to regulate the use, flow and control of all water in the Republic. The notion of public trusteeship – in this respect like the notion of the public domain – comprehensively incorporates the governance of water resources from the creation of an overall purpose, the strategic direction of decision-making, the mechanisms for implementing this purpose and these strategies, and the capacity either to control or to undertake the operational activities required to do so.

THE ADMINISTRATION OF THE PUBLIC DOMAIN

(a) Introduction

It is a function – but not the only function – of a constitution to state some of the values recognised by the society in question. A public domain regime – or a public trusteeship regime as in South Africa – ensures that the natural resources to which it applies are used and developed generally in the public interest or in the public interest as mandated by the legal system. The legal system can do this in a number of ways. It can state the principles to be observed in making decisions and undertaking activities that affect natural resources. It can indicate the direction of decision-making on the part of the executive branch of government or on the part of individual or corporate members of the community. These are paralegal rules. Or it can put in place planning and regulatory regimes which require compliance. That is, legal rules.

Let us take this analysis further by examining the detail of emerging paralegal rules. The creation of rights of state or public ownership or of a public domain regime merely creates a structure for the governance of water resources. How are these norms of competence to be exercised? There are a number of alternatives:

- the exercise of an unfettered discretion by the institution invested with these rights;
- the exercise of these rights in accordance with a set of acknowledged principles;
- their exercise in accordance with declared statements of direction or outcome;
- their exercise in accordance with a more formalised set of strategic arrangements.

These arrangements may be discretionary or normative. Whether the arrangements are discretionary or normative depends ultimately upon their status within the legal system and how they are structured. First, here are some examples of discretionary rather than normative sets of arrangements.

(b) Administration according to Principles and Directions

Articles 1, 2 and 3 of the National Water Resources Policy of Brazil state respectively a set of principles, objectives and criteria.[43] One of the principles is that water is a limited resource with economic value. Two of the objectives are:

- to ensure that both present and future generations have access to water of a quality that is adequate for their various uses;
- to ensure the rational and integrated use of water resources to achieve sustainable development.

The criteria for the implementation of the policy include the integration of water resources management with environmental concerns. There is a specific reference to the achievement of sustainable development. In addition, the trilogy of economic, social and ecological values is implicitly recognised. But each performs a somewhat different function within this framework. For example, intra-generational and inter-generational equity are elements of the objectives, while the integration of environmental concerns within water resources management is a criterion according to which the implementation of the policy is guided. Much, therefore, depends upon the relationship between these principles, objectives and criteria. On the face of it, this relationship is discretionary rather than normative.

Significantly it is the preamble to the Water Law of Morocco rather than the water law itself that declares the relevant principles and objectives.[44] One of the principles is that agricultural value is to be increased through improvements in harnessing and utilising water in agriculture. On the face of it, this is an objective rather than a principle. Nevertheless, the objectives include the optimum mobilisation and rational management of all water resources and the protection and conservation of public water resources. These objectives are achieved through a national water plan and planning at the catchment level. The inclusion of these principles and objectives in the preamble to the Water Law suggests that they are normative in structure but discretionary in operation.

The arrangements that have evolved in different parts of Africa point to significantly different values in the management of water resources. Consider the federation of Nigeria as an early example of the exploitation of water resources. The functions of river basin development authorities – federal agencies – were, according to the legal arrangements in 1979, clearly directed towards the use and development of the relevant water resources. The functions included:

- to undertake comprehensive development of both surface and underground water resources for multipurpose use;
- to undertake and develop schemes for flood control and irrigation;
- to develop infrastructure for urban and rural water supply.[45]

One function of a somewhat different character was to control pollution in rivers and lakes in the authority's area in accordance with nationally laid down standards.[46] The creation of functions is on the face of it operational rather than normative. However the purpose of the function makes clear the basis upon which the water resources are to be managed.

Then there is the law of 1981 regulating water in Senegal. While water is managed for the public good and in the public interest, what is in the public interest is not entirely a matter of discretion. Priorities are specified by the law. For example:

- water for human consumption;
- water for growing food;
- water for other purposes such as agriculture (but not food production), energy, mining, navigation and tourism.

However 'in case of conflict between various uses, the priority should be determined by the economic needs of the local area'.[47] These rules are complemented by measures for the prevention of pollution of water. These arrangements, therefore, acknowledge the economic, social and ecological values of water but priority is afforded, on the face of it, to social and economic uses.

The Water Resource Commission of Ghana was established for regulating and managing the use of water and for coordinating the relevant policies. Its functions are largely strategic and advisory rather than operational. One of its advisory functions relates to pollution. One of its wider responsibilities is 'to propose, co-ordinate and monitor plans and activities related to the development, improvement, utilization and conservation of water resources'.[48] Accordingly development and utilisation are linked with conservation. These are structured as outcomes of plans and activities. And they are outcomes to be achieved simultaneously. There is, in other words, no discretion whether water resources should be developed or conserved. The responsibility of the Water Resource Commission relates to both. The evolution of the values underlying the arrangements in Africa has moved – consistently with other parts of the world – from use and development towards conservation and in some respects sustainable development.

(c) Administration for Sustainable Development

Sustainable development is in many respects the focus of the arrangements for the management of water resources in Armenia.[49] The Water Code of Armenia was adopted in 2002. The code has a threefold strategic structure: a purpose, a set of objectives and a set of principles. These create the foundation for the formulation of a national water policy and their implementation through water basin strategies and plans. The subject matter of the code is the 'national water reserve'. This is:

> The quality and quantity of water that is required to satisfy present and future basic human needs, as well as protect aquatic ecosystems, and to secure sustainable development and use of that water resource.

The purpose of the code is the conservation of the national water reserve and the satisfaction of the water needs of the citizens of Armenia and of the economy. This is achieved through managing usable water resources and securing their ecological sustainability. The objectives and principles associated with this purpose are – not surprisingly – consistent with it. What emerges is a set of composite functions in relation to:

- the conservation and protection of water resources;

- ensuring an adequate quantity and quality of water supply;
- protection of aquatic and related ecosystems and their biological diversity.

The principles include consideration of the 'economic value' of water in managing water resources according to the purpose, the objectives and the principles. The economic value of water is 'the sum of the drinking, environmental, energetic potential and agricultural values of water'. This is – again not surprisingly – an anthropocentric approach to the value of water. It includes the environmental values of water, but in this respect the environmental values are those attributed to it by society rather than the intrinsic values of water themselves. In any event what emerges from this set of arrangements is a recognition of sustainable development and the principles that support it including intra-generational and inter-generational equity and the intrinsic values of ecosystems.

The focus of the arrangements in Kazakhstan is similarly sustainable development.[50] The Water Code of 2003 is based on the notion of the water fund which refers to all bodies of water in the territory of Kazakhstan. Sustainable development underlies the code. It seeks to achieve and maintain an ecologically and economically optimal level of water resource use together with protection of the water fund. Two of the supporting principles are equitable access to water and rational use of water. These objectives and principles find expression in a state policy 'in favour of sustainable water use and protection of the water fund'. And the administration of the water fund is guided by the sustainable use of water and the creation of optimal conditions for water use, the preservation of environmental sustainability and public health. So the three essential elements of sustainable development are implicit in these arrangements.

The approach in China has been quite different. It is essentially a public domain regime. The fundamental principles according to which the right of state ownership is exercised have been established in exercise of the executive power by the relevant Chinese agencies. There are essentially five principles:

- the principle of coordinating water resource exploitation with social development and environmental protection;
- the principle of rationalism: rational exploitation and use of water resources;
- the principle of rationalism: rational allocation of water resources;
- the principle of efficiency: efficient use of water resources;
- the effective management of water resources: the integration of water resources management.[51]

The last four of these principles may be seen as the means for bringing about sustainable development which is essentially the substance of the first principle.

The first principle has been described in more detail in this way:

> Sustainable development is the general guiding ideology of environmental protection, and is reflected in the field of water resources by coordination of exploitation and use of water resources for economic development with social development and environmental protection. This principle is the fundamental norm of water resource allocation and management. It requires that use and management of water resources be incorporated into the overall planning of the national economy. And it requires that water uses be combined with flood prevention and control and that attention is given to the social and biological benefits of water as well as its economic benefits.[52]

This principle includes the three elements of sustainable development. However in a more practical sense, perhaps, sustainable development is seen as a means of achieving environmental protection – rather than the other way round. Whether this is so or not, these principles are put into effect through the system of comprehensive planning of water resources contemplated by the Water Law of 2002. The comprehensive plan is 'legally binding'.[53] The approach in China is an example of five principles as a set of paralegal rules and a comprehensive plan containing a series of rules that are binding: namely, legal rules.

The members of the European Union are required to give effect to the directive of 2000. The main objectives of the arrangements in place in France prior to 2000 were described as 'to ensure the protection of the water and ecosystems quality and quantity, and to value water as an economic good'.[54] Water resources in France are part of the public domain – *le patrimoine commun de la nation* – and, according to the arrangements put in place in 2003 in relation to water, 'its protection, enhancement, and development as a usable resource must be managed accordingly, with due respect to natural equilibriums'.[55] The requirement to respect natural equilibriums seems to be unique – at least in the context of contemporary jurisprudence. The concept of humans in harmony with nature was a feature of a number of earlier civilisations. It has been noted that in functional terms respect for natural equilibriums means this:

> Water management must seek to reconcile the water needs associated with, *inter alia*, public health, aquatic ecosystems, protecting the quality and quantity of available water, protecting against flooding, and such human activities as agriculture, fishing, industry, energy, and transportation.[56]

The focus of this approach is anthropocentric, at least in terms of social and economic outcomes. Certainly the protection of aquatic ecosystems is included. But while the elements of sustainable development are included, sustainability in the sense of ensuring the ongoing capacity of water resources to be conserved in all respects is implicit rather than explicit.

CONCLUSION

The way in which water resources are used and developed reflects to some extent the fundamental *Grundnorms* underlying the system. Increasingly these *Grundnorms* are included in constitutions. Just as significantly, these *Grundnorms* are incorporated in what may be described as charters – often in statutory form – for the management of water resources. Much depends upon whether the system is a public domain system or a system based upon individual rights and duties. But in either case the distinction between arrangements governed by public law and those governed by private law is beginning to disappear rapidly. It is no surprise that the values underlying the arrangements in any state reflect the circumstances and the priorities of the state itself. In almost all cases it is recognised that water is used for a range of different and potentially conflicting purposes. In some cases there is emphasis upon the use of water for basic human purposes; in others for the use of water for the production of food; in some for purposes associated with industrial and commercial activity; and in others for the protection of aquatic ecosystems.

Not only is there a different focus and a different balance of values in each of the jurisdictions. There is also a range of different instruments for achieving these outcomes. These *Grundnorms* are stated differently: as constitutionally protected values; as values given priority in a statutory form of a charter; values in the form of a set of strategic directions – unenforceable – to those managing water resources; values expected to be included in the planning and operational arrangements for the management of water resources. Increasingly these *Grundnorms* are given legal status and effect. In this sense the legislatures have spoken. It is up to the executive to implement these directions and it is similarly up to the judiciary to ensure compliance. The legal arrangements in any one state are in one sense unique. The relationship between the legislative, executive and judicial functions varies. What is emerging is an increasing formalism of the way in which these *Grundnorms* are expressed and developed. In other words there is less discretion and more formality in the normative framework.

NOTES

1. Singh (2000, p. 57).
2. European Commission (2000, p. 47).
3. European Commission (2000, p. 122).
4. Upadhyay (2005, p. 197).
5. Upadhyay (2005, p. 197).
6. Salman and Bradlow (2006, p. 120).
7. Tarlock (2005, p. 41).
8. Tarlock (2005, p. 41).
9. Tarlock (1997, p. 183).
10. Salman and Bradlow (2006, pp. 112 and 113).
11. Salman and Bradlow (2006, p. 119).
12. Salman and Bradlow (2006, pp. 24 and 25).
13. Salman and Bradlow (2006, pp. 77 and 78).
14. Herrera et al. (2006, p. 224).
15. Laube and van de Giesen (2006, p. 166).
16. Hu (2006, p. 36).
17. Hu (2006, p. 37).
18. Singh (1991, pp. 39 and 40).
19. See also Fisher (2003, pp. 76–87).
20. Navia (1994, p. 124).
21. Navia (1994, p. 125).
22. Navia (1994, p. 125).
23. Navia (1994, p. 125).
24. Navia (1994, p. 127).
25. *Hartman's case* (1906, p. 688).
26. *Emmert's case* (1979, p. 1028).
27. *Day's case* (1961).
28. Dzurik (1990, p. 71).
29. Tarlock (1997, p. 183).
30. European Commission (2000, p. 262).
31. Hu (2006, pp. 68 and 78) and Salman and Bradlow (2006, pp. 36–38).
32. Salman and Bradlow (2006, p. 93).
33. Salman and Bradlow (2006, p. 97).
34. Salman and Bradlow (2006, pp. 86 and 87).
35. Salman and Bradlow (2006, p. 86).
36. European Commission (2000, p. 48) and Salman and Bradlow (2006, p. 57).
37. Salman and Bradlow (2006, p. 57).
38. Salman and Bradlow (2006, p. 61).
39. Shrubsole (1990, pp. 91 and 92).
40. Fisher (2000, pp. 101–116).
41. Fisher (2004c, 2005, 2006a).
42. *Harper's case* (1989, p. 334).
43. Salman and Bradlow (2006, p. 24).
44. Salman and Bradlow (2006, pp. 85 and 86).

45. Salau (1990, p. 195).
46. Salau (1990, p. 195).
47. Salman and Bradlow (2006, p. 97).
48. Laube and van de Giesen (2006, p. 170).
49. Salman and Bradlow (2006, pp. 14 and 15).
50. Salman and Bradlow (2006, pp. 68 and 69).
51. Xi et al. (2002, pp. 9–13).
52. Xi et al. (2002, pp. 9–10).
53. Xi et al. (2002, p. 14).
54. European Commission (2000, p. 149).
55. Salman and Bradlow (2006, p. 57).
56. Salman and Bradlow (2006, p. 57).

GENERAL REFERENCES

Dzurik, Andrew A. (1990), *Water Resources Planning* (Savage, Mld: Rowman and Littlefield Publishers Inc).

European Commission (2000), *Towards Sustainable and Strategic Management of Water Resources* (Luxembourg: European Communities Publications).

Fisher, D.E. (2000), *Water Law* (Sydney: LBC Information Services).

Fisher, D.E. (2003), *Australian Environmental Law* (Sydney: Lawbook Company).

Fisher, D.E. (2004c), 'Land, water and irrigation: hydrological and legal relationships in Australia (Part 1)', *Water Law* 15, 229–237.

Fisher, D.E. (2005), 'Land, water and irrigation: hydrological and legal relationships in Australia (Part 2)', *Water Law* 16, 14–22.

Fisher (2006a), 'Water law and policy in Australia – an overview', *Environmental Law Reporter (USA)* 36, 10264–10276.

Herrera, P. et al. (2006), 'Information use and water resources laws and policies in Ecuador', in J. Wallace and P. Wouters (eds), *Hydrology and Water Law – Bridging the Gap* (London: IWA Publishing), pp. 218–246.

Hu, Deshang (2006), *Water Rights: An International and Comparative Study* (London: IWA Publishing).

Laube, Wolfram and Nick van de Giesen (2006), 'Ghanaian water reforms: institutional and hydrological perspectives', in J. Wallace and P. Wouters (eds), *Hydrology and Water Law – Bridging the Gap* (London: IWA Publishing), pp. 157–197.

Navia, Jose M. Borrero (1994), *Environmental and Water Law in the South* (Utrecht: International Books).

Salau, Ademola T. (1990), 'Integrated water management: the Nigerian experience', in Bruce Mitchell (ed.), *Integrated Water Management: International Experiences and Perspectives* (London: Belhaven Press), pp. 188–202.

Salman, Salman M.A. and Daniel D. Bradlow (2006), *Regulatory Frameworks for Water Resources Management: A Comparative Study* (Washington D.C.: World Bank).

Shrubsole, D.A. (1990), 'Integrated water management strategies in Canada', in Bruce Mitchell (ed.), *Integrated Water Management: International Experiences and Perspectives* (London: Belhaven Press), pp. 88–118.

Singh, Chhatrapati (1991), *Water Rights and the Principles of Water Resource Management* (Bombay: N.M. Tripathi Pty Ltd).

Singh, Chhatrapati (2000), 'Water rights and the principles of water management in India', in D.E. Fisher and Noeleen McNamara (eds) (2000), *Integrated Water and Land Management* (Brisbane: Queensland University of Technology), pp. 49–79.

Tarlock, A. Dan (1997), 'Current trends in United States water law and policy: private property rights, public interest limitations and the creation of markets', in Edward H.P. Brans et al. (eds), *The Scarcity of Water: Emerging Legal and Policy Responses* (London: Kluwer Law International Ltd), pp. 183–195.

Tarlock, A. Dan (2005), 'Water transfers: a means to achieve sustainable water use', in Edith Brown Weiss et al. (eds), *Fresh Water and International Economic Law* (Oxford: Oxford University Press), pp. 35–59.

Upadhyay, Surya Nath (2005), 'The legal regime of water resources in Nepal', in Surya P. Subedi (ed.), *International Watercourses Law for the 21st Century* (Aldershot: Ashgate Publishing Limited), pp. 195–209.

Xi, Wang et al. (2002), 'Managing water resources for a sustainable futurer: law, policy and methodology of China', *Asia Pacific Journal of Environmental Law* 7, 5–41.

LEGAL INSTRUMENTS

Constitution of South Africa
National Water Act 1998 of South Africa

CASES

Day's case: Day v Armstrong (1961) *Pacific Reporter 2nd Series* 362, 137–151.
Emmert's case: People v Emmert (1979) *Pacific Reporter 2nd Series* 597, 137–151.
Harper's case: Harper v Minister for Sea Fisheries (1989) *Commonwealth Law Reports* 168, 314-.
Hartman's case: Hartman v Tresise (1906) *Pacific Reporter* 84, 685–692.

15. Planning for water resources development

INTRODUCTION

So much for principles and directions. Whether or not there is formal provision for them in legal instruments such as constitutions or statutes, they inform and guide how decisions and activities in relation to water resources are made and undertaken. Sometimes, they may be translated into statements of strategy or statements of policy which provide more specific guidance about the management of water resources. These instruments perform a range of different functions. They may operate at different levels: for example national, regional or sub-regional. They may assume the form of strategies, policies, plans or programmes. These instruments may contain enforceable rules. If not, they set parameters for decision-making and operational activities for which the public sector is largely responsible. In this event, they are not use-specific, location-specific or project-specific. Nevertheless they impact upon – as they are designed to do – existing and proposed use-specific, location-specific and project-specific activities. It is at this level of specificity that these paralegal rules of strategic planning are transformed into legal rules of use and development planning and regulation. Planning may thus be governed either by legal or by paralegal rules.

There is nothing novel about planning. It is unlikely that any project for the use and development of water resources will be designed and implemented in the absence of some kind of plan or planned arrangement. If it is a public sector project to be undertaken by a public agency in exercise of its operational functions, it will have been planned to achieve a public objective. If it is a private sector project, it is for the proponent to determine the objectives to be achieved – largely or perhaps entirely for the benefit of the proponent. Increasingly, in either event, but particularly in relation to a private sector project, the development will require approval by a public agency in the public interest. In this sense the development is directly regulated in the context of a planning system which may itself be either formal or informal. Let us consider first the function of planning by the public sector when it is incidental to either the formulation of strategy or the implementation of a public sector project: that is, strategic planning or project planning.

PLANNING TO FACILITATE WATER RESOURCES DEVELOPMENT

(a) Informal Planning

The formulation of a policy, a strategy, a programme or a plan – by whatever name – is frequently the informal function of the public agency involved in the regulatory or operational management of water resources. In the former case it is strategic planning and in the latter case project planning. There has been a variety of approaches. In summary form, here are some examples:

- a national water policy in India since 1987 mandating equitable distribution of water resources and sustained yield use;[1]
- a national water plan in the Republic of Yemen which is binding and the basis for the development of a water plan for each water basin and zone;[2]
- a national water plan for Morocco linked to master plans for integrated water resources development;[3]
- a national water policy in Armenia implementing the objectives and principles of the water code relating to the development of water basin strategies and plans for the next 10 to 15 years;[4]
- a state plan in Kazakhstan for the protection and use of the water fund designed to ensure a scientifically justified distribution of water use between users and the protection of water resources;[5]
- a set of strategies and policies in Mexico for regulating water use and a national water programme which integrates regional, basin, state and sectoral sub-programmes;[6]
- *les schémas directeur d'aménagement et de gestion des eaux* in France which represent the main tool for water resources planning at the river basin level and cover all the objectives and guidelines to ensure a quantitative and qualitative balance of surface water and groundwater as well as the protection of aquatic ecosystems;[7]
- a national water resources plan in Portugal which is linked to river basin plans and encompasses a global – that is a technical, economic, environmental and institutional – approach to the enhancement, conservation, protection and management of the relevant water resources;[8]
- a national hydrological plan in Spain and river basin plans to be established to ensure that they contain the minimum contents set out in the water law;[9]
- water management framework plans for river basins and conurbations required to be formulated in Germany by the *Länder* in accordance with the federal legislation;[10]
- river basin plans and regional plans in China which together constitute unified plans for the development, utilisation, conservation and protection of water resources and for the prevention and control of water disasters.[11]

It seems reasonably clear from these examples that strategic planning and project planning can easily be accommodated within a public domain regime, simply because the relevant public agency engages in both types of planning.

But there is no reason why strategic planning cannot be undertaken in any legal system including those that do not recognise the notion of the public domain. For example, in England and Wales during the 1970s regional water authorities were established to provide water services and to the extent of providing water services manage water resources. The areas of these authorities coincided with the boundaries of natural watersheds. Accordingly 'the new concept meant that water policy could be developed and implemented on a catchment basis'.[12] This amounted to a form of strategic planning able to be implemented because the planning and the operational functions were performed by the same institution.

The function of providing water services at the operational level was subsequently committed in England and Wales to the private sector. But the regulatory functions that related to the protection of the water environment were retained by the National Rivers Authority set up for this purpose. 'Although the new body was a national organisation, it retained a regional management structure based as before on the ten major river catchment boundaries. Thus catchment management planning could be introduced without difficulty by each of the regional offices.'[13] Regional planning was thus preserved as part of the regulatory function whereas previously the function of planning had been performed informally as an element incidental – by no means compulsory – to the provision of water services.

In the United States of America, doctrinally a common law regime, there have been many examples of interstate agreements according to which each of the states involved has agreed to cooperate in the management of their shared water resources. This has almost necessarily involved informal strategic and project planning. But the structure of the agreements may lead to more formal planning and even regulation. These arrangements have assumed the form of interstate compacts which require the approval of the Congress of the United States. Effectively these amount to norms of competence enabling the activities and the regulatory functions authorised.

An example is the Delaware River Basin Commission formed in 1961. The establishment of the commission prepared the way for 'meaningful basin-wide planning' by including the United States itself as a party to the interstate agreement. Not only that – the commission was invested 'with the sufficient authority to manage and allocate the waters of the Delaware' and any allocation was required to be based on equitable apportionment.[14] These powers to regulate water use have been exercised through licensing, through the construction and operation of projects, and through the raising of revenue for operations.[15] The commission has accordingly engaged in a relatively comprehensive process of planning and regulation. Significantly this has happened in a common law jurisdiction which has not in any strict sense acknowledged the notion of a public domain regime.

(b) Project Planning

Let us then look in a little more detail at project planning: that is, the function of planning as an incidental aspect of the provision of water services by public sector agencies. Water resources planning, it has been suggested, 'has been practised in some form since antiquity',[16] for example 'development along the Indus River and the Tigris and Euphrates rivers' more than 9,000 years ago. Similarly 'various forms of water management and engineered water systems in Egypt, Iraq, and China'.[17] Significantly these were engineered water systems: in other words artificially created mechanisms to facilitate the use of water resources for the range of purposes contemplated by the society in question. Planning was likely to have been an integral part of these arrangements.

But what is planning? The undertaking of any activity almost necessarily involves deciding what is the intended outcome of the activity; where, how and when it will be undertaken; the human and non-human resources required to do so; the management arrangements for implementing it; and finally doing it. The larger the scale of the activity, the more complex the procedures involved. These procedures may be formal in the sense that they are imposed from outside the activity, or informal in the sense that they are imposed upon those undertaking the activity by them themselves. No project is likely to be successfully implemented in the absence of some kind of planning arrangements.

The way water resources are used is a matter of interest and concern to the whole of the community. In this respect many of the ancient civilisations acknowledged a form of a public domain regime. In this context it has been suggested that 'as authoritarian sociopolitical systems declined, however, the corresponding water systems slowly degenerated'.[18] But any system requires authorisation in one form or another, either accepted or imposed: for example, compliance with acknowledged customs, compliance with enforceable rules of law or compliance with directions given by an administrative agency consistently with the relevant norms of competence.

The arrangements in modern civilisations are intrinsically no different. As we have seen, the codified arrangements in jurisdictions based upon a public domain regime provided for the construction and operation of major hydraulic works according to the rules in these codified arrangements. In jurisdictions not based upon a public domain regime the use and development of water resources through major projects were authorised directly by legislation or in accordance with statutory arrangements. These development projects were intended particularly during the 19th and 20th centuries to promote economic growth and social development. The fundamental doctrines of the common law – for example the riparian doctrine or the prior appropriation doctrine – were designed to protect individual interests and rights but not the public interest. Nevertheless major water resources projects increasingly required approval to ensure they were in the public interest from a range of perspectives, including the environmental perspective.

In the United Kingdom, for example, the use and development of water resources for public purposes such as public water supply and inland navigation received their authorisation from Parliament on an ad hoc, and consequently a fragmented, basis. Similarly in the United States of America, water resources have been developed for public purposes by public instrumentalities. Many of these activities have been undertaken or sponsored by federal agencies in exercise of the commerce clause in the United States Constitution. It enables the Congress to 'regulate commerce among the states'. The Supreme Court has indicated that:

> The power to regulate commerce necessarily included power over navigation. To make its control effective the Congress may keep the 'navigable waters of the United States' open and free and provide by sanctions against any interference with the country's water assets. It may legislate to forbid or license dams in the waters; its power over improvements for navigation in rivers is 'absolute'.[19]

Thus the 'federal government is involved in navigation, flood control, watershed development, power production, and environmental protection'.[20] So, while individual rights were protected by the common law, the use of water resources in the public interest has become the responsibility of public agencies acting in accordance with their statutory mandates.

(c) Water Resources Planning

Water resources planning emerged informally as an aspect of project planning for specific developments. This linked project planning with strategic planning so as to ensure that the use and development of water resources promoted the public interest in the widest sense. Such an informal system of planning was no doubt influenced by the wider public policy objectives current at the time, including those of the federal institutions in the United States of America. Accordingly, it has been suggested:

> The long history of integrated land and water planning seemed to die in the US in the 1980s. The Reagan Administration failed to fund the existing river basin commissions and the federal government continued its withdrawal from subsidised water development.[21]

In any event activities associated with the use and development of water resources were required to comply not only with the rules of the common law but also with those statutory rules designed to prevent pollution and protect the environment. So compliance with these rules provided the framework for planning the future use and development of water resources rather than rules for planning the use and development of water resources themselves. In other words the system became much more fragmented and much less integrated.

Water resources development projects undertaken by public agencies may be authorised generally rather than on an individual project basis. Two examples are the arrangements in India and in Australia. Custom and the rules of the common law were the foundation of arrangements in India during the 19th century when the Easement Act of 1882 was enacted. While the Act was concerned principally with the recognition and protection of individual rights, the public interest in how water resources were managed emerged as an element of these arrangements. The Act in section 2(a) specifically recognised and protected:

> Any right of the government to regulate the collection, retention and distribution of the water of rivers and streams flowing in natural channels, and of natural lakes and ponds, or of the water flowing, collected, retained or distributed in or by any channel or other work constructed at the public expense for irrigation.

This comprises what has been described as 'the power of the state over water' and as an 'absolute right'.[22] Such a right of the state in relation to water has been confirmed in later legislation of some of the states within the Indian federation. The exercise of such a right does not necessarily involve the performance of a planning function prior to the implementation of any project so authorised. If there is any planning, it is likely to be project planning rather than water resources planning.

The arrangements were much the same in Australia. Water was a matter in the first instance for the colonies and then the states within the federation of Australia. Originally the rules were those of the common law of England. The beginning of the 20th century saw the introduction of a different approach. The right to the use and flow and to the control of most water resources within the states was vested in the Crown. For what purposes was this right able to be exercised? In the state of New South Wales the purposes were:

- for the conservation and supply of water;
- for its more equal distribution and beneficial use;
- for its protection from pollution.[23]

For most of the 20th century the right was exercised in such a way that it facilitated the economic growth and social development of Australia. During this period the stated purposes for the exercise of the right were in practice no more restrictive than in the case of the unrestricted right conferred by the Indian legislation. Consequently any planning was likely to be project planning rather than water resources planning.

PLANNING FOR WATER TRANSFERS

(a) Introduction

The construction and operation of hydraulic infrastructure have been a feature of the use and development of water resources for thousands of years. Some of these projects have been very large, almost, perhaps, to the point of overwhelming the local community and the environment. Their purpose has been to move or convey water physically from one location to another, and often over vast distances. Much of the infrastructure has been concerned and is concerned with the transfer of water in this physical sense. Such a transfer of water must be distinguished from a transfer of rights in relation to water – a point for subsequent analysis. Sometimes these movements of water involve its transfer from one water basin to another water basin, or indeed from the territory of one state to that of another. It is for a state to decide whether and to what extent inter-basin transfers should be permitted within the state, and it is a matter for nation states to determine themselves the extent to which interstate transfers should take place. Such transfers involve the construction and operation of infrastructure and this requires authorisation in the same way as the construction and operation of any infrastructure. Consequently any planning function is incidental either to the formulation of strategy or to project development. Let us consider some approaches.

(b) Interstate Transfers

The transfers may be between states, between states within a federation, or within a state. The highlands water project in southern Africa 'catches water from Lesotho's highlands and ferries it via an elaborate system of dams and underground tunnels to South Africa'.[24] The project is authorised and managed, as we have seen, by an interstate agreement – a treaty – between the two states. Each state must therefore be presumed to have performed the appropriate planning functions for the development of its water resources prior to entering into the agreement. The enforceability of these arrangements depends upon the relevant rules of public international law.

Different arrangements apply when states buy and sell water as a commodity according to commercial contracts entered into between the states: for example arrangements between Turkey and Israel and between Malaysia and Singapore.[25] Presumably the doctrinal basis of these arrangements is the ownership of the water in a form in which it can be physically transferred by the state or the agency of the state in question. The nature of these arrangements suggests that they would be enforced in the normal way as a set of commercial contracts. This assumes that there is no impediment within the national legal system to the transfer of water outside the jurisdiction: for example, rules in the nature of prohibitions or restrictions upon the export of a commodity.

The water export law of Bolivia permits the export of water.[26] On the other hand, 'Canada provides an excellent example of limitations on bulk water export at both the federal and provincial level'.[27] At the federal level in Canada the removal of boundary waters outside the water basin is prohibited[28] and at the provincial level there are limitations on bulk water transfers.[29] These arrangements, whether permissive or restrictive of transfers, are reflective of policy decisions taken at the highest levels in the public interest. The creation of these permissive or restrictive regimes is on the face of it a matter for the discretion of the state in question. Whether and to what extent these decisions are consistent with the strategies and rules for water resources planning is similarly a matter for the states in question.

While Canada has prohibited or restricted cross boundary water transfers at both the federal and provincial levels, it is by no means uncommon for water to be transferred within a province. Such transfers and the infrastructure required for such transfers are governed by the law of the province. It is likely that the infrastructure required for an intra-provincial transfer of water will require an environmental impact assessment. It has been suggested that 'while environmental impact assessment is a tool which can facilitate effective water resource development', it has required improvement. For example towards the end of the 20th century it was suggested that 'important planning' was 'done behind closed doors and has made the public suspicious about governments, large industries and major water projects'.[30] While environmental impact assessment may well be part of the planning process, it merely assists in determining how water resources should be used and developed most effectively in the public interest. That ultimate decision depends upon other factors which are at least to some extent independent of any environmental impact assessment process.

(c) Interbasin Transfers

There are many examples of inter-basin transfers and interstate transfers of water in Europe, although in Europe they are relatively small compared with those in, for example, Canada, China and Egypt.[31] For example, there is the Rhône-Languedoc transfer and the Canal de Provence in France and the Tajo Segura transfer in Spain. In France the relevant water agency has financially supported a project for the transfer of water by canals to deficit zones. The objectives are:

- to bring the resources to the deficit zones;
- to save the transported water;
- to prevent saline intrusion into canals near the Flemish coast.[32]

In Spain the water law supported 'balancing water availability between catchment areas'.[33] Accordingly the relevant plans have provided for intra-basin transfers from what are described as surplus regions to deficit regions. How does this work? In this way:

> Theoretically, the volume of water transferred would depend on the water supply and needs in each respective catchment. 'Surplus' water is a flexible concept which depends upon the hydrological conditions of the catchment concerned, and in particular it refers to defining minimum flow under 'normal' and 'exceptional' conditions to ensure environmental sustainability.[34]

Clearly, then, the flexibility of this approach reflects the need to ensure that the water resources of one water basin are not detrimentally affected by the transfer of water to another basin: in other words, the need for some form of planning or regulatory control.

An inter-basin transfer of water almost inevitably involves a reallocation of water. It may well interfere with existing water rights in the water basin from which the water is being transferred. Water resources planning should in principle, then, address both issues and how each impacts on the other. Large inter-basin schemes have been constructed in South Africa. It has been suggested that 'in consideration of water allocation, inter-basin water transfers must be taken into account, and special planning requirements and implementation procedures have to be met'.[35]

There have been major inter-basin transfers of water in China. But arrangements have also been put in place in China for the diversion of water not to another water basin but to benefit other parts of the river. For example, the administrative arrangements put in place for the waters of the Yellow river in the Yellow River Basin:

> The Yellow River Basin was the first basin where the allocation of the rights of water use was conducted. Because of the serious problem of no-stream flow in the lower reaches of the Yellow River, which happened nearly every year in the 1990s, the regional ecological environment increasingly deteriorated. Under the approval of the State Council, in December 1998, the State Planning Commission and the Water Resources Ministry jointly issued a Scheme for Distribution of the Water Supply and Allocation of the Mainstream Flow of the Yellow River, which empowers the Yellow River Commission to allocate the stream flow of the Yellow River.[36]

In the drought year of 1999, diversion of water was successful. In other words, the diversion of water from one use to another by administrative decision was authorised by an instrument in the nature of a planning scheme.

In the Republic of Yemen the Water Law permits the inter-basin or inter-zone transfer of water provided certain conditions are met.[37] The power to authorise such transfers is conferred upon the National Water Resources Authority with the approval of the Council of Ministers. Specified volumes of groundwater or surface water may be pumped from one basin or zone to another. The conditions are these:

- the conveyance of the water will not have any future adverse affect on the water in the basin or zone from which it is conveyed;
- the water will be used only for drinking or domestic purposes;
- there is a shortage of water in the recipient basin or zone;
- there is coordination and consultation with all stakeholders in the water basin from which the water is conveyed.

This is one of the few – if not the only – examples of a set of codified conditions precedent to permitting inter-basin transfers.

PLANNING TO REGULATE WATER RESOURCES DEVELOPMENT

(a) Introduction

Water resources planning may be a part of the process for putting together water resources development projects which are either public or private. While the activities necessary to undertake such developments may require approval – and most jurisdictions will require approval – under laws designed to protect components of the environment, the use and development of water resources may in addition require approval within an enforceable framework of water resource planning. In other words, there will be a requirement to obtain:

- a water entitlement;
- a water use approval;
- an infrastructure construction and operation approval;
- an environmental approval.

Applications for a water entitlement and a water use approval would be determined in accordance with a water resources plan and the activities so authorised would be undertaken in accordance with the water resources plan. A planning regime thus involves a number of norms. There are norms of competence: the power – perhaps the duty – to formulate the plan. This is a function of the public sector with the involvement and participation, in most cases, of the private sector and the public at large. Then there are norms of obligation. First, there are obligations imposed upon those formulating the plan to comply with its procedural and substantive requirements. And there are obligations imposed upon those proposing to use and develop water resources and upon those actually using water resources. We have already suggested that it is in principle easier to establish and implement planning regimes where the legal arrangements acknowledge the notion of the public domain. But in other jurisdictions it is a function of legislation simply to create a planning regime.

(b) Planned Regulation in the United States of America

Let us consider, first of all, how a mandatory regime for water resources planning by regulation emerged in the state of Florida in the United States of America. 'It is the home of the Everglades, miles of beautiful beaches, and the Florida Keys. Accordingly whether Florida will be able to protect its natural treasures is largely contingent upon how it manages its water resources.'[38] Before the reforms beginning in the 1970s, the use and development of the water resources of Florida were planned – if at all – as an aspect of the formulation and implementation of water resources development projects. Thus 'these water resources were diked, dammed, dredged, and drained'.[39]

The enactment of the Water Resources Act in 1972 changed this. Water management districts were established according to the hydrologic basin boundaries in Florida. The range of functions able to be performed by the water management districts was considerable:

- to tax;
- to enter into contracts;
- to survey water resources;
- to establish saltwater barrier lines;
- to establish minimum levels and flows;
- to develop water shortage plans;
- to promulgate rules for surface water use and management;
- to develop a permit system for the consumptive use of groundwater;
- to construct works;
- to purchase land;
- to establish basin boards;
- to cooperate with federal institutions in relation to flood control, reclamation and conservation.[40]

Notwithstanding the importance of all these functions, two have been particularly critical in the context of a planning regime. The first is the power to promulgate rules for surface water use and management: in other words, the creation of enforceable rules for the use of surface water. The second is the power to develop a permit system for the consumptive use of groundwater.

What has emerged is a traditional regulatory arrangement, namely, a prohibition on the consumptive use of groundwater without a permit. The authority to engage in the relevant use would be derived not from a set of rules – as in the case of the use of surface water – but from a permit granted administratively in accordance with the statutory scheme. And the responsibility lay upon the applicant for a permit to show that:

- the consumptive use was a reasonable and beneficial use;
- that it would not interfere with any existing legal use;
- and that it would be consistent with the public interest.[41]

Implicitly, therefore, an application could not be granted unless these three conditions were satisfied.

The next step in the creation of a formal planning regime was the establishment of the Department of Environmental Regulation. Round this department there has emerged a regulatory planning regime. There are essentially four sets of plans:

- a state plan put together by the executive office of the Governor and intended as a 'direction-setting plan' which provides 'long-range policy guidance';
- a state water use plan developed by the Department of Environmental Regulation as 'an integrated, coordinated plan for the use and development of waters of the state';
- regional water plans prepared by the water management districts;
- comprehensive plans prepared by local governments providing for inter alia conservation of water, water recharge areas, wetlands, waterworks, estuaries, flood plains, rivers and lakes.[42]

Most importantly of all, 'all future development must be in accordance with the local government's comprehensive plan. If a development proposal is not consistent with the plan, a local government cannot approve the necessary development permits.'[43] There is in addition a need for consistency between and among all the plans.

The planning regime in Florida thus impacts upon first activities associated with water and second the use of water itself. The scope of this regime has been described in these two ways. First:

> Virtually any construction or other activity that alters the flow of water across the surface of the land may be regulated by the districts to ensure that water quality is not degraded, downstream areas are not flooded, and aquifers and wetlands that wildlife are dependent on are not adversely affected. If any development activity is proposed for a wetland, then it must also meet a public interest test based on a review of the impacts to fish and wildlife habitat, recreation and other wetland functions.[44]

And second:

> Another major area for regulation by the districts is the regulation of activities that affect the quantity of water available for use by humans or the environment. The districts have authority to regulate artificial recharge facilities and consumptive use of water. Consumptive use permitting is one of the most important responsibilities of the districts. Virtually any use of water that involves withdrawing or diverting it from its source can be regulated by a water management district under this authority.[45]

The comprehensive scope of such a regulatory regime brings it close in practical but not doctrinal terms to a public domain regime.

(c) Planned Regulation in Public Domain Regimes

The management of water resources in China and in Vietnam is based upon state ownership and control. The comprehensive national plan for China is intended to perform these functions:

- assessment of the present situation of water resources exploitation and use;
- drafting of plans for water saving, water resources protection and sewage treatment;
- analysis of both the potential of water resources exploitation and use and the sustainability of water resources;
- drafting of water resource allocation schemes;
- working out arrangements for exploitation, restoration, distribution, saving and protection of water resources, and an implementation scheme for all the measures;
- drawing up countermeasures and methods for water resource management, and setting up management systems for water resources appropriate to the socialist market economy.[46]

The plan provides 'the scientific basis and norms with legal effect for the macro-management of water resources in the future'.[47]

The Law on Water Resources of Vietnam provides for using and protecting water resources and for overcoming their harmful effects. This must be based on river basin planning.[48] In addition the socioeconomic development plan must include a plan to prevent and combat water pollution and to restore the quality of polluted water.[49] The implementation of water plans is monitored and enforced by a specialised inspectorate.[50] While plans are one element of the set of arrangements in Vietnam, they have effect within the broader constitutional arrangements in Vietnam, which include the strict prohibition of activities that deplete natural resources and destroy the environment.[51]

The National Water Resources Policy, a part of the legal arrangements for managing water resources in Brazil, provides for the formulation of water resources plans, and one of their functions is to implement the Policy. These water resources plans are 'long-term plans that should have a time horizon that is compatible with the period over which the proposed programmes and projects are to be implemented'.[52] More specifically, the subject matter of these plans includes:

- a diagnosis of the current status of water resources;
- an analysis of population growth, the evolution of productive activity, and land-use patterns;
- a statement of the future supply of and demand for water resources in terms of quantity and quality, including identification of potential areas of conflict;
- targets for rationalising the use of water as well as improving its quantity and quality, and measures to be taken to reach these goals;
- priorities for the awarding of water-use rights;
- guidelines and criteria for water-use fees;
- proposals for the creation of areas with the restrictions on water-use that are designed to protect to protect water resources.[53]

These plans deal comprehensively with the social and economic uses of water resources, and this includes the protection of the quality of the water but from the human rather than the ecological perspective. The issue of pollution and the protection of water more generally is governed – as is often the case – by specific provisions of the National Water Resources Policy. The water resources plans are complemented by water use plans which are developed for each river-basin, for each state within the federation of Brazil and for Brazil itself.[54] It appears to be expected but not required that rights in relation to water are allocated consistently with these water use plans.

Then there is Australia – while not strictly a public domain regime, in practice it is. Most of the states within the Commonwealth of Australia either enable or require the formulation of plans in relation to water resources. In the state of South Australia, for example, there is provision for the formulation of four types of plans:

- the state natural resources management plan;
- regional natural resource management plans;
- water management plans;
- water allocation plans.

The state and regional plans are relatively strategic in nature, while the water allocation plans deal specifically with the allocation and use of water. The provisions of the water allocation plan are to achieve an equitable balance between the environmental, social and economic needs for the water on the one hand and the sustainable rate of use of the water on the other hand. Most importantly, these three plans are linked to each other and to the objects of the legislation in general. The requirement in legal terms is for consistency between the water allocation plan, the regional plan and the state plan.[55]

Portugal, as we have noted, adopts a 'global approach' to the management of its water resources. The Water Institute has the overall responsibility for water resources management and planning. It is responsible for the preparation of two types of plan:

- the national water plan;
- river basin plans.

The national plan 'requires the integration of river objectives with national economic and social policies' and a river basin plan must include inter alia means to protect and enhance the river network and areas subject to land use restrictions.[56] These are elements of sustainable development at the planning level.

Similarly in Spain there is provision for a national hydrological plan and river basin plans. The formulation of a river basin plan is the responsibility of the relevant river basin authority.[57] However these authorities have further functions:

- administering and controlling the hydraulic public domain;
- designing, developing and operating hydraulic works according to hydrological planning criteria;
- licensing for the hydraulic public domain.[58]

These river basin authorities thus have both planning and regulatory functions together with the undertaking of hydraulic works. Significantly, however, the undertaking of hydraulic works is 'according to hydrological planning criteria'. This brings together planning by water resources regulation and planning by water resources development.

CONCLUSION

Planning has become a significant element of water resources management. It may be formal or informal. It may be voluntary or mandatory. It can operate at different levels: the formulation of strategy; the formulation of policy objectives; the implementation of policy; even operational activities. It may impact upon public agencies as well as individuals and corporations. It may comprise legal rules as part of a regulatory system or paralegal rules as part of a context for making decisions and undertaking activities. Water management, it has been suggested, involves supply management and demand management. Consequently 'planning integrates both aspects, along with environmental concerns, and provides an analytical basis for choosing between them'. It contemplates 'the need for governments to intervene in its allocation and management'.[59] But planning involves cooperation between the public sector and the private sector. Planning and the procedures for planning are particularly important in this sense because they point to and indeed are predicated upon the responsibility of society as a whole for the way water resources are managed – particularly in the context of sustainable development.

NOTES

1. Singh (2000, p. 55).
2. Salman and Bradlow (2006, pp. 122 and 124).
3. Salman and Bradlow (2006, p. 86).
4. Salman and Bradlow (2006, p. 15).
5. Salman and Bradlow (2006, p. 71).
6. Salman and Bradlow (2006, p. 77).
7. European Commission (2000, pp. 149 and 150).
8. European Commission (2000, pp. 50 and 237) and Chave (2001, p. 120).
9. European Commission (2000, pp. 50 and 262).
10. Chave (2001, p. 119).
11. Salman and Bradlow (2006, pp. 38 and 39).
12. Chave (2001, p. 46).
13. Chave (2001, p. 47).
14. Sherk (2000, p. 36).
15. Muckleston (1990, p. 35).
16. Dzurik (1990, p. 6).
17. Dzurik (1990, p. 6).
18. Dzurik (1990, p. 6).
19. *Appalachian Electric Power case* (1940, pp. 404 and 405).
20. Tarlock et al. (2002, p. 787).
21. Tarlock (2000, p. 225).
22. Singh (1991, p. 40).
23. Fisher (2000, pp. 90 and 91).
24. Brown Weiss (2005, p. 74).
25. Brown Weiss (2005, p. 75).
26. Brown Weiss (2005, p. 75).
27. Brown Weiss (2005, p. 76).
28. Brown Weiss (2005, p. 76).

29. Brown Weiss (2005, p. 77).
30. Shrubsole (1990, p. 109).
31. European Environment Agency (1999a, pp. 78 and 79) and European Commission (2000, pp. 108 and 109).
32. European Commission (2000, p. 108).
33. European Commission (2000, p. 108).
34. European Commission (2000, p. 108).
35. Hu (2006, p. 131).
36. Xi et al. (2002, p. 18).
37. Salman and Bradlow (2006, p. 122).
38. Dzurik (1990, p. 71).
39. Dzurik (1990, p. 71).
40. Dzurik (1990, pp. 72 and 73).
41. Dzurik (1990, p. 73).
42. Dzurik (1990, p. 76).
43. Dzurik (1990, pp. 76 and 77).
44. Tarlock et al. (2002, p. 281 quoting Richard Hamann).
45. Tarlock et al. (2002, pp. 281 and 282 quoting Richard Hamann).
46. Xi et al. (2002, pp. 13 and 14).
47. Xi et al. (2002, p. 14).
48. Salman and Bradlow (2006, p. 114).
49. Salman and Bradlow (2006, p. 116).
50. Salman and Bradlow (2006, p. 118).
51. Salman and Bradlow (2006, p. 113).
52. Salman and Bradlow (2006, p. 26).
53. Salman and Bradlow (2006, pp. 26 and 27).
54. Salman and Bradlow (2006, p. 27).
55. Fisher (2007, p. 117).
56. Chave (2001, p. 120). See also European Commission (2000, pp. 50 and 237–239).
57. European Commission (2000, p. 259).
58. European Commission (2000, pp. 259–263).
59. Haddadin (1996, p. 59).

GENERAL REFERENCES

Brown Weiss, E. et al. (eds) (2005), *Fresh Water and International Economic Law* (Oxford; New York: Oxford University Press).

Chave, Peter A. (2001), *The EU Water Framework Directive* (London: IWA Publishing).

Dzurik, Andrew A. (1990), *Water Resources Planning* (Savage, Mld: Rowman and Littlefield Publishers Inc).

European Environment Agency (1999a), *Sustainable Water Use in Europe: Part 1: Sectoral use of Water* (Luxembourg: European Communities Official Publications).

European Commission (2000), *Towards Sustainable and Strategic Management of Water Resources* (Luxembourg: European Communities Official Publications).

Fisher, D.E. (2000), *Water Law* (Sydney: LBC Information Services).

Fisher, D.E. (2007), 'Delivering the National Water Initiative: the emergence of innovative legal doctrine', in Karen Hussey and Stephen Dovers (eds), *Managing Water for Australia – the Social and Institutional Challenges* (Collingwood: CSIRO Publishing).

Haddadin, M.J. (1996), 'Water management: Jordanian viewpoint', in J.A. Allan (ed.), *Water, Peace and the Middle East* (London: Tauris Academic Studies), pp. 59–73.

Hu, Deshang (2006), *Water Rights: An International and Comparative Study* (London: IWA Publishing).

Muckleston, Keith W. (1990), 'Integrated water management in the United States', in Bruce Mitchell (ed.), *Integrated Water Management: International Experiences and Perspectives* (London: Belhaven Press), pp. 22–43.

Salman, Salman M.A. and Daniel D. Bradlow (2006), *Regulatory Frameworks for Water Resources Management: A Comparative Study* (Washington, DC: World Bank).

Sherk, G.W. (2000), *Dividing the Waters: the Resolution of Interstate Water Conflicts in the United States* (London: Kluwer Law International).

Shrubsole, D.A. (1990), 'Integrated water management strategies in Canada', in Bruce Mitchell (ed.), *Integrated Water Management: International Experiences and Perspectives* (London: Belhaven Press), pp. 88–118.

Singh, Chhatrapati (1991), *Water Rights and the Principles of Water Resource Management* (Bombay: N. M. Tripathi Pty Ltd).

Singh, Chhatrapati (2000), 'Water rights and the principles of water management in India', in D.E. Fisher and Noeleen McNamara (eds) (2000), *Integrated Water and Land Management* (Brisbane: Queensland University of Technology), pp. 49–79.

Tarlock, A. Dan (2000), 'Putting rivers back into the landscape: the revival of watershed management in the United States', in D.E. Fisher and Noeleen McNamara (eds) (2000), *Integrated Water and Land Management* (Brisbane: Queensland University of Technology), pp. 195–247.

Tarlock, A. Dan et al. (2002), *Water Resource Management: A Casebook in Law and Public Policy* (5th edn, New York: Foundation Press).

Xi, Wang et al. (2002), 'Managing water resources for a sustainable futurer: law, policy and methodology of China', *Asia Pacific Journal of Environmental Law* 7, 5–41.

CASES

Appalachian Electric Power case: United States v Appalachian Electric Power Company (1940), *United States* 311, pp. 377–434.

16. Water rights as regulatory instruments

THE SOURCES OF WATER RIGHTS

The community through its institutions is responsible for the way in which water resources are managed. But the community and each individual member of it are the beneficiaries of the way the water resources are managed. The function of the law is to promote and protect not only the public interest but also the interests of each and every member of the relevant society. These interests may differ; they may conflict; but it is the function of the law to accommodate them. The public interest is promoted and protected in accordance with constitutional arrangements, strategies, policies, the development and operation of planned infrastructure, the implementation of planning arrangements and the enforcement of legal rules. But how are the interests of individual members of society protected? The simple answer is by the recognition, creation and protection of rights. But rights perform different functions within the legal system. Our attention turns now to a review of the way water rights are continuing to play a significant role as instruments for water resources governance. The focus is not on the rights of the state or of the public sector but on those of persons – individual or corporate – in the private sector: in other words those who use or benefit from the use of the water in question. This includes – rather paradoxically – water rights designed to protect ecological values.

The role played by water rights recognises the wide variety of interests in water and water resources. The interest may be a share of a water resource; access to a water resource; use of a particular body of water; receipt of a flow of water; receipt of a reticulated supply of water; or even receipt of water as it falls to the surface of the earth. Some of these interests are connected with land, while others may have no such connection. The complexity of these arrangements is compounded by the variety of sources of these interests. An interest in water may exist simply because the person in question is a human being. It may exist because of a practice or custom within the community. It may exist simply because of its recognition by the common law or the civil law. Or it may exist only after it has been granted in accordance with a recognised process. It may consequently be an interest held by a person individually or as a member of a group. It may be in an interest in a particular water resource, in a particular body of water, in any water resource or in any body of water.

Water rights – whatever their function and substance – derive from a number of legal sources:

- they exist by reason of a practice or custom;
- they exist according to a rule of the civil law, the common law, Islamic law or the relevant system of private legal relationships;
- they are acquired according to a rule of law of the same kind;
- they are conferred directly by legislation without administrative involvement;
- they are granted administratively in accordance with legislation.

The first two and the fourth exist automatically as a matter of law. No activity or decision is required. The third requires some action on the part of the individual seeking to benefit from the water right. The fifth requires a decision on the part of a public agency. Water rights as a set of private legal relationships continue to be recognised in some jurisdictions either in their original form or modified by legislation, for example in the United States of America. Increasingly water rights have been abandoned as a set of private legal relationships and are consequently granted in accordance with regimes of public law, for example in systems based on state ownership or the notion of the public domain. In some jurisdictions there is a combination of approaches.

WATER RIGHTS AS PRIVATE LEGAL RELATIONSHIPS

The function of the civil law and of the common law has been to recognise and protect the interests and rights of individual members of the community. It was the function of the criminal law in these systems to protect the public interest. The notion of the public domain has emerged in civil law jurisdictions as the basis for the management of natural resources, including water resources, in the public interest. No such concept emerged in the common law. It has been suggested, however, that the public trust doctrine recognised by the common law has performed a function similar to that of the public domain. Perhaps, but the public trust doctrine itself found its origins in Roman law. In any event, the doctrinal pillars of the common law – the appropriation doctrine on the one hand and the riparian doctrine on the other hand – were unable to accommodate the public interest in the use and development of water resources.

In some common law jurisdictions – for example Australia and New Zealand – the doctrines of the common law increasingly gave way during the 20th century to water resource management regimes created by the legislatures. Towards the end of the 20th century the doctrinal foundations of the common law in relation to water were abandoned in England. The prior appropriation doctrine and the riparian doctrine have remained, although in many respects in substantially modified form, in the United States of America. While individual rights remain, whether vested in individual persons, corporations or public utilities, the manner of their exercise has been increasingly restricted to some extent by substantive rules of law designed to protect elements of the public interest, including the protection of aquatic ecosystems.

The rules of the common law according to which individual rights in relation to water are recognised or created remain particularly relevant in the United States of America. Three models have been suggested:

- a tort approach;
- an exclusive property rights regime;
- a managed property rights regime.[1]

The common law of riparian rights is an emanation of the relationship between rights and liability in tort. Riparian rights are rights associated with the ownership of land. The community of interest in this case is the interest of all riparian landowners in the watercourse. Accordingly 'individuals are allowed to use water without the need for any kind of prior approval to the point that another user suffers a demonstrable harm such as diminished flow or pollution'.[2] As we have seen, it is the common law of nuisance which historically has placed a limitation upon the exercise of this right. Riparian rights have been categorised as incorporeal property rights.[3] A riparian right is a right of property in the sense that it is a right incidental to the ownership of land, and in this sense capable of alienation.

The prior appropriation doctrine is different. A right arises as a consequence of taking water and using it for a beneficial purpose. It is the action of taking – irrespective of where the action takes place – that is the foundation of the existence of the right. It is not in any way associated with the ownership of land. It is an exclusive property rights regime. And 'the function of the law is to establish the ground rules for the acquisition of a relatively exclusive right and to police the exercise of the right to protect the interest of other users'.[4] It, too, is a right of property in the sense that it is alienable. The mere action of taking the water, however, is no longer sufficient. Three conditions have been imposed:

- the right must be exercised: that is, the water must be taken and used;
- the use must not be wasteful;
- the use must be for a beneficial purpose.[5]

The prior appropriation doctrine – just like the riparian doctrine – is based upon a set of rules. In theory, therefore, there is no administrative involvement and there is no judicial involvement except in the case of any disputes which the courts are called upon to settle.

In practice, however, the pristine rules of the common law have been modified. To some extent these rules-based property rights regimes have evolved into managed property rights regimes. In effect, 'an administrative scheme to create and protect rights and public interest limitations' has been superimposed upon 'a base of individual exclusive rights'.[6] Individual rights are thus modified in the public interest. Originally a right of prior appropriation was acquired 'by posting a notice on a stream and digging a ditch'.[7] The right does not now become 'perfect' without the involvement of a public agency. The public agency must be satisfied inter alia that the appropriation is in the public interest.[8] But the public agency does not grant the water right. It merely 'perfects' it.

WATER RIGHTS AUTOMATICALLY AVAILABLE

(a) Water Rights Conferred Directly by Legislation

So much for water rights as private legal relationships. Water rights may be conferred on individuals directly by the law and thus available without any administrative or judicial involvement. First, there are interests in water – inchoate rights – which exist simply as a matter of law. In France water rights exist within the framework of its public domain regime. For example, riparian landowners are entitled to use water that borders or crosses their land provided this does not modify the flow of the water. In addition an amount of water limited by volume per day may be abstracted for domestic use as a right and without any authorisation.[9] Similarly in the states in the Commonwealth of Australia riparian landowners have a right to use water in a watercourse for domestic purposes and for the watering of stock. Specifically the use of the water for commercial and industrial purposes – including irrigation – is excluded.[10] In each of these examples, the right to use the water is a rule of law declared by the relevant code in France and by the relevant legislation in Australia. In this sense the right is neither appropriated nor allocated. It is a right available to whoever falls within the category of riparian landowners so benefited.

The water fund in Kazakhstan, as we have seen, includes all bodies of water in the territory of Kazakhstan and it is managed in accordance with the water code. Uses of water are categorised in a number of different ways. One such category is the general use of water. All water bodies that are part of the water fund are for general or common use unless otherwise provided. It is specifically provided in addition that any person may use water bodies that are for general or common use. For this purpose a general use is a use of the water 'without application of facilities or technical devices having an impact on water conditions'. This right 'accrues to citizens from the moment of their birth and cannot be assigned'. This has been described as 'public water servitude'.[11] The use of water in other categories effectively requires a licence or a permit.

The arrangements in China are similar. Article 7 of the Water Law contemplates a system of licensing for the taking of water. There are exceptions created administratively. These include:

- extracting water for family living and livestock;
- extracting small quantities of water for agricultural irrigation;
- extracting small quantities of water by human force, animal force or other methods.[12]

While these may not amount to legal rights, to engage in these activities would in practice not amount to a breach of any duty. In this sense these activities are lawful.

These rights, available either to the riparian landowner or any individual, are very limited in terms of purpose of use and method of use. These rights reflect traditional practices. It is interesting to speculate about their origins: perhaps in relation to France, the civil law; for Australia, the common law; and for China, customary law. In any event these are individual rights and recognised as such through their status as rules of law. It would be surprising if their origins were not the civil law, the common law or customary law. These three sets of arrangements evolved in the context of agricultural rather than industrial economies, with the population living in rural rather than urban communities. In each of these cases it appears to be a right to use water – a usufructuary right – and not a right of ownership of water.

In each case the right has been conferred by legislation – except in China where it has been conferred administratively. These rights resemble in many ways customary rights. Why have they been in a sense retained by 21st century legislation? Despite increasing industrialisation and urbanisation during the 19th and 20th centuries, many local communities have remained rural, agricultural or pastoral. It is important to recognise, therefore, that water resources need to be managed so as to recognise agricultural and rural needs on the one hand and industrial and urban needs on the other hand. The balancing of these needs and the creation of ways and means of satisfying these needs are elements of any system of management. It is a function of strategic planning, policy formulation, development planning and regulatory planning to do so. The legal arrangements for doing so are almost as diverse as the circumstances themselves. One response has thus been to recognise the traditional ways of managing water resources.

(b) Customary Water Rights

Customary arrangements have played an important part in managing resources for thousands of years. Are they still relevant? In India, 'both the Hindu and the Muhamadan customary law recognized various types of water rights of the people'.[13] The Easement Act 1882 of India specifically provided that an easement could be acquired 'in virtue of a local custom'. So customary rights in relation to water continue to be recognised. In Africa customary laws remain relevant, and in some instances 'customary rights have been expressly recognised in legislation'.[14] For example, in Mozambique the Water Act of 1991 specifically retains traditional uses of water by local communities for domestic, personal or family purposes provided the use does not cause pollution.[15]

In Ghana, similarly, traditional rules continue to perform a function – particularly in rural and deprived urban areas.[16] The water resources are controlled and the local rules enforced by chiefs and spiritual leaders. But 'some of the traditional rules are losing their binding character as members of the communities undergo christianisation or the secularising effects of modernity'.[17] But some remain relevant – for example, prohibiting fishing and livestock watering in certain circumstances and securing resource access for everybody – and have 'important environmental, sanitary and distributional outcomes'.[18] The Water Law of the Republic of Yemen positively protects customary law. The use of water normally requires some form of authorisation. However:

> The right to use water entitles the holder of the right to use water in a way that does not conflict with the public interest or with the prevailing customs and traditions in each water zone or basin.[19]

On the face of it, this appears to give priority to prevailing customs and traditions in the event of a conflict.

The relationship between customary rules and statutory arrangements in Nepal is particularly interesting. When arrangements in relation to land and water were codified in 1964, it was declared that 'where water is shared traditionally, it should be maintained as per the share agreed upon previously'.[20] But in 1967 a system of licensing was introduced. The Supreme Court nevertheless took the view that customary rights of access to water had not been removed. In one case, a landowner prevented access to a well on the land by constructing a wall. The local municipality pulled down the wall. Although this action was unlawful, the Supreme Court decided that 'the local people should be given access to the well because they had been dependent on it for a long time'.[21]

The licensing system introduced in 1967 was continued by the Water Resources Act of 1992. This prohibited the use of water resources without a licence. However it was also declared that licences were not required for most water uses in rural areas. In the absence of any rules governing these uses, 'it is generally assumed that the previous legal rules and principles relating to control and use of water resources' would apply.[22] On the face of it, therefore, customary rules applied in such circumstances.

Customary laws, as we have noted earlier, have tended to be based upon notions such as water resources as shared resources; a community of interest in water resources; and a common responsibility for their use and the consequences of their use. While individual rights may emerge from such a set of values, these individual rights – or interests – may be perceived perhaps as a means of identifying those benefited by these arrangements and at the same time required to comply with them. Is this an effective approach to the sustainable development of water resources?

While the notion of sustainable development as such almost certainly was never in the minds of those managing and using water resources according to customary rules, it has been suggested that the values underlying sustainable development endorsed in the report of the World Commission on Environment and Development in 1987 tended to reflect the values underpinning customary law rather than those underpinning the water resources management regimes of the second half of the 20th century. Thus:

> The notions of inter-generational equity, communal title, inalienability, trusteeship, unity of mankind and nature, balance between personal desires and communal obligations, which the [World Commission on Environment and Development] report speaks of emphatically, are more ubiquitous in the customary laws than in modern laws. In fact the prescriptions of the commissions report will demand of the developed legal system to urgently limit and redefine the attributes of property.[23]

This points inevitably to an examination of the nature and function of property in relation to water resources in any management system. In other words, property is about control and not ownership.

WATER RIGHTS BY ALLOCATION

(a) Introduction

At the other end of the doctrinal spectrum from the availability of water rights as a rule of law are those allocated to their holders in accordance with a system of administrative regulation. Allocation is a function associated with state or public ownership of water resources and a public domain regime. From the comparative perspective of the law in the United States of America, a state ownership regime comprises three elements:

- water rights are limited to the privilege to use water as opposed to individual ownership of streams and aquifers;
- access to water requires state permission in the form of a permit or licence;
- access can be denied if the state determines that there is a higher or more efficient alternative use of the water.[24]

The function of state ownership or the public domain is simply the creation of rights in relation to water by a system of allocation. There are – not unexpectedly – many ways of doing so. The basis of a system of water rights by allocation is a prohibition – either explicit or implicit – of the use of water without some form of authorisation. For this purpose use may be described in various ways, including taking, abstracting, collecting in reservoirs, constructing reservoirs for this purpose, even undertaking any activities in relation to water. In the context of sustainable development the critical issue is the methodology according to which decisions are made: for example, an open discretion; a restricted discretion; or a regulated discretion. If the exercise of the discretion is regulated, is sustainable development relevant?

(b) Allocation According to Principles

The National Water Resources Policy of Brazil does not indicate how rights to use water should be allocated.[25] However these decisions are made within a framework of water use plans. These plans are part of the wider planning and policy framework created by the legal system. Accordingly, the discretion to grant a water right is at most guided by these planning and policy considerations but in no way is it required to apply them or give effect to them. The approach in Armenia seems to be much the same. The Water Code states criteria according to which water use permits are granted. The decision is additionally guided by the basic principles stated in the code.[26] These include the economic, social and ecological elements of sustainable development.[27] However these elements appear as principles which guide but do not direct the decision.

In some jurisdictions the exercise of the discretion is more directly regulated. In Senegal, for example, 'the officials must reconcile the interests of the different categories of water users, respect for the law, the rights of prior users, and protection of the nation's water resources'.[28] In the Republic of Yemen the Water Law states priorities. Water used for drinking and domestic purposes is given priority. 'Thereafter water may be allocated to any of the following purposes: watering livestock, use in public facilities, industrial purposes, and meeting environmental needs.'[29] The capacity of the legal system to ensure compliance with requirements such as these depends ultimately upon the capacity of the courts to adjudicate upon the decision-making processes of the administration.

Increasingly, water resources planning is becoming a significant function within the system. According to the Law on Water Resources of Vietnam the regulation and distribution of water resources – their allocation – must be based on the planning of the river basin. In addition there must be respect for the principles of fairness, reasonableness and priority in the quantity and quality of water for human lives. Indeed the state is required to take action to ensure that priority is given to living – that is basic human – purposes.[30] In Morocco an application for an authorisation or a concession is determined only after a public inquiry has been held and a consultation has taken place with the commission which held the public inquiry.[31] In relation specifically to the use of water resources for agricultural purposes, a permit will not be issued for a use which may result in the degradation of water resources or of arable soils.[32] It is a matter of circumstance whether water resources and arable soils are degraded or not. While the determining body must on the face of it comply with this requirement, in practice it involves a judgement based upon the evidence of particular facts and circumstances in accordance with a relatively flexible criterion.

(c) Allocation According to Procedures

These examples indicate that it is common practice for a right in relation to water to be granted by the relevant public agency in accordance with prescribed procedures and criteria. There is no reason in principle why different processes should not be adopted. In Mexico, an application for a concession is made to the National Water Commission, but an application can be granted only after a competitive bidding process has taken place. The application must be supported in the usual way by information about location, quantity and use together with an environmental impact statement.[33] While a competitive bidding process adds an additional dimension, the commission in determining the application takes into account the availability of water in accordance with the water resource programme, existing rights to use water, and restrictions on the right to use water from the relevant water resource.[34] Such a duty to take into account specific factors is an increasingly common approach. The factors to be taken into account in this case relate to the circumstances of the proposed use. It is to be expected, therefore, that a decision will take into account not only these circumstantial factors but also the underlying principles of the system.

The arrangements in Nepal follow the common practice of requiring a licence for the use of water resources except for minor uses of a domestic nature.[35] An application for a licence is determined by the relevant public agency in accordance with the arrangements in the legislation. However, in addition, provision is made for the use and development of any water resource to be undertaken in accordance with a contract between the relevant public agency on the one hand and any national or foreign company, corporate body or person on the other hand.[36] The contract specifies the relevant terms and conditions. Although on the face of it this power, although conferred by the legislation, is independent of the terms of the legislation, it would be unusual if such a contract were not expected to reflect the principles underlying the system. This is by no means clear. Otherwise, a water resource may be used and developed in accordance with the arrangements specified in the contract rather than those stated in the legislation.

(d) Allocation According to Rules

The ownership of water resources by the state in China carries with it according to the civil law a number of 'subsidiary rights'. These include the right to possess, use and dispose of the subject matter of ownership. These subsidiary rights are legally transferable. Accordingly 'the state does not have exclusive rights of use'.[37] In relation to water resources, the state exercises its subsidiary rights by administering a system of licensing for water takings. The administrative measures talk of 'water-drawing', which means drawing water directly from rivers, lakes or underground aquifers with the aid of water installations or machinery.[38] These administrative measures are implemented within the context of the Water Law. Accordingly the principles in the Water Law and the arrangements described in the range of plans contemplated by the Water Law are relevant to the implementation of the administrative measures for licensing the taking or drawing of water.

Thus a licence would be expected to further the purpose of the Water Law which is to promote the development, sustainable use, management and protection of the water resources of China in a way which contributes to economic and social development in China. In addition it would be expected that a licence would observe the priorities stated in the Water Law: first, the domestic needs of both urban and rural inhabitants and, second, 'in descending order', the needs of agriculture, industry, environment, and navigation.[39] The procedures for obtaining a licence include submission of a feasibility study report, an application with supporting documents, an inspection and a decision whether or not to grant the licence.[40] In the event of a refusal to grant the application, there is an opportunity for reconsideration by a higher authority. If the refusal is confirmed, 'the applicant may initiate a legal action against the authority'.[41] It is not clear what is the nature of this legal action.

In France the Ministry of Environment is responsible for granting abstraction and discharge permits. Permits are required for all water abstractions and in-stream uses except those considered minor in quantitative terms. The decision whether or not to grant the permit lies with *le Préfet* and the permit is a statutory instrument – *un arrêté préfectoral*.[42] Significantly a permit can be cancelled or amended without compensation for a number of reasons:

- harmful impact on drinking water supply to the population;
- immediate threat to ecosystem protection;
- bad maintenance of installations;
- threats to public safety from floods and flooding.[43]

In Spain an administrative concession is required for all water uses with minor exceptions. The responsibility for granting a concession lies with the river basin authorities and their decisions are made in accordance with the provisions in the hydrological plan and in the relevant river basin plan.[44] While the river basin plan itself establishes an order of priorities, the Water Law also creates an order of priority. Thus:

- supply to population and low consuming industries connected to the municipal network;
- irrigation and agricultural uses;
- electricity production;
- other industrial uses;
- aquaculture, recreational uses, navigation and aquatic transport;
- others.[45]

It has been noted that environmental requirements are generally placed second in the list of priorities in most of the river basin plans in force.[46]

Access to water resources and the use of water in Australia are regulated for the most part by the legal arrangements within each of the states. Although the arrangements are by no means uniform, access to water resources is generally granted in accordance with the strategies and the outcomes described in the legislation and in the rules contained in the plans. In most circumstances the strategies, the outcomes and the rules are linked, and it is this set of combined legal and paralegal rules that comprises the total legal framework within which rights in relation to water are granted.[47] Generally the outcomes contemplated by the legislation include sustainable development or elements of sustainable development. Accordingly it is a set of legal arrangements that is theoretically capable of enforcement.

TRADING IN WATER RIGHTS

(a) Introduction

The creation of rights in relation to water resources is by no means the only way to allocate water resources to their best use and development in the public interest. A study by the World Bank has identified a number of mechanisms for the allocation of water resources. There are in particular four:

- public or administrative allocation;
- marginal cost pricing;
- water markets;
- user-based allocation.

The first, considered in the preceding paragraphs, requires a public agency to decide which water resources can be used by the hydrological system as a whole and allocates and distributes water within different parts of the system. The second requires a price to be set for water which is equal to the marginal cost of supplying the last unit of that water. The third involves the exchange of water use rights between neighbouring users. The fourth involves the establishment by or on behalf of water users of collective action institutions with authority to make decisions about water rights.[48]

Each of these sets of arrangements – with the possible exception of marginal cost pricing – depends upon a set of norms of competence, norms of procedure, norms of substance and norms of obligation which brings together those creating rights and those exercising rights, and in each case in accordance with their legal responsibilities. It is a function of the legal system to recognise or to create these relationships. Water rights must have a juridical existence if they are to be exchanged in a commercial context – a market. They may exist either by the simple force of a rule of law, by appropriation or by allocation. Similarly user-based arrangements depend upon the juridical existence of water rights. What, more specifically, are the legal arrangements which support water markets?

(b) The Creation of a Market

A market traditionally has been a place where things of value are bought and sold or exchanged. A market can exist and often does exist in the absence of a place or location where transactions take place. The first step is to identify the subject matter of the transaction. In relation to water, what is it? Water may be a commodity. If there is to be a transaction, it must be physically possible to transfer the water. A precondition is for the person transferring the water to be the owner of the water. This raises many of the issues discussed earlier about ownership and property. Alternatively, the subject matter of the transaction may be an interest in water or in a water resource. What is being bought and sold is an asset in the form of an entitlement. The asset – the entitlement – has a juridical but not a physical existence.

Does this matter? For a market to operate, three conditions need to be satisfied:

- individual ownership of the asset to be traded;
- full and comprehensive information about the asset and about the rights and duties associated with ownership of the asset;
- a system of competitiveness.[49]

The first is about title: that is, the identification of the owner. And the second is about use: what can or cannot be done with the subject matter of ownership. It is the answer to this second question that gives value to the asset. In other words, 'an essential precondition for markets is the clear and unambiguous definition of property rights to the goods to be traded'.[50] Even more positively:

> Sometimes it requires the intervention of government to create the conditions necessary for markets to operate (defining water rights, creating the institutional and legal framework, investing in infrastructure to allow water transfers).[51]

Water markets may facilitate an exchange not only of water or water entitlements but also of water use rights. These water rights may exist by dint of custom, by a rule of law, by appropriation or by allocation. Their source and nature impact upon their tradeabilty. Customary and legal rights – as we have seen on many occasions – are frequently framed in relatively general and flexible language. It is often difficult to know the parameters of conduct or indeed of decision-making until a precise set of circumstances arises. This is not a very secure foundation for trading in these rights. Appropriated rights and, to an even greater extent, allocated rights are often much more specific in terms of what can be done, where it can be done and subject to what restrictions. The reason is quite simple. These entitlements are intended to be both specific and circumstantial. So it is easier to place a value upon them.

If these arguments are extrapolated to water and water resources, there are some particular challenges. When does a right of access become a right of use and when does a right of use become a right of ownership? The characteristics of water ensure that the precise subject matter of ownership cannot be identified until the water is contained within a facility from which it cannot escape. But this is almost impossible. This is the reason why in many jurisdictions rights in relation to water have traditionally been associated with rights in relation to land. So until water is physically constrained in this way, it is not a commodity or a good able to be traded in the traditional sense.

But what is capable of being traded? Quite simply – the collection of rights, duties and liabilities associated with the interest in water or in the water resource held by the owner of the entitlement. These interests vary: a right of access; a right of use; a right to receive; a right to be supplied; a right to a share. Trading in water rights is increasingly seen to be an effective mechanism for allocating and reallocating water resources. It is against this doctrinal background laced with practical difficulties that we turn now to a consideration of legal arrangements for trading in water rights.

(c) Trading in Common Law Water Rights

The *locus classicus* of market arrangements in relation to water resources is the United States of America. Even in this jurisdiction there are 'many different types of water transfer regimes'.[52] Some provide for physical conveyance or movement of water. In this case it is more often than not in accordance with statutory or interstate consensual arrangements. Sometimes these are part of a set of more comprehensive arrangements addressing a wider range of issues, once again, usually in statutory form. Alternatively physical movements of water may operate within a framework of water banking. A public agency may have the infrastructural but not the legal capacity to do so. So how was it done legally? A Drought Water Bank, for instance, was established in the first half of the 1990s in California to manage a series of successive droughts. It was authorised in this way:

> Emergency legislation was enacted to allow water suppliers the authority to enter into contracts with the bank and to provide that any temporary transfer would not affect the supplier's water rights.[53]

Importantly, statutory arrangements were devised ad hoc to circumvent the difficulties associated with the transfer of common law water rights. It is only transfers of common law water rights that have been described as 'true purchases and sales'.[54]

Whether a water right is doctrinally associated with rights in relation to land or not, it is probably impossible to exercise a right in relation to water without exercising a right in relation to land. The two may be doctrinally linked in relation to riparian rights but doctrinally separate in relation to prior appropriation rights. In riparian jurisdictions 'the most frequent method of purchasing a riparian right is to purchase riparian land, but riparian rights can be severed from the land'.[55] Rights acquired by prior appropriation, on the other hand, are separate from the land. In other words, a right of this kind can be exercised so as to benefit land other than the land where it was first exercised. In this sense water rights of this kind have always been transferable.

Even if water rights are severed from land the exercise of the right in the way that the purchaser wishes may require the exercise of rights in relation to the land over which, through which or under which the water is to be physically moved so as to enable the purchaser of the water right to benefit from it. These incidental activities may attract a requirement for an approval of one kind or another on the part either of the owner of the land or a regulatory agency in the performance of functions designed to protect a component of the public interest: protection of the environment and aquatic ecosystems.

But neither a riparian right nor an appropriated right is absolutely clear, certain and stable. On the one hand:

> A riparian does not have a right to a fixed quantity until the right has been adjudicated. Even an adjudicated riparian right is theoretically vulnerable to the subsequent claims of other riparians who were not subject to the adjudication. A riparian cannot bind other riparians. Thus, a riparian can convey the right to use its share of a watercourse, but it is an inchoate or undefined right, which may be challenged by other riparians not party to the agreement.[56]

And on the other hand:

> An appropriative water right cannot be transferred unless there is no injury to junior water right holders. This means that the senior right holder can only transfer the amount of water that is consumed rather than the amount of water that it originally diverted.[57]

And consequently:

> In appropriation jurisdictions, transfers are subject to administrative scrutiny. In brief, the amount of the transferable right must first be determined. Some states limit transfers to the historically beneficial use rather than the decreed amount; others allow the amount of water presently being put to beneficial use to be transferred.[58]

These are examples, once again, of aligning legal doctrines relating to rights, duties and liabilities with the physical use – in this case the transfer or conveyance – of water. Neither a riparian nor an appropriation jurisdiction manifests rules based arrangements that provide the clear definition of usufructuary rights – usually called property rights – that is a condition for an effective trading system.

(d) Trading in Statutory Water Rights

So much for trading in water rights that exist as a matter of law or through the application of legal rules. What about trading in water rights that are granted or allocated according to legal arrangements based upon ownership or control by the state or upon a public domain regime. What is being traded and then transferred under these arrangements is the interest in the water granted to the holder of the right in accordance with each and every aspect of the system: that is, from statements of value, principle and direction, through strategic plans and programmes, to the detailed rules in water basin and water allocation plans. The rights and duties are administratively rather than judicially created. The strength of a market in common law water rights has often been attributed to their characterisation as property rights or rights like property rights. Can rights granted administratively to their holder be similarly characterised? Certainly, the holder of the right is the owner of an asset. In one sense the asset is property and the holder of the right is its owner. Beyond that, it is difficult to argue that these rights are property rights within the contemplation of the common law simply because these rights are the creatures of legislation.

The use of economic instruments – including market arrangements – has received international approval. How have nation states responded in the context of their own legal arrangements? Some prohibit. Most permit. In Kazakhstan the Water Code prohibits the disposition of a right to use a water body by the holder of the right.[59] In Senegal a water authorisation is held by an individual and it cannot be transferred accept to the heirs of that person.[60] The holder of a water use permit in Armenia is authorised to transfer a portion of the water right in accordance with the relevant procedures. However the permit itself may prohibit any transfer.[61] In the Republic of Yemen a water use permit may be assigned only with the permission of the National Water Resources Authority.[62] Similarly in Nepal, a licence holder may transfer the licence only with the approval of the relevant administrative officer.[63]

The arrangements in Mexico and Morocco are more complicated. The National Water Act of Mexico authorises the transfer of concessions and grants. A distinction is drawn between a transfer which is a change of title holder and a transfer which affects third parties or 'the conditions of the pertinent water basin'.[64] In the first case the National Water Commission approves the transfer and lodges it in the public register of water rights. In the second case the prior permission of the commission is required and, in addition in some cases, the transfer is permitted only 'for specified causes' and 'with substantiated justification'.[65] In Morocco the arrangements focus upon the use of water for irrigation. Irrigation water permits are linked to a specific land holding. A new permit is required if the water is to be used to benefit another holding. A transfer of the title to the land carries with it an irrigation water permit.[66] This system makes it possible for the holder of the title to the land to be different from the holder of the water permit. Accordingly it is provided:

> Agricultural land that uses water for irrigation where the water rights belong to a third party can only be leased or transferred if the acquirer of the land obtains a contract to use the water from the third party owner of those water rights.[67]

A recognition, in other words, that the arrangements for the use of land need to be accommodated within the arrangements for the use of water. Title may be separate but uses impact upon each other.

Although the Water Law of China does not specifically provide for trading in water or in water rights, transactions are beginning to take place in relation to them. It has been suggested that water markets can be divided into two levels. The first contemplates a market within which water resources are transferred between their owners, and the second contemplates a market for the trade of water between the users of water resources.[68] A transfer of water rights seems to be a part of the arrangements in either of these two contexts. In any event these transactions take place informally rather than formally.

According to an agreement entered into in 2000 the city of Yiwu agreed to buy annually a certain volume of water from the city of Dongyang.[69] The city of Dongyang was an upper riparian and the city of Yiwu a lower riparian of the river Jinhua. The city of Dongyang constructed the Hengjin reservoir and part of the water stored in this reservoir was the source of the water sold to the city of Yiwu. These arrangements did not apparently change in any way the ownership of the reservoir and presumably the waters in the reservoir. It was the responsibility of the city of Yiwu to construct and operate the infrastructure necessary to divert the waters from the reservoir to the city. The relevant works were completed and water has been delivered to the city of Yiwu since 2005. These arrangements have been described as 'the first case of trans-city water right transfer in China, breaking the state monopoly over the right to use water resources'.[70] Significantly, it was an arrangement between two cities and not between individuals.

More formal arrangements have been put in place to enable transactions between individual members of the relevant community. Not surprisingly, these arrangements involve the use of water for irrigation. An example is the set of arrangements in the Liyuanhe irrigation area of the city of Zhangye.[71] Every user of water receives an allocation according to a water use quota which differentiates between the various purposes for which the water is used. The water users are organised into water user groups, and these are included in a number of water user associations. Each water user receives a certificate. The information on the certificate includes:

- the name of the holder;
- details of those needing water;
- the relevant water use quota;
- the annual quantity of water.

A certificate enables its holder to buy water tickets. These authorise the abstraction and use of the water. They are also 'negotiable under the regulated management'.[72] The way in which these transactions are managed depends upon the amount of water involved. Those responsible for managing the transactions include:

- the relevant water user association;
- the relevant water management station;
- the Liyuanhe Water Management Office;
- the relevant department of water administration.

It is for the parties to the transaction to negotiate the price. However a maximum price of three times the original price is prescribed. This is an example of individual members of a community receiving a negotiable certificate of water use: in other words a transferable water right. These arrangements are administered locally: a significant task for the local community which originally comprised 10,680 water users.

In Australia rights in relation to water are granted by a public agency in accordance with the relevant legislation. In the first instance this is a matter for the states and territories. Traditionally a right to use water has been granted only to the owner of the land. However to facilitate trading in water rights in a market context, rights in relation to water have in some cases been separated from rights in relation to land. While a right in relation to water may be transferred, it may be difficult and perhaps impossible for the new holder of the right to exercise it in a way and for the purposes intended by the new holder. For example, the new holder may wish to use the water for a different purpose, in a different location and perhaps in different quantities.

These issues have been addressed to some extent in the state of Queensland in Australia.[73] While a water licence remains associated with land, a water allocation is not so linked. A water allocation states the nominal volume of water for the allocation in accordance with the volumetric limit, the location from which the water may be taken and the purpose for which it may be used. It in effect represents a statement of the share of the water available to the holder of the allocation. Any use of the water must be in accordance with the relevant resource operations plan and a resource operations licence.

A water allocation may be transferred, and the transfer is simply registered in the water allocations register. In this respect the transfer of a water allocation is different from a change to a water allocation. A change to a water allocation is a reconfiguration of elements of the holder's entitlement under the allocation: for example, the nominal volume, the location, the purpose of use, the priority grouping, the maximum rate for taking water or the flow conditions under which it may be taken. A change to a water allocation must comply with the relevant rules in the resource operations plan or receive the approval of the relevant public agency. In effect an application to change a water allocation may not be approved unless the change:

- is compatible with the objectives of the water resource plan;
- is in the public interest;
- will not significantly affect water entitlement holders, resource operations licence holders or natural ecosystems in an adverse way.

Accordingly, in principle a change to a water allocation may not take place unless interests in the water resource other than those of the holder of the water allocation are protected. In other words, these arrangements disclose a subtle accommodation between the public interest on the one hand and a range of private interests on the other hand. The implication is that unregulated market mechanisms are unlikely to protect natural ecosystems.

WATER RIGHTS FOR ECOLOGICAL PURPOSES

(a) Introduction

The protection of water quality and the prevention of water pollution have been functions of the legal system for many years. The protection of public health in these ways is clearly an anthropocentric function of the legal system. The protection of the intrinsic ecological values of water resources has been a much more recent phenomenon. These values are protected as an element of the management of water resources in general, by provisions of a more prescriptive nature included in plans and regulations, as factors to be taken into account in making decisions about the use and development of water resources, by the administration of environmental water rights on behalf of the environment or a combination of any of these.

(b) Protection by Administrative Integration

In South Africa, the constitution establishes the *Grundnorm* of the environmentally sustainable use and development of natural resources. In relation to water resources, one of the principles upon which the legal arrangements have been constructed is this:

> The quantity, quality and reliability of water required to maintain the ecological functions on which humans depend shall be reserved so that the human use of water does not individually or cumulatively compromise the long term sustainability of aquatic and associated ecosystems.[74]

This brings together the economic, social and environmental values of water, although perhaps the emphasis is anthropocentric rather than ecocentric. The legal arrangements have specifically incorporated, as we shall see, the mechanism of a 'reserve' one of the functions of which is the protection of ecological values.

The legal arrangements in France and Germany are a reflection of their public domain regimes, and in England and Wales a reflection increasingly of a public domain regime. In France the purposes of water resources management include:

- the preservation of aquatic ecosystems including wetlands;
- protection against pollution;
- the conservation and free flow of water.[75]

In Germany water resources are to be managed so that inter alia 'any avoidable damage to their ecological functions does not occur'.[76] In England and Wales, the environmental duties imposed upon the public sector protect a range of specific environmental values rather than ecosystems and the environment in general. Accordingly the ecological values of water resources are not directly included.

One of the purposes of water resources management in England and Wales stated in section 19(1)(a) of the Water Resources Act of 1991 is the conservation of water resources. While there is provision in section 21 for determining and implementing minimum acceptable flows in relation to inland waters, the focus of these protectable interests are the economic and social rather than the ecological values of water. This is not to suggest that environmental values are irrelevant – far from it. Protection of the ecological values of water resources and the need to set minimum acceptable flows for ecological purposes are however implicit rather than explicit in the responsibilities of the public sector. Consequently, neither in France nor in England and Wales do the arrangements specifically and explicitly contemplate the protection of the environmental values of water resources. It is by implication that these values are recognised. On the other hand in Germany the ecological functions performed by water resources are explicitly recognised.

The conservation and protection of water resources are a part of the arrangements in the Water Law of China. The principles for their use and development acknowledge the need to protect ecological values. According to Article 21, while first priority must be given to satisfying human needs, the need of water in 'the ecological environment' must be taken into consideration. Even more significantly, it is further provided by Article 21:

> In developing and utilising water resources in arid and semi-arid areas, full consideration shall be given to the need of water in the ecological environment.

While this does not on the face of it create any priorities, by implication it suggests that in these environments it is particularly important that ecological values should be protected. In addition, according to Article 15, the protection of ecological values is one of the purposes to be achieved in the formulation of plans. For example, according to the plan for wetland protection for the period between 2006 and 2020 'degraded and lost wetlands are to be restored and wetlands and their biodiversity are to effectively protected'.[77] Collectively, it has been suggested, these arrangements amount to 'the duty of the government to fulfil the environmental right to water'.[78]

Ecological values may be protected not only in the context of planning but also as a function of administration. An example is the Tarim River Basin. The river Tarim in North West China is the longest inland river in China. The basin covers 1.02 million square kilometres and is administered by five prefectures.[79] Its water resources are managed in accordance with a set of regulations first adopted in 1997.[80] These regulations are simply one part of the overall planning and regulatory framework for their management created by the Water Law. One of the principles stated in the regulations is joint consideration of economic development and ecosystem protection.[81] One of the responsibilities of the Tarim River Basin Water Resources Commission, established by the regulations, is water environment protection.[82] The Water Law, as we have seen, acknowledges the need to protect ecosystems. Water allocation is critical in this context. The regulations for the river Tarim effectively require all the prefectures to agree on the annual water use ration for each prefecture. Once this agreement has been approved by the commission, it becomes effective as a matter of law. The principal objective of water allocation in this way is 'to ensure the required mean annual flow and ecological water use of the mainstream Tarim'.[83] In China, therefore, ecological values of water resources are protected administratively in accordance with relevant regulations, plans and the Water Law itself.

While the ecological values of water are able to be protected by an allocation system operated administratively, they are protected in certain jurisdictions in accordance with the principles and rules incorporated within the strategic framework for water resources management. In the Republic of Yemen, the Water Law establishes a list of priorities. The first is the use of water for drinking and domestic purposes. Water may then be allocated for use for a number of purposes including meeting environmental needs.[84] The National Water Resources Policy of Brazil makes it clear that one of the objectives of the law 'is to protect the quality, quantity, and flow of water'.[85] Clearly this has anthropocentric as well as ecocentric objectives. The objectives in Brazil include the rational integrated use of water resources to achieve sustainable development.[86] The integration of environmental concerns is one way of doing so. So the protection of ecological values represents one of the elements in an overall balancing approach. In Armenia the Water Code requires water resources to be protected.

Accordingly the use of water is permitted only 'in conditions that allow for water resource protection and restoration'.[87] In addition, since water is an element of an ecosystem, 'its protection demands the maintenance of balance within a given ecosystem'.[88] Under arrangements such as these, therefore, ecological values are protected not only in general in the context of the macro legal system but also specifically when the rules of the micro legal system are applied in particular sets of circumstances.

(c) Protection by Water Rights

The ecological values of water resources are protected somewhat differently in the United States of America. Rights in relation to water resources inhere fundamentally in individuals within the community. The exercise of these rights is in certain circumstances constrained by elements of the legal system designed to protect aspects of the public interest. These include restrictions designed to prevent pollution, to ensure the quality of water, to protect the environment of which the water resources are a part, and to protect specific values associated with water such as their cultural and aesthetic values.

The arrangements for achieving this in the state of Colorado have been institutionalised. The statutory objectives of the Colorado Water Conservation Board are to assist in the development and protection of the state's water rights for the benefit of present and future inhabitants.[89] These include intra-generational and inter-generational equity as principles of sustainable development. The arrangements described as the instream flow programme were initiated in 1973. The programme's objectives were to correlate the activities of mankind with some reasonable preservation of the natural environment.[90] How does the board do this? Essentially by these means:

- the exclusive ownership of instream flow and natural lake-level water rights in Colorado intended for environmental purposes;
- the authority to obtain instream flow and lake-level rights by appropriating, leasing, acquiring and receiving donated water rights;
- the authority to develop insteam flow and lake-level rules in coordination with other departments;
- the authority to protect instream flow and lake-level rights in water court proceedings.[91]

In other words it is an institution mandated to preserve the natural environment of water resources and invested with water rights as the mechanism for achieving this outcome.

The board has no monopoly of water rights. Instream water rights may be appropriated by any person for the purposes of recreation and power generation. But only the board may hold instream water rights for environmental purposes.[92] It is open to any person to apply to the board for a reach of a stream or a water body to be included in the programme. An application is determined after a process which involves public participation. If a water right is necessary, it cannot be granted unless:

- there is a natural environment to be preserved in whole or in part by the acquisition;
- the acquisition can occur without material injury to other water rights.[93]

Once these tests have been satisfied, the matter proceeds in the normal course of events to the water court for the right to be perfected.

Once the process is complete, the effect is that instream water rights are defined in cubic feet per second, for a river length and for a duration. The title that evidences the existence of these water rights is transferable. By way to example:

> The instream flow right for Buckskin Creek (Middle Fork South Platte River, Park County), is for 5 cubic-feet per second over 6.1 miles reach of the river between 1 January and 31 December. Most instream flow rights occur in unregulated streams in mountainous areas but also occur downstream of storages and in urban areas.[94]

Once in existence, these instream flow water rights operate and have effect in the same way as any other water rights. The essential difference is that they are held not by an individual but by an institution with public responsibilities.

(d) Protection by Reserves

While the arrangements in the United States of America reflect to some extent the doctrines of the common law, there is no reason why the protection of ecological values in water resources should not be based upon a set of different doctrinal foundations. South Africa has introduced the doctrine of the reserve. According to section 1(1)(xviii) of the National Water Act of 1998 the reserve means the quantity and quality of water required to satisfy basic human needs and simultaneously to protect aquatic ecosystems in order to secure ecologically sustainable development and use of the relevant water resource. The National Water Resource Strategy must provide for the requirements of the reserve and identify water resources from which particular requirements must be met. In addition there is a duty to determine for every significant water resource a class for that water resource in accordance with a classification system and resource quality objectives based on that class. The objectives include those which relate to the reserve. Finally there is an obligation to determine the reserve for all or part of that water resource in accordance with the class so determined. Once the reserve has been determined in this way, it prescribes the quantity and quality of water required to protect aquatic ecosystems. And finally, according to section 18:

> The Minister, the Director-General, an organ of state and a water management institution must give effect to the Reserve as determined in terms of this Part [of the Act] when exercising any power or performing any duty in terms of this Act.

The reserve thus constitutes one of the doctrinal foundations of the arrangements in South Africa. Establishing the reserve, implementing it and complying with it are legally binding obligations.

In Australia ecological values are protected in a number of different ways: in plans, as conditions in licences or as environmental water entitlements. The legal arrangements put in place by the Water (Resource Management) Act of 2005 in the state of Victoria have introduced the concept of a reserve. The environmental water reserve comprises water which is set aside for the environment and maintained so as to preserve the environmental values and health of water ecosystems, including their biodiversity, ecological functioning and quality of water and the other uses which depend on environmental condition. The objective of the environmental water reserve is achieved in a number of ways:

- as an environmental entitlement;
- as a condition placed on a bulk entitlement or other licence or authority;
- as an element in a management plan.

The water in respect of which an environmental entitlement may be allocated includes:

- water in a waterway;
- groundwater;
- water – but not recycled water – in the works of certain public supply agencies.

The purpose for which the environmental entitlement is allocated is either maintaining the environmental water reserve in accordance with its objective or improving the environmental values and health of water ecosystems. The environmental entitlement is held by the Minister responsible for the environment. Although an allocation of water under an environmental entitlement may be assigned to the owner or occupier of land specified in a water use licence or a water use registration, the allocation must not be assigned unless the Minister responsible for the environment is of the opinion that the assignment would benefit the environmental water reserve.

While the objective of the environmental water reserve is achieved through these various mechanisms, there appears to be no obligation to formalise the reserve in any particular way. However, the power to grant rights in relation to water is constrained by the requirement to have regard to the need to maintain the environmental water reserve in accordance with the environmental water reserve objective. Accordingly there is no obligation to maintain the reserve: only an obligation to take the reserve into account in performing the function of allocation. Mechanisms thus exist for the recognition and protection of the reserve: for example, the environmental entitlement. It is a matter for administrative discretion within the framework of the Act to determine whether to protect the reserve and how to do it.

WATER RIGHTS ADMINISTRATION AND ADJUDICATION

(a) Introduction

The diversity of arrangements for managing water resources within nation states is a reflection not only of their cultural diversity but also of the range of interests in water and of the uses to which it is put. Individuals have an interest in water. So have those with a community of interests. It is in the public interest on the one hand to recognise and protect these interests and on the other hand to recognise and protect other interests in water not capable of direct protection through a system of rights, duties and liabilities. For example, the interests of those outside the group with a community of interest and the interest of individual members of the public and of the public at large in ensuring their development, the conservation of their resources, the protection of their environment and ultimately sustainable development. A range of institutions contribute to balancing these interests, including the legislatures, the executive and the judiciary.

For reasons which are cultural as well as historical, the development of legal doctrine has been approached from one of two perspectives: the promotion and protection of private interests on the one hand and the promotion and protection of the public interest on the other hand. But they have become quintessentially related in the context of water resources management. Individual rights exist and are protected, but their exercise is controlled so as to encapsulate the values of society at large. Alternatively the values of a society at large are realised through the existence and protection of individual rights. The emergence over thousands of years of customary arrangements, legal arrangements or administrative arrangements simply emphasises how private interests and individual rights on the one hand and the public interest and its realisation on the other hand are mutually accommodated within the one system. Individuals are the ultimate beneficiaries of how water resources are managed. The benefits accruing to the individual depend upon how each individual behaves, how every other individual behaves and how the environment itself behaves.

The systems for managing water resources across the diversity of jurisdictions comprise a matrix of legal and paralegal rules: a set of macro legal arrangements and a set of micro legal arrangements. There are norms in the form of constitutional statements and norms in the form of principles, directions and outcomes. And there are norms in the form of protectable rights and enforceable obligations. The relationship between the macro legal system and the micro legal system is critical. The strength or weakness of this relationship points to its effectiveness particularly in the context of its enforceability. It is important to consider:

- who decides how water resources in general should be used and developed;
- who decides how a particular water resource should be used;
- who is responsible for administering these decisions;
- how the rules for managing water resources are enforced;
- how disputes are resolved.

(b) The Creation and Enforcement of Rights

A legislature may make rules of general application for the management of the water resources in its jurisdiction. Their application is in the first instance a matter for the persons concerned. There is no judicial involvement except in the case of a dispute. This is so for those protecting the public interest or a private interest. Frequently, the general rules provide for rights to be granted to individuals in particular circumstances. It is important who does so and how. In a public domain regime it is the function of a public agency acting administratively to plan for the use and development of water resources, to undertake the relevant activities and to grant any individual rights in relation to water accordingly. In some jurisdictions these administrative processes are subject to some form of judicial supervision to ensure their legality. Less commonly, a court or a tribunal may be asked to review the decision on its merits. This provides the judiciary with an opportunity to analyse more comprehensively and in greater depth the substantive norms according to which decisions are made. Significantly, whether the substantive decision is made by a public agency acting administratively or a court or tribunal acting similarly in an administrative capacity, their function is not only to apply rules but also to create rights and duties that do not otherwise exist. In this sense the function is positive, proactive and creative.

In a rules based system for managing water resources, the function of the judiciary is quite different. The rules – such as those underpinning the riparian doctrine or the prior appropriation doctrine – exist as a matter of law. The judiciary may have played a part in developing these rules. But the function of the judiciary is essentially adjudicative: in other words to settle disputes about the substance and application of these rules. It is in addition a function of the judiciary to enforce these rules by providing a remedy for a breach of a duty or a violation of a right. In either case, the adjudicative function involves settling disputes which have arisen in the past or determining liability in relation to events which have taken place in the past. In this respect it is a reactive function.

Whatever the doctrinal basis of a water resources management system, there are rules of law. A failure to comply with a rule of law normally attracts a sanction. In almost all cases the ultimate sanction is the imposition of a penalty when a breach of the duty constitutes a criminal offence: for example, taking water without an authorisation, failure to comply with a condition in an authorisation or discharging polluting matter into water. In addition, a failure to comply with a condition in an authorisation is a ground for suspension or termination of the authorisation. In some cases the circumstances may give rise not only to criminal liability but also to civil liability. But in the case of a failure by a public agency to comply with a rule of law, the sanction is different. A decision of a public agency may be declared invalid and the decision-making process abandoned or commenced a second time according to the rules. The remedy for a breach of a rule is thus very specific to the jurisdiction.

(c) The Adjudication of Rights

While it is the traditional function of a court to ensure compliance with a rule of law by providing a remedy for a failure to do so, disputes about the use and management of water resources cannot always be accommodated comfortably within the traditional function of a court. Many of the issues are technically complex in hydrological terms. The issues are thus determined by a body of experts or a judicial body with expert assessors. It is not uncommon, therefore, for dispute settlement or dispute resolution procedures to be put in place specifically for disputes about water resources. For example, in Armenia the Dispute Resolution Commission is responsible for resolving disputes relating to water use permits. The method for doing so may be either mediation or binding arbitration.[95] In Brazil the National Water Resources Management System settles conflicts by arbitration subject to an appeal to either the State Council on Water Resources or the National Council on Water Resources.[96] In South Africa it is the Water Tribunal – a specialist body – which hears appeals against decisions made administratively. An appeal lies from the Water Tribunal to the High Court.[97] The Environment Court in New Zealand determines appeals against administrative decisions on their permits subject to an appeal on matters of law to the High Court. Under the Resource Management Act of 1991 the Environment Court is a judicial body assisted by expert assessors.

The range of issues – both technical and legal – which may be in dispute is potentially extensive. They may impact upon private as well as public interests. The courts may have an inherent jurisdiction to deal with any disputes as they arise. In some cases there are no special arrangements for the settlement of disputes about water resources. In the absence of any such arrangements, as in Morocco, disputes will be adjudicated in accordance with the civil code.[98] Similarly in the Republic of Yemen, the Water Law says nothing about dispute settlement. However, if the Water Law does not deal with an issue, the civil law and the principles of Islamic jurisprudence apply.[99]

The use of water is a location-specific and a purpose-specific activity. Disputes are, accordingly, often very local in nature. A number of jurisdictions recognise the relevance of local practice and custom. Again, depending on circumstances, there may be a clear community of interest in how a particular water resource is managed. It is not uncommon for water resources to be managed locally and for disputes about water resources to be resolved locally. These functions are undertaken in a number of jurisdictions by water users' associations specifically established for these purposes. In Armenia disputes are resolved among their members by water users' associations.[100] In the state of Colorado in the United States of America water users' associations – described locally as ditch companies – include within their functions monitoring and enforcing water consumption within the ditch area.[101] Whatever is the doctrinal foundation of the system, it is not uncommon for local disputes to be settled locally.

The Water Law of China deals comprehensively with enforcement and dispute resolution. According to Articles 56 and 57, disputes about water resources are expected to be resolved through consultation. This applies to disputes between different administrative regions. If consultation is ineffective, the dispute is resolved by an administrative agency at a higher level in the hierarchy of administrative responsibility. In the case of disputes between units or individuals or between units and individuals, if consultation is not successful the dispute may be resolved either administratively or by a civil action in the People's Court.

The range of sanctions for failure to comply with a duty is extensive. Criminal or administrative sanctions may be imposed upon officers in administrative departments in a range of circumstances. In terms of Article 64:

- for neglect of their duties;
- for the issue of licences that do not meet the statutory requirements;
- for failure to allocate water in accordance with the water allocation plans.

A fine may be imposed upon a unit or an individual for a very extensive and detailed range of activities proscribed by Article 65. These relate to the construction of buildings, the construction of water infrastructure as well as interfering with water resources and using water in unauthorised ways. The focus in China is thus upon the personal responsibility of individuals and administrative officers rather than any institutional responsibility associated with the formulation and implementation of strategies and plans and their implementation in specific sets of circumstances.

CONCLUSION

It is trite to observe that rights in relation to water are a phenomenon of any set of arrangements for the management of water resources. This is so whether it is a public domain regime or one based upon individual rights. Rights in relation to water, for example, may inhere fundamentally in the state or arise through the application of the appropriation doctrine or the riparian doctrine. While rights such as these have probably always been transferable – at least in theory – the transfer of these rights in the sense of trade has become a phenomenon of regulation over recent years in many jurisdictions: in other words, the existence or the creation of a market in water rights. Once again there is a close and increasingly closer relationship between public and private rights, duties and liabilities. The state, for example, may have the fundamental right to water. In this case the right of the individual to water arises as a result of allocation. Or – at the other end of the spectrum – the individual right in relation to water may be acquired by appropriation or in exercise of the riparian right. In some jurisdictions a right in relation to water may arise as a result of custom or practice.

While these rights are and always have been transferable in theory, their effectiveness as a regulatory mechanism has been more controversial. One source of controversy has been the impact of a transfer upon third parties and another source of controversy has been the impact of a transfer upon the environment. The law has been concerned not only to recognise the transferability of rights in relation to water but also to ensure that any transfer does not inadvertently impact upon how water resources are managed in the public interest. The matter has been made more complicated by the recognition in some jurisdictions of environmental or ecological rights in relation to water. The management of rights in relation to water is and always has been an aspect of water resources management. The way in which these rights are exercised similarly is and always has been governed by sets of legal rules. These rules are increasingly being complemented by paralegal rules: another example of the interrelationship between legal rules and paralegal rules. Rights in relation to water, their exercise and their transferability are an integral part of the emerging set of arrangements for the management of water resources.

NOTES

1. Tarlock (1997, p. 184).
2. Tarlock (1997, p. 185).
3. Tarlock (1997, p. 185).
4. Tarlock (1997, p. 187).
5. Tarlock (1997, p. 188).
6. Tarlock (1997, p. 188).
7. Tarlock (1997, p. 189).
8. Tarlock (1997, p. 189).
9. Salman and Bradlow (2006, p. 57).
10. Fisher (2000, pp. 93–100).
11. Salman and Bradlow (2006, p. 70).
12. Xi et al. (2002, p. 17) and Hu (2006, p. 73).
13. Singh (1991, p. 56).
14. Navia (1994 pp. 289 and 290).
15. Navia (1994, p. 311).
16. Laube and van de Giesen (2006, p. 167).
17. Laube and van de Giesen (2006, p. 167).
18. Laube and van de Giesen (2006, p. 167).
19. Salman and Bradlow (2006, p. 121).
20. Upadhyay (2005, p. 196).
21. Upadhyay (2005, p. 198).
22. von Benda-Beckmann et al. (1997, p. 231).
23. Singh (1991, p. 55).
24. Tarlock (2005, p. 44).
25. Salman and Bradlow (2006, p. 27).
26. Salman and Bradlow (2006, p. 16).
27. Salman and Bradlow (2006, pp. 14 and 15).
28. Salman and Bradlow (2006, p. 98).
29. Salman and Bradlow (2006, p. 120).
30. Salman and Bradlow (2006, p. 114).
31. Salman and Bradlow (2006, p. 88).
32. Salman and Bradlow (2006, p. 88).

33. Salman and Bradlow (2006, p. 78).
34. Salman and Bradlow (2006, p. 79).
35. Salman and Bradlow (2006, p. 93).
36. Salman and Bradlow (2006, p. 94).
37. Xi et al. (2002, p. 16).
38. Hu (2006, p. 73).
39. Salman and Bradlow (2006, p. 38).
40. Hu (2006, p. 74).
41. Hu (2006, pp. 74 and 75).
42. European Commission (2000, pp. 153 and 154).
43. European Commission (2000, p. 154).
44. European Commission (2000, p. 262).
45. European Commission (2000, p. 262).
46. European Commission (2000, p. 262).
47. Fisher (2007, pp. 115–120).
48. European Environment Agency (1999b, pp. 15 and 16).
49. Fisher (2006b, pp. 104 and 105).
50. Luterbacher and Wiegandt (2005, p. 33).
51. European Environment Agency (1999b, p. 16).
52. Tarlock (2005, p. 47).
53. Tarlock (2005, p. 53).
54. Tarlock (2005, p. 51).
55. Tarlock (2005, pp. 51 and 52).
56. Tarlock (2005, p. 46).
57. Tarlock (2005, p. 46).
58. Tarlock (2005, p. 52).
59. Salman and Bradlow (2006, p. 71).
60. Salman and Bradlow (2006, p. 98).
61. Salman and Bradlow (2006, p. 17).
62. Salman and Bradlow (2006, p. 122).
63. Salman and Bradlow (2006, p. 94).
64. Salman and Bradlow (2006, p. 79).
65. Salman and Bradlow (2006, p. 79).
66. Salman and Bradlow (2006, p. 88).
67. Salman and Bradlow (2006, p. 88).
68. Xi et al. (2002, p. 18).
69. Hu (2006, p. 70) and Xi et al. (2002, p. 18).
70. Xi et al. (2002, p. 18).
71. Hu (2006, pp. 76 and 77).
72. Hu (2006, p. 77).
73. Fisher (2004a, pp. 37 and 38).
74. Hu (2006, p. 142).
75. Hu (2006, p. 44).
76. Hu (2006, p. 45).
77. Xi et al. (2002, p. 24).
78. Hu (2006, p. 87).
79. Zhang (2006, p. 108).
80. Zhang (2006, p. 126).
81. Zhang (2006, p. 127).
82. Zhang (2006, p. 128).
83. Zhang (2006, p. 135).
84. Salman and Bradlow (2006, p. 120).
85. Salman and Bradlow (2006, p. 27).
86. Salman and Bradlow (2006, p. 25).
87. Salman and Bradlow (2006, p. 18).
88. Salman and Bradlow (2006, p. 19).
89. Productivity Commission (2003a, p. 12).
90. Productivity Commission (2003a, p. 13).
91. Productivity Commission (2003a, p. 13).
92. Productivity Commission (2003a, p. 23).
93. Productivity Commission (2003a, p. 38).

94. Productivity Commission (2003a, p. 45).
95. Salman and Bradlow (2006, p. 23).
96. Salman and Bradlow (2006, p. 32).
97. Salman and Bradlow (2006, p. 112).
98. Salman and Bradlow (2006, p. 92).
99. Salman and Bradlow (2006, p. 126).
100. Salman and Bradlow (2006, p. 23).
101. Productivity Commission (2003a, p. 53).

REFERENCES

European Commission (2000), *Towards Sustainable and Strategic Management of Water Resources* (Luxembourg: European Communities Official Publications).

European Environment Agency (1999b), *Sustainable Water Use in Europe: Part 2: Demand Management* (Luxembourg: European Communities Official Publications).

Fisher, D.E. (2000), *Water Law* (Sydney: LBC Information Services).

Fisher, D.E. (2004a), *Trading in Water Rights – towards a National Approach* (Melbourne: Queensland University of Technology and Phillips Fox).

Fisher, D.E. (2006b), 'Markets, water rights and sustainable development', *Environmental and Planning Law Journal* 23, 100–112.

Fisher, D.E. (2007), 'Delivering the National Water Initiative: the emergence of innovative legal doctrine', in Karen Hussey and Stephen Dovers (eds), *Managing Water for Australia – the Social and Institutional Challenges* (Collingwood: CSIRO Publishing).

Hu, Deshang (2006), *Water Rights: An International and Comparative Study* (London: IWA Publishing).

Laube, Wolfram and Nick van de Giesen (2006), 'Ghanaian water reforms: institutional and hydrological perspectives', in J. Wallace and P. Wouters (eds), *Hydrology and Water Law – Bridging the Gap* (London: IWA Publishing), pp. 157–197.

Luterbacher, Urs and Ellen Wiegandt (2005), 'Cooperation or confrontation: sustainable water use in an international context', in Edith Brown Weiss et al. (eds), *Fresh Water and International Economic Law* (Oxford: Oxford University Press), pp. 1–34.

Navia, Jose M. Borrero (1994), *Environmental and Water Law in the South* (Utrecht: International Books).

Productivity Commission (2003a), *Water Rights Arrangements in Australia and Overseas: Annex I, Colorado* (Melbourne: Productivity Commission).

Salman, Salman M.A. and Daniel D. Bradlow (2006), *Regulatory Frameworks for Water Resources Management: A Comparative Study* (Washington DC: World Bank).

Singh, Chhatrapati (1991), *Water Rights and the Principles of Water Resource Management* (Bombay: N. M. Tripathi Pty Ltd).

Tarlock, A. Dan (1997), 'Current trends in United States water law and policy: private property rights, public interest limitations and the creation of markets', in Edward H.P. Brans et al. (eds), *The Scarcity of Water: Emerging Legal and Policy Responses* (London: Kluwer Law International Ltd), pp. 183–195.

Tarlock, A. Dan (2005), 'Water transfers: a means to achieve sustainable water use', in Edith Brown Weiss et al. (eds), *Fresh Water and International Economic Law* (Oxford: Oxford University Press), pp. 35–59.

Upadhyay, Surya Nath (2005), 'The legal regime on water resources in Nepal', in Surya P. Subedi (ed.), *International Watercourses Law for the 21st Century* (Aldershot: Ashgate Publishing Limited), pp. 195–209.

von Benda-Beckmann, K. et al. (1997), 'Disputing water rights: scarcity of water in Nepal hill irrigation', in Edward H.P. Brans et al. (eds), *The Scarcity of Water: Emerging Legal and Policy Responses* (London: Kluwer Law International Ltd), pp. 224–242.

Xi, Wang et al. (2002), 'Managing water resources for a sustainable futurer: law, policy and methodology of China', *Asia Pacific Journal of Environmental Law* 7, 5–41.

Zhang, Jiebin (2006), 'Water management issues and legal framework development of the Tarim River Basin', in J. Wallace and P. Wouters (eds), *Hydrology and Water Law – Bridging the Gap* (London: IWA Publishing), pp. 108–142.

PART VII

Towards sustainable water resources governance

So far we have considered the conceptual, doctrinal and instrumental arrangements that have been evolving over recent years for the management of water resources at international, national and local levels. International law is beginning to recognise sustainable development as an emerging principle. A number of interstate agreements are moving in the same direction. It would be premature to suggest that the function to be performed by sustainable development in the international arrangements for the management of water resources has emerged clearly and unequivocally. This is for the most part true of national legal arrangements. The concept of sustainability has been emerging in one form or another in a number of constitutional and statutory instruments. The function to be performed by sustainability depends upon the structure and the language of the instrument in question.

In this Part we return to an analysis of how sustainable development has been incorporated over recent years in a number of statutes which control the use of natural resources – including water resources – and their impact on the environment. This suggests that it is possible to structure legislation so that it encourages the achievement of sustainable development but does not strictly require it. There have over recent years been suggestions indicating how sustainable development might effectively be included within the legal arrangements for the management of water resources – particularly at the international level. After reviewing these suggestions, we confirm the challenges facing the legal system and suggest models which might be appropriate for the achievement of sustainable water resources governance.

Chapter 17 presents a synopsis of the constitutional, institutional and statutory arrangements in Chile for the management of its water resources. It then looks in detail at the structure and terminology of the statutory arrangements in a number of jurisdictions before assessing how the judiciary in one jurisdiction has responded to statutory arrangements such as these. On the basis of these evolving statutory structures Chapter 18 returns to the doctrinal foundations of the legal system. Are they capable of supporting a system for the sustainable use and development of water resources or is doctrinal innovation necessary? Chapter 19 predicts the trends for the future and reviews some suggestions for doctrinal innovation. Chapter 19 concludes by itself suggesting in some structural and terminological detail what the elements of an integrated legal structure might look like.

17. Evolving statutory structures for water resources governance

INTRODUCTION

A wide variety of approaches to the governance of water resources across the globe has emerged towards the end of the 20th century and at the beginning of the 21st century. Whatever the doctrinal foundations of their systems and whatever their formal structures, there are many common features. The effectiveness of these systems depends upon their detailed form and structure, and this in turn depends upon the language that expresses them. Before we turn to a review of some examples of these precise details, the arrangements that have evolved over recent years in Chile represent in many ways a model of what has been happening elsewhere. This, both concisely yet comprehensively, provides a synopsis of a contemporary national system of water resources governance.

The principal features of the system in Chile are these:[1]

- according to the constitution water is a national good of public use as it is essential for life, economic development, social objectives and environmental requirements;
- the constitutional function of the state as a trustee is to regulate water use so as to achieve these objectives;
- every citizen has a constitutional right to live in an environment free of contamination;
- there is a constitutional right to seek a remedy in a court for a violation of a right;
- individual water users sharing a common source of water comprise water communities with a community of interest in the water;
- the relevant public agency is required to manage water resources and the environment in a manner which maintains the volume of surface and groundwater, maintains landscape values, and protects species which are under threat of extinction, vulnerable, rare or insufficiently known;
- there is a recognition of the principle of administrative integration;
- there is a recognition of the principle of river basin level water resources management;
- there is provision for the construction and operation of water resources infrastructure by the public sector and the private sector particularly in relation to major and minor irrigation works;
- there is a recognition for the planning function to be based upon the detailed information about water resources and detailed assessment of these resources leading to a set of overall objectives, mid-level objectives and specific objectives in relation to particular areas;
- individual water rights are granted administratively;
- the holder of a water right is entitled to use, enjoy and hold the right as if it were a privately appropriated good;

- water rights for the benefit of the indigenous community are the property of the indigenous community;
- water rights are rights to use water and to the benefits of that use;
- water rights are divided into consumptive and non-consumptive categories;
- water rights are either permanent or eventual and either continuous, non-continuous or alternate;
- these categories determine the priority of water rights;
- an application for a water right is a public process and an application must be granted if water is available and the exercise of the water right would not injure third parties – the rights of other water users;
- a water right rests permanently in its holder;
- a water right is governed by the civil law;
- a water right is protected as private property;
- a holder of a water right is subject to a number of obligations including an obligation not to injure third parties;
- the relevant public agencies are required to determine minimum ecological volumes of water for each river and sustainable recharge rates for each aquifer;
- it may be that the environment is a third party the interests of which must not be injured by the exercise of a water right;
- water is allocated to the environment through trading in water rights in the market;
- water rights are not attached to land title;
- water rights are transferrable;
- a transfer of a water right requires approval in certain circumstances;
- the holder of a water right has a right to enter into negotiations with the owner of private land to gain access to that land;
- the holder of a water right has a right to construct the necessary infrastructure;
- water users associations are responsible for managing water distribution infrastructure; water distribution, monitoring and enforcing water rights in the area of the association, and mediating disputes between members of the association who use water;
- administrative decisions are reviewable on appeal by the regional appellate courts;
- appeals from regional appellate courts are heard by supreme courts;
- the decisions of administrative agencies and of water users' associations can be overridden by decisions of the courts.

The arrangements in Chile thus disclose all of the features which are typically part of a system of water resources governance at the beginning of the 21st century: in effect: constitutional values, guiding principles and objectives, planning functions, allocated water rights, protection of indigenous communities, protection of the ecological values of water resources, market mechanisms, local administration, local enforcement and dispute resolution, central administration, and supervisory jurisdiction of the courts. The arrangements in Chile comprise a collection of integrated public rights and duties; and private rights and duties. Each of these is protected in its own way by the relevant rules of private law, of public law and of the constitution. It comprises a blend of legislative, executive and judicial functions. Ultimately it is the way these arrangements have been structured and articulated which determines their enforceability. That in turn depends upon the way the fundamental *Grundnorm* of the rule of law underpins the whole system. Against this background, we turn now to a detailed analysis of the form, structure and language of a number of emerging statutory structures. The question is the extent to which they demonstrate a commitment to the sustainable development of natural resources, and water resources in particular.

Nation states manage their natural resources within the limits of international law and within the limits – if there are any – of their constitution and the fundamental doctrines of their legal system. The particular functions of the legislature, the executive and the judiciary are in theory quite clear. While in some jurisdictions each is independent of the other, in other jurisdictions the functions to some extent overlap. In any event there is usually a capacity within a state to modify existing legal arrangements. It is the primary function of the executive to administer the law and of the judiciary to settle disputes and to enforce it. Whether or not the executive is technically bound by the law, the expectation is that the executive will abide by the legal and paralegal rules which comprise the totality of the system. And so it is for the judiciary. Accordingly it is a function of both the executive and the judiciary to interpret and apply the law as it is.

It is important for the executive – in practice the relevant public agency and its administrators – to understand the law and to administer it accordingly. In some cases their administration is subject to judicial supervision and in other cases not. It is ultimately the responsibility of the judiciary – within the limits of its own jurisdiction – to explain in the context of particular circumstances what are the relevant rules and how they are applied. Notwithstanding customary rules and the rules of the civil law and the common law, it is legislation, in whatever form is relevant to the jurisdiction in question, which has tended to dominate how natural resources are managed and the environment protected. A great deal depends upon the structure and the language of the legislation. It is ultimately for the courts to determine what it means as the final arbiters within the system.

WATER RESOURCES DEVELOPMENT AND PROTECTION IN IRAN

Prior to the reception of sustainable development, the protection of the environment had emerged as a counterpoint to the development of the environment as a natural resource. It is no surprise that the consequential statutory arrangements proceeded on the basis of a sectoral or fragmented approach which accordingly focussed upon either the development or the protection of the environment. Increasingly, the one began to complement the other in the sense that environmental values began to be incorporated in development processes and developmental values incorporated in environmental processes: for example, the arrangements introduced for the management of water resources in Iran by the legislation of 1968.

The purpose of the legislation in general appears from its title: namely the Iran Water Law and the Manner of Water Nationalisation. It formalised a public domain regime based upon a concept of shared water resources with considerable local responsibility for their management. Individual rights were created as rules of law directly by the legislation or granted in accordance with a permit system. According to Article 1 all waters were considered as the national wealth and belonged to the public. The responsibility for safeguarding and utilising this national wealth was given to the relevant ministry. Water use permits were granted administratively. According to Article 4 a water use permit meant a right which was to be given to real or legal entities for the purpose of beneficial and reasonable use of water through the issue of a permit in accordance with the regulations. So the administrative discretion was governed by the provisions of the law. One such provision was that a water use permit according to Article 14 related to a 'piece of land' . The water law recognised that circumstances change. If there had been an over-allocation of water in the sense that the total of water allocated by water rights was greater than the actual and usual capacity of the rivers, then in accordance with Article 5 the matter could be reviewed.

One of the consequences of a public domain regime has been imposing responsibility for water supply upon a public agency. According to Article 19, the Ministry of Water and Power was bound to take the measures necessary for the purpose of supplying the amount of water required for the country. These included:

- to harness floods and reserve the water of rivers in surface and underground reservoirs;
- to regulate and distribute the water;
- to exploit and use beneficially underground and mineral waters;
- to desalinate saline water;
- to create irrigational systems.

Linked to the public supply of water was the quality of water. According to Article 27, holders of well permits were responsible for the prevention of pollution of the water in the well and were required to comply with health regulations. More generally Article 55 prohibited the pollution of water and Article 60 included within its range of offences the intentional pollution of water.

There was no provision for formal water resources planning in the legislation. Some form of planning no doubt took place in the context of the performance of the public supply functions by the ministry. Thus, according to Article 17, the ministry was under a duty to determine the amount of water use permitted for each region in view of local conditions and circumstances. This arose in the context of the allocation of water by water use permits. Articles 40 and 41 contemplated the joint management of the infrastructure associated with shared water resources. According to Article 46, in the event of any disputes arising in connection with priority, quality or quantity of water, these had to be settled in a 'syndic friendly manner by chief water masters and water masters'. Otherwise the matter was dealt with by the regional or district administrator. The water law made no specific reference to the functions of the courts other than obliquely through provisions for the imposition of criminal sanctions.

The Water Law of Iran of 1968 reflected the values of that time. It was a code concerned essentially with the social and economic development of the water resources in that jurisdiction. While the quantity and the quality of water were equally important, the quality of water was perceived from the human perspective rather than the ecological perspective. The administration of the law was the responsibility of the executive, with relatively little local involvement in terms either of local management or the recognition of local customs. The creation of an essentially public domain regime was clearly the intended outcome of the system, not only because of the way the legislation was structured but also because the law included within its title the reference to water nationalisation.

PUBLIC WATER RESOURCES MANAGEMENT IN SINGAPORE

The focus of the arrangements for water resources management in Singapore is the performance of functions by the relevant public agency. The way in which water resources are used and developed – probably more accurately managed – is governed by a trilogy of statutes:

- the Public Utilities Act;
- the Sewerage and Drainage Act;
- the Environmental Pollution Control Act.

The first establishes the Public Utilities Board and states its functions and duties. These include the supply of water for human consumption. The second provides for the construction and maintenance of public sewerage systems and the construction of stormwater drainage systems. The third creates a licensing system for the discharge of trade effluent, oil, chemicals, sewage and other polluting matters. There is no provision in the legislation for individual water rights. Customary rights and common law rights exist, therefore, only to the extent that they have not been taken away by the legislation.

What is the effect of section 31 of the Sewerage and Drainage Act? Sub-section (1) provides:

> No person shall, without the approval of the Board, construct any works for taking or intercepting water from any place or sea within the territorial limits of Singapore.

It is an offence to contravene this provision. But the provision does not apply to the lands, buildings and installations of the Board. Significantly the prohibition relates to the construction of works. It does not relate to the taking or intercepting of water. Accordingly, on the face of it, if water is taken or intercepted without 'works', then it does not require the approval of the Board. Whether it is lawful or not depends on whether there is a rule of law permitting such taking or intercepting. Probably not: in light of the responsibilities placed by the legislation upon the Board.

Sub-section 6(1) of the Public Utilities Act states in mandatory terms the function and duty of the Board. Despite the use of two words, it appears to comprise a unified responsibility. In the context of the duty, it includes duties:

- to provide, construct and maintain such catchment areas, reservoirs and other works as may be required or necessary for the collection, supply and use of water for public and private purposes;
- to secure and provide an adequate supply of water at reasonable prices;
- to promote water conservation.

Clearly there is a duty to construct works, and the purpose is the supply and use of water for public and private purposes. This comes close to a correlative right to be supplied with water able to be used for a private purpose. In this sense the use of the water is exclusive to the person so entitled to be supplied. This is reinforced by the duty to provide an adequate supply of water at reasonable prices. While there is a duty to supply and hence a right to receive, it is coupled with a duty to pay the 'reasonable' price for the supply. And it is limited to an adequate supply. What is adequate is a matter of circumstances. Then there is the duty to promote water conservation. It is not a duty to conserve water but merely to promote its conservation. It contemplates putting in place arrangements for conservation. There is no indication of the meaning of conservation. Nor is there any reference to sustainable development. Accordingly, how Singapore's water resources are managed is for the Board to determine. It is within its discretion to do so sustainably.

The Public Utilities Board in effect has a monopoly over the supply of water services in Singapore in relation to human consumption, drainage services and sewerage services. Any such services are provided by the Board or with the approval of the Board. Linked to this is the prohibition on the construction of works for taking and intercepting water. In return for this prohibition upon individual activities, there has been conferred a right to a supply of water for private purposes. There is thus a duty to this effect imposed upon the Board and a right conferred upon individuals. But the right is not protected and the duty not capable of enforcement. This is simply because of the provisions of section 6(3) of the Public Utilities Act. It provides:

> Nothing in this section shall be construed as imposing on the Board, directly or indirectly, any form of duty or liability enforceable by proceedings before any court.

This seems clear, but the judiciary has traditionally interpreted such immunity rules strictly. In any event in Singapore there appears to be an unenforceable right to an adequate supply of water. Accordingly, therefore, it may be more accurate to describe the arrangements in section 6(1) of the Act as functional rather than mandatory.

ARRANGEMENTS IN INDIA

(a) Public Water Resources Management

A number of statutory arrangements in the states in India involve the performance of functions rather than the implementation of duties. The arrangements for water conservation in the state of Maharashtra have involved the establishment of a water conservation corporation. Section 18 of the Maharashtra Water Conservation Corporation Act of 2000 sets out the functions of the corporation. These include:

- to prepare and execute schemes for water conservation, irrigation, and watershed management in conformity with the published policies of the state government;
- to prepare draft state plans for these purposes in conformity with the directions of the state government;
- to plan, investigate, design, construct and manage the water conservation projects in the watersheds.

The powers conferred by sections 19 and 20 in support of the performance of these functions include the acquisition of property and the compulsory acquisition of land. In addition the corporation may assist in the establishment of water users' associations, for the better use of facilities made available by the corporation. The corporation is accordingly a public agency equipped with all the planning and operational powers necessary to achieve its objectives. In addition the performance of these functions must be in conformity not only with the published policies but also with the directions of the state government. On the face of it, the powers of the corporation are discretionary but linked only to the functions stated by the legislation. These are directed at the social and economic development of the state. In this context, therefore, the conservation of water is more likely than not to be its conservation for human rather than ecological purposes.

The purpose of the Rajasthan Farmers' Participation in Management of Irrigation System Act of 2000 is to make formal arrangements for the participation by farmers in the management of their irrigations systems. It is structured as a set of functions, but functions to be performed at local rather than state level. Section 3 provides for the delineation of water users' areas and section 4 for the establishment of water users' associations. A water users' association according to section 4(2) consists of all water users' who are land owners in a water users' area. According to section 16 the objects of the Farmers' Organisation are threefold:

- to promote and secure distribution of water among its users, adequate maintenance of the irrigation system, efficient and economical utilisation of water to optimise agricultural production;
- to protect the environment;
- to ensure ecological balance.

This is achieved by involving the farmers and by inculcating a sense of ownership of the irrigation system in accordance with the water budget and the operational plan. There is provision for operational plans, irrigation schedules, and regulation of water use. These are linked. Water may be distributed, for example, either by the *osrabandi* system or the *warabandi* system. By the former system water is allocated to water users in proportion to the area of land irrigated or capable of irrigation, and by the latter system water is allocated to water users by turn according to the day, the duration and the time of supply. The irrigation system is accordingly managed by the water users' association in accordance with the plans and schedules made specifically for it. If there are any disputes, section 28 provides that they are determined in the first instance by a committee of the water users' association with provision for appeals and ultimately an appeal to the Apex Committee whose decision is final. While environmental protection and ecological balance are two of the objects to be achieved by water users' associations, the focus appears to be the use of water for irrigation to optimise agricultural production. Environmental protection and ecological balance are elements, but no more than elements in the performance of these functions. It is, once again, essentially a functional approach with stated but not enforceable objectives.

(b) Constraints upon Public Management

In India, the statutory arrangements in states, whether the approach is functional or mandatory, have effect within the framework established not only by the Constitution of India but also by legislation which relevantly impacts upon the use and development of water resources. In the *Tirupathi* case,[2] for example, the Supreme Court of India acknowledged the relevance of a range of international, constitutional and fundamental doctrines and principles. The town of Tirupathi is located in an area prone to drought. There were located in the town two 500-year-old tanks used for irrigation purposes but also as lakes which improved groundwater recharge. This area of the town was proposed for redevelopment for housing. Construction works had already commenced, and the land where these tanks were located had been transferred to the relevant public agencies to enable the developments to take place. It was claimed that protection of the environment and protection of 'valuable and most cherished freshwater resources'[3] had been ignored in favour of economic growth. The issues fell ultimately for determination by the Supreme Court of India.

The court accepted that the need to protect the environment and the freshwater resources in question had been largely ignored. There had, however, been a review of many of these issues by a committee whose recommendations, it was determined, should be taken into account. The remedy sought in this case was the preservation of the two tanks and their restoration to their former condition. The court pointed out that 'the right to shelter does not seem to be so pressing under the present circumstances so as to outweigh all environmental considerations'.[4] But what could the court do about it? Merely make a declaration to that effect or be more proactive. The response was this:

> The court should do the most it can to safeguard the two tanks in question. However, due to the persistent developmental activities over a long time, much of the natural resources of the lakes has been lost, and considered irreparable. This, though regrettable, is beyond the power of this court to rectify.[5]

But that was not the end of the matter. The expert committee appointed by the court had considered the factual situation and the feasibility of the revival of the two tanks. The court then proceeded to make a number of specific orders about the two tanks based upon the findings of the committee. These included:

- a prohibition on further construction;
- bore wells and tube wells to be prohibited;
- any development leading to concretisation of the ground surface to be prohibited;
- stormwater to be recharged to groundwater;
- recharge structures to be constructed for rainwater harvesting;
- roof top rainwater harvesting to be provided for each house constructed.[6]

A set of specific orders quite clearly capable of being enforced.

But what were the legal principles for the approach adopted by the Supreme Court of India? These included:

- the tanks were communal property and to be managed for the benefit of the community but without infringing the rights of the community;[7]
- the right to a healthy environment derived inter alia from the fundamental duties and the directive principles of state policy declared in the constitution;[8]
- shelter as a basic human need;[9]
- the principles declared in the Stockholm Declaration 1972, including the principle of intergenerational equity, that 'must be applied in full force for protecting the natural resources of this country';[10]
- the application of the public trust doctrine to the extent that 'the state as a trustee is under the legal duty to protect the natural resources';[11]

- the responsibility of a nation state to protect the environment according to the principles of international law leading to 'a responsibility bestowed upon the government to protect and preserve the tanks which are an important part of the environment of the area'.[12]

This approach amounts in effect to an endorsement of the substantive principles for water resources governance which have been evolving since the last decades of the 20th century. The court went on to embrace the principle of sustainable development. The Supreme Court noted the origins of the principle and its juridical development by the Supreme Court of India itself. In this context the court held that:

> Merely asserting an intention for development will not be enough to sanction the destruction of local ecological resources. What this court should follow is a principle of sustainable development and find a balance between the developmental needs which the respondents assert and the environmental degradation which the appellants allege.[13]

The court proceeded to do so. It could not prevent the development but it placed substantial restrictions upon future activities related to the development. The statutory functions of the state must therefore be performed in accordance with the fundamental doctrines stated in the constitution and with those incorporated into the jurisprudence of India by the Supreme Court. This included sustainable development.

SUSTAINABILITY IN INDONESIA

As early as 1982 the legislature in Indonesia incorporated the notion of sustainability in the Act concerning Basic Provisions for the Management of the Living Environment. It was concerned as much with the use and development of natural resources – including water resources – as with the protection of the environment. The approach is comprehensive. The structure of the legislation includes:

- a basic principle;
- a set of objectives;
- a set of rights vested in individuals and in the state;
- a set of obligations imposed upon individuals and the state;
- the creation of supporting management regimes;
- the creation of supporting legal liabilities.

The *Grundnorm* of these arrangements lies in Article 3 which states the basic principle, namely:

> The management of the living environment is based upon the sustenance of the capability of the harmonious and balanced environment to support continued development for the improvement of human welfare.

This incorporates the three elements of sustainable development: economic, social and ecological. According to Article 4, which states the objectives of this management system, one objective is the development of the Indonesian individual 'in his totality'. Clearly this involves the economic as well as the social development of the individual in the widest sense to include cultural and aesthetic perspectives. It includes the development of the individual from the perspective of the living environment, for Article 4(c) seeks to develop the individual as a proponent of the living environment. In addition, according to Article 4(b), development embraces economic development in the sense of the wise use of natural resources. Development in this sense is not only to be wise but also to include 'environmental consideration for the interest of present and future generations'. While there is no declared priority among these provisions, there appears to be a focus on social development in the context of environmental values rather than an integration of the three sets of values that comprise sustainable development.

This perspective is consistent with Article 5(1) which gives to every person the right to a good and healthy living environment. Correlatively Article 5(2) places an obligation upon every person to maintain the living environment and to prevent and abate environmental damage and pollution. This is itself supported by the creation in Article 21 of absolute civil liability for causing such damage or pollution and in Article 22 of criminal liability for causing such damage or pollution intentionally or negligently. Rather uniquely, Article 7 places a specific obligation upon persons 'engaged in an enterprise'. This includes an enterprise which uses and develops natural resources including water resources. This is the obligation:

> To maintain the sustenance of the capability of the harmonious and balanced living environment to support continued development.

This comes close indeed to an obligation to ensure that any development is a sustainable development. Particularly important is that this duty is placed upon every person so engaged.

This set of individual responsibilities is complemented by an equivalent set of state responsibilities. According to Article 10(1) natural resources are controlled by the state and utilised for the maximum welfare of the people. This is realised through the authority granted to the state to regulate the allocation, development, use, reuse, recycling, provision, management and supervision of resources, very much a public domain regime. But – very significantly – a public domain regime in accordance with the principles and objectives stated in Articles 3 and 4. While there is no textual link between the control of natural resources by the state and the principles and objectives stated for the management of the living environment, there is little doubt that they are linked contextually. Accordingly, the legislation establishes a framework within which the management of the living environment and its natural resources takes place. While neither the state nor individuals are strictly bound by these provisions, these legal and paralegal rules function as elements of the interstitial normativity of this statutory regime.

SUSTAINABLE DEVELOPMENT GOALS IN THE PHILIPPINES

The enactment by the legislature of the Philippines in 1991 of the Strategic Environment Plan for Palawan Act was a response to perceived environmental degradations in that part of the Philippines. The legislation was concerned with natural resources in general, but clearly this included water resources. The legislation:

> Reiterates the policy of the state to protect, develop and conserve the natural resources. It lays down the comprehensive framework for the sustainable development of the island province of Palawan, often referred to as the country's last frontier.[14]

It was intended to address some issues including:

- the unsustainable harvesting of forest;
- agricultural encroachment on steep lands;
- erosion and pollution caused by inadequate control over a range of development activities.[15]

As the title of the Act suggests, it effectively creates a strategic environmental plan. Indeed section 6 specifically indicates that its function is to guide the government agencies concerned in the formulation and implementation of plans, programmes and projects affecting the environment and natural resources of Palawan. Accordingly all relevant public agencies – national, regional and local – are under an obligation not only to coordinate but also to align their projects with the strategic environmental plan itself. One of the mechanisms for implementing the plan is the establishment according to section 7 of a graded system of protection and development control over the whole of Palawan: namely the Environmentally Critical Areas Network. The network is obliged to ensure a number of outcomes, and one of these is the protection of watersheds. The network deals generally with protection and development control. For this purpose – particularly development control – a zoning system is prescribed to regulate the nature, scope and intensity of the use of the relevant natural resources. The zones in section 9 are a maximum protection or core zone; a buffer zone and a multiple or manipulative use area.

The implementation of the plan and indeed the plan itself are guided by the principles and the objectives incorporated in the legislation. According to section 2 the policy of the state is to protect, develop and conserve its natural resources. The purpose of the plan and of its implementation is to preserve and enhance the environment and at the same time to pursue the socio-economic development goals of the country: in other words, the three elements of sustainable development. It goes on to state that one of the purposes of the plan is to support and promote the sustainable development goals for the provinces through proper conservation, utilisation and development of natural resources to provide optimum yields on a continuing basis. The focus of this seems to be sustainable economic development rather than sustainable development in the wider context of sustainability. But sustainability in this wider sense is itself incorporated in the legislation.

Section 3(2) defines sustainable development as the improvement in the quality of life of the present and future generations through the complementation of development and environmental protection activities. This clearly incorporates social, economic and environmental values. Consistently with this, section 5 states the philosophy underpinning the plan. It is the sustainable development of Palawan. This in turn is:

> The improvement in the quality of life of its people in the present and future generations through the use of complementary activities of development and conservation that protect life support ecosystems and rehabilitate exploited areas to allow upcoming generations to sustain development growth.

This juxtaposes sustainable economic development, sustainable social development and sustainable ecosystems.

Section 5 goes on to indicate that the plan shall have three particular features: ecological viability; social acceptability; and an integrated approach. Ecological viability focuses upon the productivity of natural ecosystems. Social acceptability requires public participation and equality of access to resources and their benefits. An integrated approach incorporates inter alia the environmental dimension. While sustainable development is the intended outcome of the implementation of the plan, the legislation approaches it in a somewhat fragmented way by emphasising one or another of its aspects in the different sets of provisions in the legislation. This approach highlights specific issues in particular contexts. To some extent, however, it detracts from the integrity of sustainable development as the unifying and unified principle around which the more detailed objectives, mechanisms and instruments are organised.

INTEGRATED WATER RESOURCES MANAGEMENT IN FLEMISH BELGIUM

Our focus turns to the arrangements for the management of water resources in the European Union prescribed by the Directive 2000/60/EC Establishing a Framework for Community Action in the Field of Water Policy of 2000. It contains a number of duties in the context of a statement of purpose which prescribes an overarching outcome supported by a set of additional outcomes directed at more specific sets of circumstances. One of these critical duties is the formulation of river basin management plans and another is a statement of the environmental outcomes to be achieved in making operational the programmes of measures specified in the river basin management plans. It is a matter for each state within the European Union to decide how to implement the duties and perform the functions contemplated by the directive.

In 2003 statutory arrangements were put in place to give effect to the directive in the Flemish region of Belgium, namely, *le Décret relatif à la politique intégrée de l'eau*. These arrangements included the formulation of river basin management plans, their supporting institutional arrangements, ecological objectives to be achieved in relation to surface and groundwater, and a range of supporting instruments. Most important, however, are the statements of object, objective and principles in Articles 4–7. The fundamental object of these arrangements is *la politique intégrée de l'eau*. This is described in this way:

> *La politique intégrée de l'eau est la politique orientée sur le développement coordonné et intégré, la gestion et la restauration de systèmes d'eau afin d'atteindre les conditions connexes nécessaires à conserver ce système d'eau en tant que tel, et en vue de l'utilisation polyvalente, compte tenu des besoins des générations actuelles et futures.*

This incorporates the coordinated and integrated development of water resources, their management and their restoration with a view ultimately to their conservation by taking account, in recognition of their beneficial uses, of the needs of present and future generations.

The objectives to be achieved include:

- the protection, improvement and restoration of surface water and groundwater so as to reach the prescribed condition by 2015;
- the prevention and reduction of surface water and underground water pollution;

- the sustainable management of surface and underground water resources;
- the prevention of all degradation of aquatic ecosystems;
- the improvement and restoration of aquatic and terrestrial ecosystems that depend directly on water.

In doing this there must be taken into account the relationships between:

- the water and other elements of the environment, particularly the ecosystem linked with water;
- underground water, surface water and rainwater;
- the quality and quantity of water.

Then there must be similarly taken into account a number of principles. These include:

- prevention of any deterioration;
- intervention to prevent harmful effects;
- preventative measures taken at the source;
- the precautionary principle;
- the polluter pays principle;
- the user pays principle so as to include the environmental costs and the resource costs in providing the water services.

It is not only the statements of object, objectives and principles that are important. It is the relationship between and among them. The three are effectively interlinked. There is no weight attributed to one or the other. The protection of the environment, the protection of aquatic and terrestrial ecosystems, the use and development of water resources for social and economic purposes – all are equally part of the arrangements leading to their integration.

SUSTAINABLE WATER RESOURCES DEVELOPMENT IN SCOTLAND

(a) The Statutory Structure

The Parliament of Scotland in exercise of its devolved legislative powers enacted the Water Environment and Water Services (Scotland) Act in 2003 to implement the directive of the European Union of 2000. In 2002 the responsibility for the provision of public water supply, drainage and sewerage services had been given to Scottish Water by the Water Industry (Scotland) Act of 2002. According to schedule 3(1) Scottish Water is a statutory public agency which is not the agent of the Crown and does not have any of the immunities or privileges of the Crown. It is subject in some respects to directions by the relevant ministers. The performance of its functions is governed directly by the Act of 2002. Its primary duty is to provide a supply of wholesome water where a supply of water is required for domestic purposes and can be provided at a reasonable cost. There is by implication a correlative right to receive a supply of water in these circumstances. In other circumstances, arrangements are put in place between the supplier and the consumer essentially by agreement.

The way in which Scottish Water performs its functions is controlled in two ways. Section 50 of the Act of 2002 states:

> Scottish Water must, in exercising its functions, seek to ensure that its resources are used economically, efficiently and effectively.

Essentially this requires Scottish Water to undertake its operations as if it were a commercial entity notwithstanding its status as a statutory public agency. Even more importantly, section 51(1) provides:

> Scottish Water must, in exercising its functions, act in the way best calculated to contribute to the achievement of sustainable development.

But sub-section (2) limits this to the extent that it is consistent with the purposes of any enactment relating to its functions. The application of the test of consistency is not easy. What is important, however, is how the reference to sustainable development is structured. Clearly sustainable development is an outcome to be achieved in the performance of functions. Its achievement is not cast as an absolute outcome. It is limited, first, by the reference to 'the way best calculated'. While there is an element of discretion in this, clearly it is contemplated that, of the various alternatives, there will be one more likely to achieve sustainable development than any other. This is a judgement based upon the circumstances but not an unfettered discretion. Then, second, there is the reference to its contribution to the achievement of sustainable development. The implication is that sustainable development may be achieved in other ways. There is, however, nothing in the Act of 2002 to indicate what sustainable development means in this context.

The Act of 2003 goes some way to filling this gap. The Act discloses this structure:

- a purpose;
- how this is to be achieved;
- the subsidiary aims to be achieved in doing so;
- a range of duties restricting the performance of these functions;
- a duty to designate river basin districts;
- a duty to formulate river basin management plans;
- a duty to include specified matters in river basin management plans;
- a prescribed public process for preparing river basin management plans;
- a duty to review river basin management plans within prescribed periods;
- a duty to prepare sub basin plans;
- a duty to have regard to river basin management plans and sub basin plans in performing functions affected by them;
- a power by regulations to regulate any activity for the purpose of the protection of the water environment;
- a duty to include in every river basin management plan a list of the environmental objectives to be set for every body of water in the district.

(b) The Statutory Details

So much for the structure. Just as important are the details in section 1. The purpose is to make provision for the protection of the water environment rather than to protect it. Its protection includes preventing further deterioration of aquatic ecosystems and their related terrestrial ecosystems; promoting sustainable water use; enhancing protection of the aquatic environment; progressively reducing pollution of groundwater; and mitigating the effects of floods and droughts. Linked to these is the provision of a sufficient supply of good quality surface water and groundwater as needed for sustainable, balanced and equitable water use. These are essentially an indication of the ways and means for achieving the purpose. The purpose is the protection of the water environment. The water environment is the source of a sufficient supply of good quality surface water and groundwater. The criterion for sufficiency is what is needed for sustainable, balanced and equitable water use. The provision of a sufficient supply of water is a factor, but it is not the dominating factor in the context of this structure.

Section 2 specifies the duties imposed upon those performing statutory functions. This includes specifically the relevant ministers and the Scottish Environment Protection Agency, and by implication Scottish Water. Indeed sub-section (5) specifically requires every public body and office holder 'to have regard to the desirability of protecting the water environment'. This is a deliberative obligation – a duty to consider or have regard – and the substance of the duty is simply the desirability of protecting the water environment rather than, again, actually protecting it. Sitting side by side with this duty is an additional duty to have regard to the social and economic impact of the performance of these functions. Obviously elements of sustainable development. Linked to this is a duty in identical terms to the duty imposed upon Scottish Water by the 2002 Act, namely a duty – so far as consistent with the purposes of the relevant legislation – 'to act in the way best calculated to contribute to the achievement of sustainable development'. Overall, therefore, these statutory arrangements are about protecting the water environment and not about achieving sustainable development. The achievement of sustainable development is nevertheless an element – a subsidiary element – in these arrangements.

River basin management plans and sub basin plans are the principal instruments for protecting the water environment. Section 9(1) requires the Scottish Environment Protection Agency, first, to set environmental objectives for each body of water in the district covered by the river basin management plan and, second, to prepare a programme of measures to be applied to achieve those objectives. For this purpose environmental objectives are those required to comply with Article 4 and Article 7(2) and (3) of the directive of the European Union. Environmental objectives are prescribed for each body of water and must reflect particular circumstances. While the principal function of a river basin management plan is the protection of the water environment, two of the matters to be included are significant pressures on the water environment and the impact of human activity on it. To that extent social and economic factors are relevant – consistently with the general duty to take them into account.

(c) The Techniques of Regulation

There is, in addition to plans and environmental objectives, a third mechanism for protecting the water environment, namely the regulation of controlled activities in accordance with section 20. But these three mechanisms are linked. One of the purposes stated in paragraph 2(2) of schedule 2 of regulating controlled activities is to secure that the functions of the relevant public agencies are performed with a view to achieving the environmental objectives set out in river basin management plans. Effectively, therefore, the plans state the objectives and the objectives are achieved by regulating controlled activities.

The function of regulation of controlled activities is performed in four ways. These represent in essence the function of the law in achieving objectives. They are set out in paragraph 3 of schedule 2 to the Act. These are, in paraphrased form:

- prohibition of an activity unless authorised by a rule;
- prohibition of an activity unless authorised by an administrative decision – for example the grant of a water use licence;
- engaging in an activity authorised by a rule of law;
- engaging in an activity in a way authorised by a rule of law.

In other words, it is the function of the law to prohibit or permit activities in one of two ways:

- generally as a matter of law;
- specifically in particular sets of circumstances by an administrative decision made in accordance with a rule of law.

These represent the legal means according to which environmental objectives can be achieved. Such a classification of legal rules is fundamental to the formulation of a legal structure for doing so in more general terms.

CONSERVATION OF WATER RESOURCES IN SOUTH AFRICA

The focus of the National Water Act of 1998 of South Africa is the use, development and conservation of the water resources of South Africa rather than the protection of the water environment as such. Accordingly the statement of the purpose of the Act in section 2 indicates that the nation's water resources are to be conserved in ways which take into account a range of specific factors. There is clearly a duty to take these into account, and this process is the means by which the water resources are to be conserved. The matters to be taken into account include:

- meeting the basic human needs of present and future generations;
- promoting equitable access to water;
- promoting the efficient, sustainable and beneficial use of water in the public interest;
- facilitating social and economic development;
- protecting aquatic and associated ecosystems and their biological diversity;
- reducing and preventing pollution and degradation of water resources.

As declared in section 3, the national government is the public trustee of the nation's water resources. The responsibility to conserve water is part of this fiduciary duty, which also includes a responsibility to ensure that the water is used and developed in a sustainable and equitable manner and for the benefit of all persons in accordance with the mandate in the constitution. Thus, subject to the constitution, the concepts of sustainability and equity are specifically included in these arrangements. Essentially the purpose of the Act is structured not as one single overriding purpose but as a series of linked purposes which are themselves expanded by the reference to a range of – possibly subsidiary – factors to which there is a duty to have regard.

So much for the purpose of these statutory arrangements. The foundation of the regulatory scheme created by the Act is section 151(1)(a). It states quite simply that no person may use water otherwise than as permitted under this Act. Use for this purpose is given a very wide meaning by section 21. It includes not only taking but also storing water, impeding or diverting the flow of water, discharging waste into water and storing water. The use of water requires a licence unless it is permissible as a scheduled use. These permissible uses are associated with domestic use or a continuation of an existing lawful use or are permissible by way of a general authorisation.

The licensing and authorisation of water uses have effect within the national water resource strategy required under section 5 and catchment management strategies established for water management areas under section 8. The national strategy is designed to achieve the purpose of the Act and it must provide for the requirements of the reserve. These are, first, the satisfaction of basic human needs and, second, the protection of aquatic ecosystems in order to secure ecologically sustainable development and use of the relevant water resource. A catchment strategy must not be in conflict with the national strategy. In addition there is a general obligation to give effect to both the national strategy and a catchment strategy.

A water use licence may be granted either if the applicant wishes to apply for one or if the applicant is required to apply for one in accordance with section 43. Applications are compulsory if it has been decided that an allocation regime is needed to protect the public interest in a number of ways. These include to achieve a fair allocation of water to promote its beneficial use and to protect its quality. In these circumstances an allocation schedule is prepared and this indicates the quantity of water to be assigned inter alia to the reserve and to be allocated to any applicants. There is no obligation to allocate all available water. In any event the grant of a licence does not imply a guarantee relating to the statistical probability of supply; the availability of water; or the quality of water.

Although the overall management of South Africa's water resources is the responsibility of the state acting through its several public agencies, individual users of water have an interest not only in how water is allocated but also in how it is used in particular ways and in particular locations. Provision is accordingly made in Chapter 8 of the Act for the establishment of water users' associations. Although the functions of an association are prescribed by its constitution, it is an association of individual water users acting cooperatively to engage in activities related to the use of the water. It is in a sense recognition of their community of interest in the water resources in question.

The Act of 1998 incorporates in schedule 5 a model constitution of a water users' association. This is indicative of the functions of an association. These functions include:

- to protect water resources;
- to prevent unlawful activities in relation to water;
- to regulate the flow of a watercourse;
- to construct and operate waterworks.

One of the functions is to supervise and regulate the distribution and use of water from a water resource according to the relevant water use entitlements. The obligation to do so according to the relevant water use entitlements is critical. These entitlements reflect the national water resource strategy, the relevant catchment management strategy, the allocation schedule, the water incorporated in the reserve, permissible water uses and water use licences. In this respect it is quite different from a group of water users ensuring the management of the water resource according to acknowledged practices and customs. To this extent their functions are formal rather than informal.

WATER RESOURCES USE PRIORITIES IN AUSTRALIA

(a) The Structure of the Legislation

The arrangements for the management of water resources in the state of New South Wales in Australia comprise a comprehensive arrangement of objects, principles, rights and duties. In accordance with sections 341–343 and 392 of the Water Management Act of 2000 the system is predicated upon the vesting in the state of the exclusive right to the control, use and flow of more or less all water resources in the state, coupled with a prohibition on taking and using water and on constructing infrastructure for doing so. In addition common law riparian rights are abolished by section 393 without any reference to the range of common law rights that otherwise exist. The rights of individuals as distinct from the rights of the state are either the basic land holder rights conferred directly by the legislation or access and use rights conferred by licences. Licences are granted and have effect within the wider framework of these statutory arrangements.

An access licence, for example, must not be granted unless the management plan so permits and adequate arrangements are in force to ensure that minimal harm will be done to any water resource as a consequence of water being taken from the water source the subject of the licence: section 63(1). In addition, a management plan must be consistent inter alia with the state water management outcomes plan and with government policy including government policy in relation to the environmental objectives for water quality and river flow: section 16(1). The state water management outcomes plan is in furtherance of the objects of the Act. At the apex of these arrangements are the objects of the Act, and these are expanded by the water management principles stated in considerable detail. This strategic framework is implemented by the duty imposed by section 9.

The objects of the Act stated in section 3 include one general object and eight particular objects. The general object is to provide for the sustainable and integrated management of the water resources of the state for the benefit of both present and future generations. Cleary aspects of sustainable development are included: the reference to sustainable management; the reference to present and future generations; and the reference to integration. However the relevant verb is provide but not achieve, and the relevant noun is management but not use and development. Consequently the object of the Act does not go so far as to incorporate the achievement of the sustainable development of the water resources of the state.

Similarly the eight particular objects incorporate aspects of sustainable development. These include the protection of water resources and their associated ecosystems; the recognition of the significant social and economic benefits accruing from the sustainable and efficient use of water; provision for the orderly, efficient and equitable sharing of water; and the integration of water resources management with the management of the environment generally. One of the particular objects is to apply the principles of ecologically sustainable development and these principles are described in related legislation.

The water management principles stated in section 5 are generally consistent with these objects and in some cases described in more detail. Significantly, in relation to water sharing, there are three principles:

- sharing of water from a water source must protect the water source and its dependent ecosystems;
- sharing of water from a water source must protect basic landholder rights;
- sharing or extraction of water under any other right must not prejudice these two principles.

Even more important is the duty imposed by section 9(1). It provides:

> It is the duty of all persons exercising functions under this Act:
> (a) to take all reasonable steps to do so in accordance with, and so as to promote, the water management principles of this Act, and
> (b) as between the principles for water sharing set out in section 5(3), to give priority to those principles in the order in which they are set out in that sub-section.

The first duty is not absolute but limited to taking all reasonable steps. At most it is a duty to do so by observing the water management principles and by promoting these principles. Some of the water management principles are structured in the form of outcomes. There is accordingly no duty to achieve any of these outcomes. The second duty is to give priority to these principles in the order in which they are set out. Accordingly, on the face of it, priority must be given to protecting the water source and its dependent ecosystems. Next is the protection of basic landholder's rights. Rights arising under licences come last. But has the legislation been interpreted in this way?

(b) Judicial Interpretation of the Legislation

The Court of Appeal in New South Wales has had an opportunity to express a view on this duty creating priorities in a case brought by the Nature Conservation Council of New South Wales.[16] Before the matter reached the Court of Appeal, it was judicially commented that 'in determining the priority between extraction and the environment, priority must be given to the environment'.[17] The Court of Appeal was less positive. 'Decision-makers should give priority in relation to water sharing to protecting the water source and its dependent ecosystem'.[18] The first comment expresses the statement of priority as an enforceable rule of law, while the second comment expresses it as a principle – perhaps even a principle to be observed. In any event, the issue in this case was the validity of the relevant water management plan. According to section 8(2) of the Act a management plan must contain what are described as environmental water rules. These are rules for the identification, establishment and maintenance of planned environmental water. And environmental water is water committed for fundamental ecosystem health or other specified environmental purposes.

In this case[19] the issue was the validity of the relevant clause 14(a) in the plan. It stated:

> Water volume in excess of the long term extraction limit established in clause 30 of this plan may not be taken from this water source and be used for any purpose.

This comprises an obligation not to take water in excess of a stated limit: an obligation of behaviour or conduct. But what was the stated limit? The application of clauses 14(a) and 30 together meant that the long term extraction limit became the long term average annual extraction. Did the provision in the plan commit water for fundamental ecosystem health? According to the Court of Appeal, it did not do so. This was the reason:

> The idea of 'commitment' is quite inconsistent with any abstract or theoretical concept of water. What must be identified is actual water. The words 'establish' and 'maintain' are not suggestive of anything other than actual water. What is required is water that is consistently provided for and which, absent acute drought conditions, will in fact be available to protect 'fundamental ecosystem health'. To the extent water is present at all; priority is to be given to fundamental ecosystem health.[20]

Accordingly actual water was not committed to fundamental ecosystem health. This amounted to a failure to observe the statutory requirements. The plan, however, remained valid. Thus:

> The objects of the Act set out in section 3(a) and section 3(b), the water management principles sets out in section 5(2)(a) to (d) and section 5(3)(a), together with a priority established by section 9(1)(b) strongly suggest that the Parliament was concerned with matters of substance rather than form when it required the establishment of environmental water rules.[21]

And the court proceeded to deal with it on this basis. While the rules were formally invalid, they nevertheless achieved the substantive outcome expected of them. In effect, water was committed to be used for fundamental ecosystem health purposes even though the actual water so committed could not be specifically identified at all times. This acknowledges – if nothing else – the difficulty in creating formal arrangements for ensuring that the ecological values of water resources are protected and conserved.

A duty to commit water to the environment is easy enough to state at a strategic level. It is altogether different to implement such a duty in particular sets of circumstances. In the New South Wales case it emerged as an obligation of conduct. Alternatively, the duty may emerge as a duty linked to an outcome – in this case ecological or environmental objectives. The values to be protected will vary from water resource to water resource. In the state of Queensland in Australia – similar to the approach in New South Wales – provision has been made for environmental flow objectives in the relevant water resource plan. According to section 46(1) of the Water Act of 2000 there is an obligation for the plan to state the outcomes, including the ecological outcomes, for the sustainable management of the water.

The water resource plan for the Boyne River Basin in Queensland is a useful example. It contains a requirement for river flows to be managed to achieve seven purposes. The natural flow of the water through the river has been affected by the construction of one large dam and several weirs. The seven ecological outcomes to be achieved are synoptically these:

- to maintain existing habitats which sustain native plants and animals;
- to provide water for fine sediment removal and long term water quality suitable for riverine ecosystems upstream of the dam and ecosystems downstream of the dam;

- to allow movement by marine and estuarine fish over and around the weir;
- to provide water to stimulate reproductive processes for estuarine animals downstream of the dam;
- to provide water for delivery of catchment based nutrients from upstream of the dam to riverine and estuarine reaches downstream of the dam;
- to allow river forming flows upstream of the dam full supply level and in the riverine and estuarine reaches downstream of the dam;
- to allow for an increase in the frequency and duration of marine conditions in the estuarine reach downstream of the dam leading to a shift towards plant and animal species which favour or tolerate the increase.

The plan goes on to specify environmental flow objectives such as the monthly low flows for each named month required to be between 94% and 102%; performance indicators such as the number of days of flow to inundate the riparian zone; and strategies for achieving ecological outcomes such as the release downstream from the dam of water from the upper or oxygenated layers of water in the dam and the volumes of water to be released from the dam for the benefit of the downstream environment. To be effective these arrangements require monitoring and reporting and the plan specifically includes requirements to this effect. A water resource plan in Queensland – just like a water management plan in New South Wales – is an enforceable legal instrument. The substantive obligations in the Queensland plan are essentially obligations of outcome or result, and in the New South Wales plan obligation of behaviour or conduct. In either case their implementation is the responsibility of those managing the water resources on an operational basis. The obligations are intrinsically technical and their enforcement that much more difficult. While the management of water resources is a challenge in any event, the achievement of ecological outcomes – however expressed structurally – is for these reasons particularly challenging.

CONCLUSION

Despite the wide variety of doctrinal foundations of national legal systems and despite the complexity of the values underlying these systems, the end of the 20th century and the beginning of the 21st century have seen the introduction of a range of different statutory forms and structures to respond to the challenges facing the management of water resources across the globe. While the detail of the arrangements varies tremendously, there is for the most part a desire and an attempt to introduce sustainability in one form or another into these arrangements. In this sense the sustainable use and development of water resources have become part of the normative structure for the management of water resources. How it is implemented and whether it is enforced are matters for the national legal system in question. In some cases responsibility lies primarily with the executive, while in other cases the judiciary has an important function to perform. Accordingly the constitutional and institutional arrangements are critical. There is no single model which can ensure success. But in all cases it seems that the law performs a significant function. The key appears to be coherence within the overall normative structure of these arrangements. The examples discussed in some detail in this chapter tend to suggest that sustainable development is a goal and a process but not a legally enforceable duty.

NOTES

1. Productivity Commission (2003b, pp. 5–10, 17, 18, 23–30, 35–40 and 46).
2. *Tirupathi case* (2006).
3. *Tirupathi case* (2006, p. 302).
4. *Tirupathi case* (2006, p. 324).
5. *Tirupathi case* (2006, p. 324).
6. *Tirupathi case* (2006, pp. 324 and 325).
7. *Tirupathi case* (2006, p. 324).
8. *Tirupathi case* (2006, pp. 322 and 323).
9. *Tirupathi case* (2006, pp. 321 and 322).
10. *Tirupathi case* (2006, p. 321).
11. *Tirupathi case* (2006, p. 320).
12. *Tirupathi case* (2006, pp. 317 and 318).
13. *Tirupathi case* (2006, p. 319).
14. Tolentino (2000, p. 89).
15. Tolentino (2000, p. 89).
16. *Nature Conservation Council case* (2005). See Fisher (2006c, pp. 29–32).
17. Fisher (2006c, pp. 29–32).
18. *Nature Conservation Council case* (2005, para 54).
19. *Nature Conservation Council case* (2005).
20. *Nature Conservation Council case* (2005, paras 65 and 66).
21. *Nature Conservation Council case* (2005, para 95).

REFERENCES

Fisher D.E. (2006c), 'Water resources governance and the law', *Australasian Journal of Natural Resources Law and Policy*' 11, 1–41.

Productivity Commission (2003b), *Water Rights Arrangements in Australia and Overseas: Annex J, Chile* (Melbourne: Productivity Commission).

Tolentino, Amado (2000), 'Integrated water and land management in the Philippines', in D.E. Fisher and Noeleen McNamara (eds), *Integrated Water and Land Management* (Brisbane: Queensland University of Technology), pp. 81–109.

LEGAL INSTRUMENTS

Australian States

Water Management Act, being Act No. 92 of 2000 of New South Wales.
Water Act, being Act No. 34 of 2000 of Queensland.

Belgium

Décret relatif à la politique intégrée de l'eau, adopted by the Flemish Parliament on 18 July 2003.

European Union

Directive 2000/60/EC Establishing a Framework for Community Action in the Field of Water Policy 2000 [2000] OJ L327/1–69.

India

Maharashtra Water Conservation Corporation Act, being Maharashtra Act No. III of 2001.
Rajasthan Farmers' Participation in Management of Irrigation System Act, being Rajasthan Act No. 21 of 2000.

Indonesia

Law Concerning Basic Provisions for the Management of the Living Environment, being Law No. 4 of 1982 of Indonesia, in Katō, Ichirō et al. (eds) (1983), *Water Management and Environmental Protection in Asia and the Pacific* (Tokyo: University of Tokyo Press), pp. 163–169.

Iran

Water Law and the Manner of Water Nationalisation Law approved by the Parliament of Iran on 18 July 1968, in Economic Commission for Asia and the Far East (1968), *Water Legislation in Asia and the Far East*: Water Resources Series No. 35 (New York: United Nations), pp. 217–226.

Philippines

Strategic Environment Plan for Palawan Act approved on 19 June 1992, being Act No. 7611 of the Philippines, in D.E. Fisher and Noeleen McNamara (eds) (2000), *Integrated Water and Land Management* (Brisbane: Queensland University of Technology).

Scotland

Water Environment and Water Services (Scotland) Act, being Act No. 3 of 2003 of the Scottish Parliament.

Singapore

Public Utilities Act 2001 of Singapore.
Sewerage and Drainage Act 1999 of Singapore.
Environmental Pollution Control Act 1999 of Singapore.

South Africa

National Water Act being Act No. 36 of 1998 of South Africa.

CASES

Nature Conservation Council case: Nature Conservation Council of New South Wales Inc. v Minister Administering the Water Management Act 2000 (2005) *New South Wales Court of Appeal* 9.
Tirupathi case: Intellectuals Forum Tirupathi v State of Andhra Pradesh (2006) *Supreme Court Journal* 2, 293-325.

18. The need for doctrinal innovation

TOWARDS LEGAL INTEGRATION

(a) The Challenge

Water is a critical issue for each and every community across the globe. There is a diversity of interests and concerns. There is a diversity of cultural, institutional and legal arrangements. Despite this diversity, the characteristics of water are much the same across all communities. Five have been identified:

- water is an essential of life;
- its availability is variable and uncertain;
- it is supplied or available locally;
- it is used for a wide variety of independent purposes;
- there is no substitute for water.[1]

These characteristics:

> Combine to make water a 'commodity' of special significance to government, industry and the general populace and a subject which can arouse great passion. Command over water confers power especially in arid climates while, conversely, lack of government control over water leads to the risk of social instability.[2]

One of the most important questions in establishing a system for water resources governance is to determine who has control of water resources and of water at the several points of the hydrological cycle. Just as important is how this power of control should be exercised: in other words a combination of structural and normative issues. Although each community over the millennia may have responded somewhat differently to these questions, it is remarkable how much emphasis has been placed upon the need to conserve rather than waste water in the context of the arrangements for its use and development. In some communities over the last 200 or 300 years the pendulum began to swing more towards the use and development of water resources for economic purposes. However the last decades of the 20th century have witnessed a swing of the pendulum back towards sustainability.

The achievement of sustainability depends upon the existence of a set of appropriate and effective governance arrangements. Governance can be described as the totality of the processes according to which natural resources and the environment are conserved, used and developed. In relation to water resources, it has been defined as:

> The range of political, organizational and administrative processes through which communities articulate their interests, their input is absorbed, decisions are made and implemented, and decision

> makers are held accountable in the development and management of water resources and delivery of water services.[3]

But such a process achieves nothing unless it is directed towards clear and specific outcomes. What is emerging internationally, regionally and nationally is an acknowledgement that sustainable development is the outcome of such a system of governance. But how can this be achieved against a historical background – particularly over the last 100 years in some communities – of economic growth, industrialisation and urbanisation? In many respects the achievement of sustainable development necessarily involves a shift of cultural values and paradigms. One of these is the law.

If the use and development of water resources are to be sustainable, then each and every institution in the relevant community has a discrete function to perform. These include the institutions of the state, of the corporate sector and of the private sector. Some operate formally: others less so. The law plays a significant role in this. The law may facilitate; it may impede; it may be neutral. Much depends upon how the relevant legal arrangements are structured; upon whether they are descriptive or prescriptive; upon whether they are exhortatory or mandatory; whether they are flexible or rigid; whether and to what extent the rules are ultimately enforceable. A set of effective legal arrangements is not a goal in itself. The law performs an instrumental function and no more. In other words, the formal normative arrangements acknowledged by international custom, included in international and regional agreements, constitutions, regulatory regimes and rule-based arrangements of rights and duties must follow the function allotted to them. This must in turn reflect the outcomes to be achieved by this process of governance. How is sustainable development to be accommodated within such a set of arrangements as a matter of law?

(b) Integrated Water Resources Management

The process for achieving sustainable development of water resources is often said to involve integration, coherence and harmonisation. Alternatively it is described as integrated water resources management. In this respect the concepts of integration and management are important. It is a system of management rather than one of development, protection or conservation. It is integrated not only in the sense of an integrated set of processes but also as an integrated set of outcomes. The concept of integrated water resources management is relevant at all levels of governance: international, regional, national and local. It reflects the hydrological cycle. It is based not only upon the river basin as the management unit but also – perhaps more difficult – upon underground aquifers as the management unit. But each of these two management units must themselves be integrated with each other. It has been proposed as a principle of international law and it is a principle which has emerged recently in many national arrangements.

Integrated water resources management has been described as 'an attempt to manage water in an integrated fashion' and this requires a number of matters to be taken into account. These include:

- the physical, ecological and chemical characteristics of the water;
- the interests of all different users as they may collide or intercept;
- the environmental components of water use (ie the management of pollution);

- the participation of stakeholders in the decision-making process;
- the intersectoral management of water resources through the coordination of different central and local departments and agencies.[4]

Consistently with this the regulatory measures included within integrated water resources management are:

- the strengthening of formal administrative water rights systems;
- cost recovery and water pricing (the 'user pays' principle);
- the creation of new basin institutions;
- better consideration of the environment (the 'polluter pays' principle).[5]

This notion of integrated water resources management is clearly an element of sustainable water resources governance. But only one element. A comprehensive system of water resources governance brings within it a whole range of separate but related systems, each of which is ultimately directed at the achievement of the single overriding objective of sustainable development. The elements that comprise the overall system are these:

- the hydrological cycle – the inter-connections between surface water and groundwater and between evaporation, transpiration and precipitation;
- the ecological systems of which water is part;
- the individual members of the community;
- the corporate members of the community;
- the global, international, national and local communities;
- the information system including registers, cadastres and databases;
- the bureaucratic system which investigates, informs, consults, advises and makes decisions;
- the institutional system which creates strategies, policies, goals and objectives and implements them;
- the legal system which provides for the acknowledgement, recognition or creation of rules describing powers, rights, duties and liabilities.

Some of these systems are natural; some human; and some artificial. Sustainable development is concerned with all of these. An effective system of integrated water resources management incorporates each of them in ways which are consistent, coherent and harmonious. It is an integrated governance system such as this that brings together the ecological, social and economic perspectives of the use and development of water resources. While the law is a means to an end, the legal system provides the integrity, the security and the coherence that are necessary for integrated water resources management and for sustainable water resources governance.

(c) The Relevance of Legal Arrangements

International legal arrangements have a part to play in this. The nature of the environmental legal system effectively precludes the performance of legislative and to a large extent regulatory functions. It does however provide a relatively generalised and flexible set of normative principles and processes according to which more detailed arrangements in specific sets of circumstances can be put in place by agreement or by adjudication. The international community has taken the lead in identifying sustainable development as its goal for the future. The International Court of Justice itself has a role to perform. Thus the Court 'has served notice that it will hold states to newly developing norms and standards of environmental law, and will even require them to take such new rules into account with regard to existing projects that were, at their inception, not subject to the rules'.[6] Assuming that sustainable development is one such 'newly developing norm', the function to be performed by sustainable development raises critical issues concerning its relationship with the existing principles of international law.

There is the principle of the equitable utilisation of international water resources and the principle of environmental protection. Is sustainable development to be subsumed within one or both of these principles or are these principles to be subsumed within the wider concept of sustainable development? At its most fundamental level, the resolution of this issue is not simply a question of integration or even of consistency or coherence. The critical question is the point of commencement in the methodology related to making decisions and undertaking activities that involve the use and development of water resources. If the outcome is environmental protection, then equitable utilisation is subsidiary. If the outcome is equitable utilisation, then environmental protection is subsidiary. If the outcome is sustainable development, how are the competing ecological, social and economic perspectives to be satisfied?

Integration is as much a challenge for national communities as it is for the international community. Arrangements within national states necessarily reflect how water resources should be used and developed in particular locations and for particular purposes. There is a need for consistency among national, regional and local perspectives. An example is the approach adopted in some of the Mediterranean countries of the European Union. They have 'recently attempted to create a coherent framework addressing water resources management, usually characterised by one predominant framework law'.[7] Significantly it is a framework law. It provides for the management of water resources. It does not manage them. Management is undertaken regionally and locally, but in accordance with the framework so created. This approach has revealed a number of common features of water resources management laws:

- considering water as a public domain;
- protecting the resource through ensuring a minimum flow;
- controlling use and pollution;
- increasing participation by users;
- dealing with water rights and markets.[8]

These are the mechanisms or the instruments according to which water resources are managed with the capacity to achieve sustainable development by linking the ecological, the social and the economic perspectives in managing the water resources.

These instruments are integrated institutionally in accordance with three concepts:

- deconcentration;
- devolution;
- hydrological segmentation.[9]

The first transfers competences for water resources planning to different units. The second grants autonomy and independence to regional governments or to regional and local entities. The third creates river basins as planning units together with river basin authorities or committees. These institutional arrangements align with the hydrology of water resources and at the same time distribute the power to manage water resources across the community. To this extent there is a degree of fragmentation. But integrated water resources management seeks to render fragmented systems coherent. How is done? Through the concept of concertation. Thus:

> The developing concept of integrated water resources management implies coordination and the introduction of concertation mechanisms – in France, concertation is a legally binding concept introduced in the 1992 law – between the different policy, implementation and operational levels.[10]

In this sense there is expected to be a close relationship between what you are trying to do, how you intend to do it and how you actually do it. If sustainable development of water resources is assumed to be the outcome, then it is function of the law within a national state to ensure coherence among these instruments of governance.

(d) The Coherence of Legal Instruments

It is the water resources management regime that overall indicates who controls water resources and the criteria or rules according to which it is done. It may be based upon public or private rights of ownership; public or private rights of use; the uncontrolled exercise of these rights; or the exercise of these rights controlled by the law. At one extreme there is an uncontrolled right to use and at the other extreme a totally controlled right to use. Most lie within the two extremes. In this case most regimes include sets of rights in relation to water and more or less extensive regulation of their exercise. Despite the ambiguity of the word property, a water resources management regime has been stated to comprise 'a public governance component and a property rights component'.[11] Whatever the appropriate nomenclature, it is the relationship between these two components that is important.

This takes us back to the concept of coherence. Using the same terminology, 'regime coherence' has been described as 'the degree of consistency within and between the property rights structure and the governance structure of a resource regime'.[12] Four regime variables were then distinguished:

- the extent of the regime;
- the internal coherence of the property rights structure of the regime;
- the internal coherence of the governance structure of the regime;
- the external coherence between the property rights structure and the governance structure of the regime.[13]

Not only that: the internal coherence of the governance structure of the regime was analysed from five perspectives of coherence, namely:

- levels and scales of governance;
- actors in the policy network;

- problem perception and objectives;
- strategy and instruments;
- responsibilities and resources for implementation.[14]

What this means is that a water resources governance regime involves a number of institutional levels, a range of actors, a number of potentially ambiguous objectives, a number of instruments, and a range of responsibilities and resources.

Particularly important from the legal perspective is the function to be performed by instruments in such a set of regime arrangements. Each instrument is to be internally coherent. And each instrument has to be coherent with every other instrument within the overall regime. Let us be more practical. The range of rights in relation to water must be internally coherent. The range of instruments included in the governance structure must be internally coherent. The range of rights in relation to water must be coherent with the range of instruments in the governance structure of the regime. A daunting task in any sets of circumstances. It is the law – and probably only the law – that creates the formal structure which gives to the elements of the regime not only their internal but also their external coherence.

To create a coherent regime for water resources governance within a nation state is difficult by itself. To ensure that it is externally coherent with international law is even more difficult. But this is an element of integration. For how water resources are used is of global as well as of national concern. Similarly the impact of the use of water resources is a matter of global concern. This points to a community of interest at the global as well as at the national level. The nations of the world place different priorities upon the use and development of water resources. Some emphasise the need of society for water as a human and social right. Others see water resources as the engine of economic growth. Some wish to protect water and its ecosystems as a matter of priority. The recognition of a right of equitable use or utilisation – either nationally or internationally – is not by itself an adequate response. The need to protect the environment is similarly not an adequate response. The need to use and develop water resources to ensure economic growth is not an adequate response by itself. The task is to devise a set of coherent legal arrangements across the whole of the legal system – international, national, regional and local – to ensure the achievement of the sustainable development of water resources.

(e) The Function of Rules of Law

This takes us back to the fundamental function to be performed by rules of law which are intrinsically enforceable, by rules of law which inform the application of enforceable rules of law, and by the rules which comprise the framework within which these other rules have effect. In other words the sets of legal and paralegal rules in the sense already discussed. These rules – in the generic sense – perform different functions and they are structured and formalised accordingly. The rules of customary international law are for the most part relatively general and relatively flexible in their application – be they rules or principles of procedure or of substance. In some cases the rules included in interstate agreements relating to water resources are more specific and potentially more capable of enforcement. Significantly multilateral international agreements frequently create a framework of objectives and principles designed to guide the creation of more specific and potentially enforceable rules in later negotiations leading to formalised agreements. Similarly in national legal systems there has been emerging a framework of objectives and principles – either incorporated in a constitution or in legislation – which guide and inform the creation of enforceable rights and duties in more particular sets of circumstances.

Each of these various elements at either international or national level may be seen as separate and independent elements of the legal system and enforceable or not in their own terms. Or they may be seen as elements of an overall system which embraces them all in a coherent fashion. The former is intrinsically fragmented, while the latter is essentially integrated. Whether or not sustainable development is a principle of international law, it cannot be asserted currently that a development which is unsustainable is unlawful. If that were so, it would mean that the nation state was under an obligation neither to permit nor to undertake an unsustainable development of its water resources. While the same may well be true in general terms of national legal systems, in some jurisdictions sustainable development has been beginning to perform a function, but probably not yet as a standard according to which the lawfulness of a development is determined. A legal system incorporates legal rules of various kinds. Some are flexible, while others are rigid. Many afford a degree of interpretative and applicational discretion. But there is nothing new about this. The challenge now is to strike an appropriate balance between flexibility and rigidity in the structural design of rules of law.

Standards of human conduct and behaviour are derived from a variety of inspirational sources: religious or secular; spiritual or moral; aesthetic or utilitarian. Sustainability discloses some or perhaps all of these. These standards manifest themselves within the legal system through arrangements of rights, duties and liabilities. As we have seen, these may be human rights, natural rights, constitutional rights, individual rights, common rights, public rights. These rights correlate with sets of obligations such as obligations of conduct or behaviour, obligations of performance, obligations of methodology, obligations of outcome or result. The legal system becomes a mosaic comprising all of these differently shaped and functionally distinct pieces which together it is hoped form a meaningful and coherent pattern. This normative structure contains rules which are directly enforceable as well as those which are not so enforceable. So when does a rule become a legal rule?

According to one view – drawing on the analysis of Thomas Reid, the 17th century philosopher – the task of government based on a contract with society is:

> To secure the natural rights to life, liberty and property by means of laws prescribing the proper procedures that turn rules into legal rules or valid laws that direct and regulate the natural rights. The natural rights are prior to the making of positive laws and provide the moral standard that inform the content of the valid law on the one hand and provide the standard for the evaluation of the law on the other.[15]

In this sense a human or natural right to water becomes a legal rule when arrangements are formally put in place to realise this right through a set of legal liabilities which protect this right and ensure compliance with the duties that correlate with it. These rules are enforceable, but their inspiration is the human or natural right. The existence of this human or natural right – in itself not enforceable – nevertheless remains part of the pattern created by these pieces of mosaic.

Let us pursue this analysis further. The human or natural right to water is flexible in the sense that it responds and adapts to changing circumstances and needs. Moreover, the way in which natural resources and water resources in particular are managed necessarily responds to changing circumstances and needs – whether they are ecological, social or economic. Once the appropriate responses have been identified – for example the need for water as a right in particular circumstances – then enforceable rules are put in place to do so. These rules themselves may be relatively flexible or relatively rigid. Either way, they can be enforced. A rule which comprises a standard such as what is reasonable or what is equitable is enforceable in particular sets of circumstances notwithstanding its lack of absolute certainty or precision. It is simply no more and no less the function of the judiciary to adjudicate upon these matters. It has become the function of the law – certainly in the context of sustainable development – not only to state these rules but also to make clear the context in which they operate. Once again: the relationship between text and context.

In effect, therefore, the rules underpinning not only sustainable development but also integrated water resources management as an aspect of it should reflect the need not only for integrated management but also adaptive management. More specifically:

> Legal scholars have pointed to a number of legal issues that need to be addressed if resource management systems are to be managed adaptively. In particular, it may be important to reduce the law's traditional concern for predictability and certainty in favor of more flexibility to respond to changing conditions.[16]

In this respect rights of property in land have been distinguished from those in relation to wildlife, water or oil. Thus:

> These kinds of property are more fluid; they recognise the need to adapt to changes in the environment, while property rights in land have typically emphasised the value of finality.[17]

Accordingly it may be appropriate for some rules in relation to water to reflect 'the law's traditional concern for predictability and certainty' – particularly in relation to matters of process. And at the same time it may be equally appropriate for some rules to be designed to respond to changing circumstances, for example new scientific understanding, and to set contextually the *Grundnorms* underpinning the system.

(f) The Issue of Enforceability

This essentially raises the status and enforceability within the legal system of the rules about strategies, polices and plans. This is critical, for these have become integral elements of water resources governance. It has been suggested that:

> Legal enforceability of broad policy documents which simply set goals and objectives to be implemented by operational plans at a lower level in the hierarchy would be quite inappropriate. But once we begin to operationalize these objectives in component plans, then enforceability becomes a live issue.[18]

At which end of the spectrum does each rule sit? Strategic or operational? Instruments such as policy statements and the plans giving effect to them may well have been formulated in the context of enforceable rules. For example:

- an obligation upon the relevant public agency to formulate a policy or a plan;
- an obligation upon the public agency to include certain matters within such an instrument;
- a range of procedural obligations in formulating these instruments;
- an obligation upon the public agency to ensure that the plan gives effect to the policy or is consistent with the policy.

The instrument itself, if so authorised and so structured, may create enforceable rules which impose obligations or confer rights upon the persons to whom the rules apply – for example the holders of water rights.

These rights and these obligations may be structured – depending on their function – more or less flexibly or rigidly. While their capacity to be enforced depends to some extent upon the function they are designed to perform, ultimately their enforceability depends upon how these instruments are structured and the language used in the formulation of the rule. To express it slightly differently: the word used to describe the instrument – strategy, policy, plan – does not by itself attribute to it legal status or enforceability. These are matters for the substance of the instrument.

A SYNOPTIC REVIEW OF INTERNATIONAL APPROACHES

(a) The Principles of International Law

Let us now review synoptically the substance of the legal and paralegal rules supporting water resources governance and potentially their sustainable development. As we have seen, the international community has increasingly committed itself to the idea of sustainable development generally and in relation to water resources. The Stockholm Declaration of 1972 was concerned principally with the protection of the human environment. The Rio Declaration of 1992 linked development more directly to protection of the environment. The Johannesburg Declaration of 2002 affirmed the international community's commitment to sustainable development, but perhaps more in the context of social values as well as economic and ecological values. Although the expression sustainable development appeared in the 1992 and 2002 declarations, its meaning was assumed rather than explained. Subsequently the expression has been subjected to much philosophical, technical and juridical analysis.

The New Delhi Declaration of 2002 affirmed the principle that all states have, in accordance with international law, the sovereign right to manage their own natural resources pursuant to their environmental and developmental policies. However this right was complemented by the duty of states to manage natural resources, including natural resources solely within their own territory or jurisdiction, in a rational, sustainable and safe way. This is not a duty to achieve sustainable development. Neither does it have the status of a rule of law. Nevertheless it suggests that the international legal community – as distinct from the international community – is committed to the development of rules for sustainable development. The New Delhi Declaration went on to describe the supporting principles of substance and of procedure linked to sustainable development. These included the principle of good governance and the principle of integration.

Subsequent analysis by international institutions and commentators has developed and expanded the principles associated with sustainable development including the sustainable development of water resources. For instance, the report of the International Conference on Water and the Environment held in Dublin in 1992 suggested four principles to guide action at local, national and international levels. The first principle endorses a holistic approach which links social and economic development with protection of natural ecosystems. The second principle involves public participation and decision-making at the lowest appropriate level. The third principle – rather uniquely – acknowledges the specific needs of women and the importance of their participation at all levels of governance. The fourth principle emphasises the importance of managing water as an economic good. This contributes to the achievement of efficient and equitable use and to encouragement of conservation and protection of water resources. Each of the elements of sustainable development is implicit in these principles. It is the river basin – including surface water and groundwater – which was regarded by the Dublin report as 'the most appropriate geographical entity for the planning and management of water resources'. This applies as much to river basins within states as to international river basins. Consequently a high priority should be given 'to the preparation and implementation of integrated management plans'.

(b) Sustainable Development in International Law

The expression sustainable development appears in the Rio Declaration of 1992 but it is neither defined nor described. Principles 3, 4 and 8 together point to the three perspectives of sustainable development. Principle 3 brings together the developmental and environmental needs of present and future generations in the context of the right to development. On the assumption that the right to development is partly social and partly economic, its fulfilment is linked to the environmental needs of present and future generations. Principle 4 mandates that environmental protection is an integral part of the development process, and principle 8 essentially limits economic development to development which is sustainable. But sustainable development does not emerge as a single or overarching goal, objective or criterion.

The Draft International Covenant on Environment and Development develops further the function of sustainable development within the legal system. The objective of the covenant is the achievement of environmental conservation as an indispensable component of sustainable development. This is done by establishing an integrated set of rights and obligations. Some of these rights are conferred upon states and some – unusual for international arrangements – upon persons. Similarly obligations are imposed upon states and persons. The rights relate principally to the use and development of natural resources and the obligations for the protection of the environment. A set of fundamental principles guides the achievement of the objective.

According to Article 13(2) environmental conservation incorporates environmental, economic, social and cultural factors in planning and implementing activities in general. Article 13(1) requires the pursuit of what may be paraphrased as integrated sustainable development policies aimed at the relevant economic, social and ecological outcomes. The duty however is restricted to the pursuit of such policies. There is no duty to pursue sustainable development, nor is there a duty to achieve sustainable development. Article 19 creates obligations specifically in relation to water. The satisfaction of basic human needs and the maintenance of aquatic systems are the objective of the duty to take appropriate measures in relation to the quantity and quality of water available. There is, in other words, no duty to use and develop water resources sustainably. So, although sustainable development may thus be seen to be the ultimate objective of how natural resources are managed, the arrangements for doing so described in the covenant relate rather to process than to substance.

Although the covenant refers to sustainable development in these ways, it seems to be largely predicated upon the sovereign right of states to use their natural resources on the one hand and their responsibility to protect the environment on the other hand. This is complemented by the more limited duty to take appropriate measures to ensure the sustainable use of renewable resources. Does this mean that the right to use and the duty to protect remain somewhat polarised despite the process obligations in relation to integration? Does this reflect the position of international law generally or has it developed beyond these polarities?

Let us ask this question more specifically in relation to international water resources. A summary of the rights and obligations of states with a common interest in shared water resources has elicited a number of principles here expressed by way of a series of paraphrases:

- the right of use of a state is a corollary of its sovereignty;
- this right of use is not unlimited;
- there is no general obligation not to interfere with the natural flow of water;
- the exercise of the right to use is governed by the principle of equitable utilisation;
- all relevant factors and circumstances must be taken into consideration in determining whether a use is equitable;
- there is no priority among uses;
- there is a general obligation to cooperate;
- there is a general obligation to notify;
- there is a general obligation to consult and negotiate in good faith;
- the principle of equitable utilisation prevails over the principle of no harm;
- equitable utilisation is a step towards sustainable utilisation;
- an integrated management system is desirable not mandatory.[19]

Accordingly, it is equitable utilisation rather than sustainable development which lies at the foundation of this set of principles. Equitable utilisation may produce a sustainable outcome. In this event it is quite fortuitous and not the consequence of a legal obligation.

Let us be more specific and turn to the United Nations Convention on the Law of the Non-navigational Uses of International Watercourses of 1997. This is often taken to represent a codified version of the general principles of international law which apply to shared international water resources. Does it embrace sustainable development? The preamble to the convention speaks of the utilisation, development, conservation, management and protection of international watercourses and the promotion of their optimal and sustainable utilisation for present and future generations. A watercourse state is under a number of obligations:

- to use the international watercourse in an equitable and reasonable manner;
- to take all appropriate measures to prevent the causing of significant harm to other watercourse states;
- to protect and preserve the ecosystems of international watercourses;
- to prevent, reduce and control the pollution of international watercourses.

These substantive obligations are complemented by procedural obligations relating to cooperation, the exchange of data and information, notification and consultation. In addition there is a duty to take into account all relevant factors and circumstances in determining whether an international watercourse is being utilised in an equitable and reasonable manner. These factors include those relating to ecology, society and the economy. The principles of sustainable development are clearly implicit in the convention. But there is no obligation related directly either to sustainable development or the principles of sustainable development. Accordingly sustainable development itself has no formal function in these conventional arrangements.

Assuming that the principle of equitable utilisation is the *Grundnorm* of the convention, it may be that sustainable development is relevant in the application of the principle or, more particularly in the implementation of the duty to utilise an international watercourse in an equitable and reasonable manner. These perspectives sit side by side with the duties in relation to the protection of ecosystems and the prevention of pollution. The question is the relationship between them. This has been suggested:

> Factors which are relevant to the establishment of an equitable regime for the utilization of international watercourses might also be relevant to the attainment of sustainable development. Such would be the case of conservation (sustainable use) of the water resources of international watercourses. Despite this, equitable utilization and sustainable development differ in that the first is an existing principle of general international law while the second is a goal to be achieved at the national and international levels, and which aims at the harmonization of existing and emerging principles of international law. This means that, in the context of sustainable development, equitable utilization is one of the principles that have to be taken into consideration. The inclusion of equitable utilization in the process of attaining sustainable development might entail the modification of the equitable result arrived at on the basis of equitable criteria. Given the flexibility which characterises the establishment of equitable regimes for the utilization of international watercourses, the modification of the equitable result might not change the essential nature of the principle of equitable use.[20]

This analysis highlights the distinction between the application of a principle and the implementation of an obligation on the one hand and the achievement of a goal on the other hand. In this case sustainable development is treated as a goal to be achieved. One of the principles to be taken into consideration in doing so is equitable utilisation, or perhaps even more broadly equity. But equity and sustainability are intrinsically different. They may lead to the same result but equally they may not. If equitable utilisation remains the principle or the duty, then sustainable development performs and always will perform a subsidiary function. If that is the policy, so be it. But if sustainable development is to perform a different function within the legal system, then a different approach needs to be adopted. Consequently, on the assumption that the convention of 1997 reflects the approach of the international legal system, sustainable development – as a matter of process or of substance or as a principle or a goal – is relevant in determining whether a particular use of a particular international watercourse is equitable or not. Clearly no more.

(c) Sustainable Development in Interstate Agreements

While customary international law and the general principles of international law no doubt influence the arrangements entered into between and among states, it is the terms of interstate agreements which are ultimately critical. Do these agreements take our analysis of sustainable development any further? Until the last decade of the 20th century the focus of these agreements was upon the actual allocation or the criteria for allocation of water between the states or the authorisation of infrastructure to enable the relevant water resources to be used and developed. More recently sustainable development has been increasingly incorporated within these arrangements. For example, the arrangements agreed for the Southern African Development Community Region acknowledged a function for sustainable development in the Protocol on Shared Watercourse Systems. The principal obligation reflected the general principles of international law: namely an obligation to utilise a shared watercourse in an equitable and reasonable manner. In determining what is an equitable and reasonable utilisation, there was an obligation to take all relevant factors and circumstances into account including those of a technical, social, economic and environmental nature. In addition there was a recognition of the principles of intragenerational equity and intergenerational equity. These principles and what may be paraphrased as the principle of sustainable development emerged as factors of relevance in a determination of an equitable and reasonable utilisation of the shared water resources.

The Convention on Cooperation for the Protection and Sustainable Use of the Danube River of 1994 acknowledged in its preamble the importance of water management cooperation in the field of water protection and water use. Water management cooperation was to be oriented on sustainable water management. Much of the convention was concerned with preventing, controlling and reducing the transboundary impact of the discharge of polluting materials. Environmental protection of the river was clearly within the scope of the convention. Sustainable development was linked to environmental protection in the statement of objectives in the convention. While sustainable development was not defined, aspects of it were certainly acknowledged.

The critical function described in the convention emerges as management. The fundamental obligation in the convention is for the parties to cooperate on fundamental water management issues. In addition the parties are required to 'strive at achieving the goals of a sustainable and equitable water management'. This includes the conservation, the improvement and the rational use of surface waters and groundwater in the catchment area. Water management characterised as sustainable and equitable is thus the goal. The ways and means for achieving this goal include an obligation to take into account 'the urgency of water pollution abatement measures and of rational, sustainable water use'. While the goal is sustainable and equitable water management, this is achieved inter alia by taking into account rational and sustainable water use. For the most part, therefore, the adjective sustainable is linked with other adjectives – equitable and rational – and does not stand alone. The arrangements for water management are thus directed principally at environmental protection rather than sustainable development notwithstanding the relevance of sustainability.

The parties to the Agreement on Cooperation for the Sustainable Development of the Mekong River Basin of 1995 undertook a number of obligations. These included:

- to protect the environment and the ecological balance of the river basin from pollution;
- to utilise the waters of the river system in a reasonable and equitable manner pursuant to all relevant factors and circumstances;
- to make every effort to avoid harmful effects on the environment especially the quantity and quality of the water and its ecosystems;
- to promote the development of the full potential of sustainable benefits to all riparian states.

In effect, these obligations acknowledged the three elements of sustainable development: namely the relevance of social, economic and ecological considerations. Sustainable development was incorporated specifically in the preamble and in the obligation to cooperate. The parties agreed to cooperate 'in all fields of sustainable development, utilization, management and conservation of the water and related resources of the Mekong River Basin'. Sustainable development, utilisation, management and conservation were thus linked, but their relationship was not clear.

Management – unlike the other three – is a function without substantive content. The direction of management comes from other elements of these arrangements. The utilisation of water resources and the conservation of water resources are clearly relevant areas for cooperation, and so is sustainable development. The substantive obligations include the protection of the environment and the utilisation of the water resources in a reasonable and equitable manner. It is accordingly these two obligations and the relationship between them that comprise the fundamental *Grundnorms* of these arrangements. Sustainable development – once again – is relevant but subsidiary to these obligations.

The approach adopted by the Convention on the Protection of the Rhine of 1999 is somewhat different. The obligations undertaken by the parties – in some respects reasonably specific – are linked specifically to the achievement of the aims of the convention and the principles set out in it. The most significant aim of the convention is the sustainable development of the Rhine ecosystem. The parties are under a duty to 'pursue' this aim. The means for doing so include:

- maintaining and improving the quality of the waters;
- protecting populations of organisms and species diversity;
- maintaining, improving and restoring the natural function of the waters;
- conserving, improving and restoring the most natural habitats for wild fauna and flora in the water and adjacent areas;
- ensuring environmentally sound and rational management of water resources;
- taking ecological requirements into account in making decisions about development.

Significantly in this case management is qualified by reference to two criteria. And these are probably methodological rather than substantive criteria: namely management which is environmentally sound and rational. The convention then goes on to state the principles that guide the pursuit of this aim. These include not only the principles associated with the protection of the environment but also specifically 'the principle of sustainable development'. Sustainable development thus appears twice in the convention: first as an aim to be pursued and second as a principle that guides the pursuit of this aim. Clearly, sustainable development is relevant. Equally clearly, there is a difference between an aim and a principle. But the relationship between the aim of sustainable development and the principles of sustainable development is not taken any further.

(d) Conclusion

Sustainable development has been emerging as a principle recognised increasingly by the international legal system. Interstate agreements on the management of water resources in similar fashion have increasingly incorporated sustainable development in one form or another within them. The relevance of sustainable development cannot be denied, but the function it performs in any particular set of circumstances remains equivocal. Clearly there is no obligation in international law to ensure that the use and development of water resources are sustainable. Sometimes the adjective sustainable is linked to an otherwise neutral term such as management; sometimes to a less neutral term such as use. The adjective sustainable is often used conjunctively with another adjective such as equitable or rational. None of these approaches clarifies either the status or the function of sustainable development or sustainability within the international arrangements. Sustainability therefore remains something of an enigma in international law.

A SYNOPTIC REVIEW OF NATIONAL APPROACHES

(a) Constitutional Foundations

The international community has directed its attention to sustainability. So have national communities. Is the approach adopted by national legal systems any less enigmatic? The constitutional framework of the European Union places considerable emphasis upon the environment. While there is no specific reference to sustainable development in the sense under discussion, sustainability has become a significant value which has influenced the development of the arrangements in the European Union for the management of water resources, in particular, the Directive 2000/60/EC Establishing a Framework for Community Action in the Field of Water Policy of 2000. The *Grundnorm* of these arrangements is the integrated management of water resources based upon river basin planning.

These arrangements have been transposed into the law of the Flemish region of Belgium by *le Décret relatif à la politique intégrée de l'eau*. The objectives to be achieved by this integrated water resources management system include the protection, improvement and restoration of water resources; the prevention and reduction of water pollution; the protection of related ecosystems; and the sustainable management of surface and underground water resources. It is therefore a sustainable management system which is contemplated by these arrangements. In support of these, there are a number of principles, and these include those commonly associated with sustainable development. All of these provisions are linked to each other. It is consequently an integrated system. Sustainable management is one element, but only one element, of it.

In South Africa the *Grundnorm* of the arrangements for water resources management is their conservation. The matters to be taken into account in performing this function include the use and development of water resources for social and economic purposes and the protection of the environment and the ecosystems of which they are a part. The conservation of water is linked to a responsibility to ensure that the water is used and developed in a sustainable and equitable manner and for the benefit of all persons in accordance with the mandate in the constitution. These arrangements include the three elements of sustainable development. But it is conservation, not sustainable development, that lies at the foundation of the system.

(b) Statutory Foundations

One of the first manifestations of sustainable development was the approach adopted by the

Resource Management Act of 19991 of New Zealand. Its purpose was to promote the sustainable management of natural and physical resources, and sustainable management was defined to mean managing the use, development and protection of natural and physical resources. Accordingly management is given substance by reference to the use, development and protection of natural and physical resources. The substantive and procedural arrangements that comprise the details of the Act are related to this statement of purpose. The Environment Court has applied the Act and in particular the statement of purpose in such a way that sustainable development is more likely than not to be achieved. It is not guaranteed in any legal sense in the absence of an enforceable obligation to achieve the purpose of the Act.

The Directive of 2000 of the European Union has been transposed into the law of Scotland by the Water Environment and Water Services (Scotland) Act of 2003. Water resources in Scotland are managed by a public agency. Particularly significant is the obligation placed upon this agency, in exercising its functions, to 'act in the way best calculated to contribute to the achievement of sustainable development'. This is limited to the extent that it is consistent with the purposes of the legislation relating to its functions. While it is an obligation, it is not an obligation to achieve sustainable development. It is not an obligation that development must be sustainable. It is an obligation of performance linked to an obligation of outcome. It recognises that sustainable development is an outcome. The achievement of this outcome is governed by the other provisions of the legislation. It is consequently linked and linked directly to the way the public agency exercises its functions in particular sets of circumstances. It is an example, therefore, of a particular function to be performed by sustainable development in the context of the governance system of which it is a part. It is clearly a rule of law but a paralegal rule with 'interstitial normativity'.

NOTES

1. Storer (1995, pp. 261 and 262).
2. Storer (1995, pp. 262 and 263).
3. Grover (2006, p. 221).
4. Louka (2006, p. 182).
5. Van Koppen et al. (2007, p. 2).
6. McCaffrey (2007, p. 461).
7. European Commission (2000, p. 47).
8. European Commission (2000, p. 47).
9. European Commission (2000, p. 27).
10. European Commission (2000, p. 28).
11. Bressers and Kuks (2006, p. 239).
12. Bressers and Kuks (2006, pp. 238 and 239).
13. Bressers and Kuks (2006, p. 239).
14. Bressers and Kuks (2006, p. 239).
15. Bjarup (2005, p. 133).
16. Bosselman (2005, pp. 248 and 249).
17. Bosselman (2005, p. 249).
18. Farrier et al. (2000, p. 324).
19. Kaya (2003, pp. 203–214).
20. Fuentes (1998, p. 199).

GENERAL REFERENCES

Bjarup, Jes (2005) 'Social interaction: the foundation of customary law', in Peter Ørebech et al. (eds),

The Role of Customary Law in Sustainable Development (Cambridge: Cambridge University Press), pp. 89–157.

Bosselman, Fred (2005), 'Adaptive resource management through customary law' in Peter Ørebech et al. (eds), *The Role of Customary Law in Sustainable Development* (Cambridge: Cambridge University Press), pp. 245–281.

Bressers, Hans and Stefan Kuks (2006), 'Water basin regimes in Europe and institutional conditions for their sustainability', in V.I. Grover (ed.), *Water: Global Common And Global Problems* (Enfield: Science Publishers), pp. 235–268.

European Commission (2000), *Towards Sustainable and Strategic Management of Water Resources* (Luxembourg: European Communities Official Publications).

Farrier, David et al. (2000), 'Integrated water and land management in New South Wales: plans, problems and possibilities', in D.E. Fisher and Noeleen McNamara (eds), *Integrated Water and Land Management* (Brisbane: Queensland University of Technology), pp. 291–336.

Fuentes, Ximena (1998), 'Sustainable development in the equitable utilization of watercourses', *British Yearbook of International Law* 69, 119–200.

Grover, V.I. (2006), 'Water governance and water economics', in V.I. Grover (ed.), *Water: Global Common and Global Problems* (Enfield: Science Publishers), pp. 221–234.

Kaya, Ibrahim (2003), *Equitable Utilization: the Law of Non-navigational Uses of International Watercourses* (Aldershot: Ashgate Publishing Limited).

Louka, Elli (2006), *International Environmental Law* (Cambridge: Cambridge University Press).

McCaffrey, Stephen C. (2007), *The Law of International Watercourses: Non-navigational Uses* (2nd edn, Oxford: Oxford University Press).

Storer, David (1995), 'The potential role of privatizatoin in the management of water resources in the Middle East', in J.A. Allan and Chibli Mallat (eds), *Water in the Middle East: Legal, Political and Commercial Implications* (London: Tauris Academic Studies), pp. 261–269.

Van Koppen, B. et al. (eds) (2007), *Community-Based Water Law and Water Resource Management Reform in Developing Countries* (Wallingford: CAB International).

LEGAL INSTRUMENTS

Agreement on Cooperation for the Sustainable Development of the Mekong River Basin 1995 (1995), *International Legal Materials* 34, 865–880.

Convention on Cooperation for the Protection and Sustainable Use of the Danube River 1994, *Yearbook of International Environmental Law* 5, doc.16.

Convention for the Protection of the Rhine 1999 (2000), [2000] OJ L288/1.

Décret relatif à la politique intégrée de l'eau adopted by the Flemish Parliament on 18 July 2003.

Directive 2000/60/EC Establishing a Framework for Community Action in the Field of Water Policy 2000 [2000] OJ L327/1.

International Union for the Conservation of Nature (2004), *Draft International Covenant on Environment and Development* (3rd edn, Gland: International Union for the Conservation of Nature).

Dublin Statement on Water and Sustainable Development (1992), in Salman M.A. Salman and Daniel D. Bradlow (2006), *Regulatory Frameworks for Water Resources Management: A Comparative Study* (Washington, DC: World Bank).

New Delhi Declaration (2002): Declaration of Principles of International Law relating to Sustainable Development (2002), in Alistair Rieu-Clarke (2005), *International Law and Sustainable Development* (London: IWA Publishing), pp. 181–187.

Protocol on Shared Watercourse Systems in the Southern African Development Community (SADC) Region 1995, in Food and Agriculture Organisation of the United Nations (1997), *Treaties Concerning the Non-navigational Uses of International Watercourses Africa* (Rome: Food and Agriculture Organisation).

Resource Management Act 1991 of New Zealand.

Rio Declaration (1992): Declaration of the United Nations Conference on Environment and Development (1992), *International Legal Materials* 31, 874–880.

Stockholm Declaration (1972): Declaration of the United nations Conference on the Human Environment (1972), *International Legal Materials* 11, 1416–1421.

United Nations Convention on the Law of Non-navigational Uses of International Watercourses (1997), *International Legal Materials* 36, 700–718.

Water Environment and Water Services (Scotland) Act, being Act No. 3 of 2003 of the Scottish Parliament.

19. Models of doctrinal innovation

NORMATIVE STRUCTURES FOR INTEGRATED MANAGEMENT

(a) Introduction

Neither the international legal system nor national legal systems have yet incorporated sustainable development as an enforceable rule of law. Aspects of sustainable development and its associated principles are in some circumstances emerging as a relevant but subsidiary element of the overall system. The absence of an international legislature means that the introduction of new normative arrangements into international law is a difficult, controversial and lengthy process of diplomacy and negotiation. The development of new principles by the International Court of Justice is similarly a complex process. Much, then, depends upon what states do, not only in relation to each other but also within their own jurisdictions – particularly when their activities and decision-making processes impact upon the interests of other states. Similarly what happens locally impacts upon what happens nationally, and what happens nationally impacts upon what happens internationally and also the other way round. Integrated water resources management requires not only hydrological, technical, institutional and governmental integration but also legal integration by harmonising the norms of international, national and local regimes.

(b) Trends for the Future

Before looking at some legal models, let us focus upon what they are trying to achieve. We begin with the three goals that have been identified for a 'sustainable water policy' for the arid western areas of the United States of America. These are:

- efficiency;
- respect for and protection of nature;
- equity and social justice.[1]

Such a policy 'balances efficiency, respect for nature and equity as well as economic profit' and in addition 'it is necessary to create a fair and efficient decision-making process that facilitates sustainability'.[2] There are thus incorporated the twin elements of substance and process. The challenge again is to determine the extent to which the law can contribute to substance as well as process.

Seven ways of doing so were suggested:

- a holistic approach to deal with the relationships within the water cycle and between water and other resources;

- the remediation of environmental harm;
- a more extensive but careful use of water markets and water transfers;
- the adoption of conjunctive management and adaptive management;
- the adoption of technological solutions to help conserve water and use it more efficiently;
- decentralised collaboration in managing water;
- the integration of the effects of long-term climate change into water planning and management.[3]

Implicit in these suggestions are the three perspectives of sustainable development and the principle of integration. Significantly there have been reservations about the use of water markets and transfers. This was the conclusion in this respect:

> The limitations and complications of using market methods are still substantial. These and the potential damage that sales and markets can inflict suggest that market reforms will come gradually and should be carefully crafted to balance their benefits and damage.[4]

The use of economic instruments, as we have seen, has been on the international agenda for some time. Nevertheless it is clear that market arrangements have not been universally embraced.

A more juridical response has been a set of principles and a structure for an ideal water rights system. Eight principles have been suggested.[5] These are:

- the resource of water is not only a social good but also an economic good;
- the resource of water is a thing – *res* – necessarily subject to legal regulation;
- the concept of ownership is the mechanism according to which the resource of water in its natural state should be regulated;
- the law must recognise explicitly the human right to water and the environmental right to water;
- the state is under a duty to allocate water resources in a sustainable, equitable and reasonable way and to ensure that water is utilised equitably, reasonably, effectively and efficiently;
- the human right to water and the environmental right to water should be protected through the mechanism of rights of property in relation to water as a resource while the use and development of water resources should be promoted through rights of property in relation to water as a resource;
- water rights should be administered in accordance with the legal regime;
- the design of a national water rights system should be consistent with the relevant rules of international law.

While the emphasis of this model is upon rights – including rights of ownership – it is the implicit constraint upon the exercise of these rights that is particularly significant. One constraint – in principle 8 – is the need for the national system to be consistent with the international system. Even more important is the fifth principle which imposes two specific obligations upon the state. These are a restriction on the way water resources are allocated and a restriction upon the way in which they are used. Interestingly, sustainability constrains allocation but not use. Then there is the juxtaposition of the standards of sustainability, equity, reasonableness, effectiveness and efficiency. These standards are used conjunctively rather than disjunctively. To apply these standards simultaneously or conjunctively is a difficult task: simply because the application of each of these standards or criteria may in particular sets of circumstances lead to a different conclusion. For instance, what is equitable is not necessarily effective and what is sustainable is not necessarily reasonable. The juxtaposition of such criteria is, as we have seen, not uncommon. But does such an approach encapsulate sustainable development? Does such an approach amount to integration?

(c) A Conflict of Interest Rule

In relation to modern international water law, it has been suggested that:

> While being solidly founded on such well-established principles as equity, fairness and distributive justice, good neighbourliness, prevention of harm to co-riparians and the environment, [it] has also embraced the concepts of sustainable development, river basin approach, integrated water resources management, and stakeholder participation.[6]

Clearly so. But a river basin approach, integrated water resources management and stakeholder participation have been incorporated to an increasing extent within the system. While the concept of sustainable development may well have been embraced, it has scarcely been integrated within the system beyond its emergence as an inchoate principle. But how to integrate it effectively within the international legal system?

A novel approach[7] has been suggested: the incorporation of a conflict of interest rule 'as the primary rule of international law in the field of sustainable development'.[8] According to this approach, there is a conflict of interest where the economic, social or environmental needs of the state are not being met due to the activities of another state or a group of states.[9] One example given is where one state pollutes the environment of a neighbouring state, and another example is where the activities of one state adversely affect the global climate system. The existence of a conflict of interest is determined by the criterion of what is reasonable. Accordingly, 'only reasonable economic, social and environmental needs will be taken into account'[10] in determining whether or not there is a conflict of interest. If there is a conflict of interest, the issue is resolved on the basis of equity taking into account all relevant factors and circumstances. So the needs of a state for its sustainable development are determined by what is reasonable and any dispute about what is reasonable is determined by what is equitable. What is equitable is determined by taking into account – inter alia – the protection of the environment and the sustainable use of natural resources.

One of the advantages suggested for such an approach is that it is based on existing principles of international law. In this case – what is reasonable and equitable. This approach, it has been suggested, does not derogate from the existing right of a state to manage its own natural resources pursuant to its own environmental and developmental policies and its duty not to damage the environment of other states.[11] This approach would be supported by the existing range of procedural obligations relating to cooperation, notification, consultation and the exchange of data and information.[12] There are, in addition, the emerging principles such as the precautionary principle and the principles of intragenerational equity and intergenerational equity. All of these normative arrangements would in the course of time develop into 'more specific rights and obligations relating to sustainable development'.[13]

The significance of this approach is the existence of 'a primary rule relating to sustainable development'. This rule determines whether there is a conflict of interest according to what is reasonable and resolves a conflict of interest according to what is equitable. It is a primary rule – apparently – in the sense that it will be developed and expanded by a range of secondary rules which reflect existing rules and principles relating to sustainable development. Such an approach – as it itself acknowledges – is inspired by and based upon existing principles of international law. It deals with the economic, social and environmental needs of a state, the satisfaction of these needs, and the resolution of disputes about these needs. Although it deals with these elements, indeed these essential elements, of sustainable development, it does not appear directly to address the function to be performed by sustainable development and its normative status within the system.

(d) A Framework for Guardianship

A wider approach[14] emerges as a Draft Declaration on Guardianship over Water. The concept of guardianship over water incorporates directly the social, economic and ecological pillars of sustainable development. These are described essentially as access to water, control of water, and protection of water.[15] Each is linked to the others to constitute a coherent framework for water resources governance. It is directed at the international community, but the international community includes 'people, states, companies and organisations'.[16] It comprises a set of rights, duties and responsibilities affecting each and every element of this international community. It is based upon what are described as three fundamental principles:

- a set of human rights and duties;
- the qualified sovereignty of states;
- the principle of equitable and reasonable utilization.[17]

The fundamental rationale for this approach has been described in this way:

> In order to promote the sustainable development of freshwater resources, international law must construe the role of humankind as that of guardian over water by applying the three principles [above]. This implies that people, states and the international community as a whole must have access to water to meet their basic needs, and that they have the right to use water for other purposes – such as industrial and cultural uses – as long as the quantity and quality of water required for basic human needs and environmental protection is assured. Access and use of water are required to achieve the development of people, states and the world community. The protection of water at all levels is needed for it to support the life of humans and the environment. Water must be used sustainably by reducing waste and responding to problems of scarcity as well as maintaining ecosystems. The principles are to be viewed in their interrelationship and interdependence.[18]

Each of the three principles already mentioned reflects a balance between a set of rights and a set of duties or responsibilities. In other words the rights of people in relation to water are linked to their duty to take responsibility for the consequences of their actions. Similarly the rights of states are linked to their responsibilities. The application of the principle of equitable and reasonable utilisation is itself linked to the responsibilities associated with its exercise. The relationships between the rights and duties of people and the rights and duties of states reflect the relationship between the three pillars of sustainable development.

The significance of this approach is the coherence of all of those relationships and responsibilities. It is in a sense summed up by the last statement that comprises the operational rules sustaining the application of the principle of equitable and reasonable utilisation, namely:

> The international community shall perceive the hydrological cycle as a common heritage of humankind, in order to promote the sustainable use of water and the preservation of the environment at large and safeguard equitable and reasonable utilization of the world's waters.[19]

This aligns the characteristics of water and water resources with the social, economic and ecological needs of the international community from an ecocentric as well as an anthropocentric perspective so as to promote the sustainable use of water and the preservation of the environment. Once again, sustainable development emerges as a goal or outcome to be achieved in accordance with this set of coherent rights and duties in a system which reflects and aligns with the hydrological cycle. Clearly there is a function for sustainable development to perform in this way within the international legal system. The concept of guardianship over water according to this model goes some way towards formulating a set of international arrangements for sustainable water resources governance.

THE ELEMENTS OF AN INTEGRATED LEGAL STRUCTURE

(a) Integration of Normative Arrangements

International law may well recognise human rights and duties. But these rights and duties need to be realised within national legal systems. It has been suggested that 'national water codes are the cornerstones of sustainable water use. Without an effective water code, it is difficult for a nation to prevent the "tragedy of the commons", the over-use of limited supplies'.[20] A matrix of effective national legal arrangements is likely to include correlated sets of public rights and duties and private rights and duties that are managed, protected and enforced within a set of coherent institutional structures. It is no longer simply a system of allocation or of appropriation: rather a system of coherent governance. From a juridical perspective, it has been suggested that this involves two steps, namely:

> The first step is to realize that water, in contrast to other resources, is a shared resource. It must be shared both with other similarly-situated users and with the state as the representative of extra-market interests. The second step is the creation of an effective, comprehensive national water law regime that implements the idea that water is a commodity subject to the public interest and that all entitlements must be linked to the achievement of fair and sustainable water use.[21]

It is the detail of a national water law regime that is just as important as the detail of an international water law regime. The challenges of water resources governance are now recognised. There is substantial agreement about the strategies, policies, objectives, principles and mechanisms needed to respond to these challenges. Each nation state has responded in its own way. Despite the doctrinal differences in their legal systems, it is remarkable how similar many of the responses have been. Sustainability is emerging as a significant issue. It is achieving credibility as a concept. It may have the status of 'interstitial normativity' in some jurisdictions but it has not yet become a clearly defined and legally enforceable rule of law.

The achievement of sustainable development perhaps requires a cultural shift in the values of all communities and societies. Perhaps it also requires a different approach to legal doctrine. What might a set of detailed arrangements for the sustainable development of water resources look like?

The nature of water points to water resources as shared resources. This is certainly true of water in watercourses which flow across international boundaries and across boundaries that separate land in different ownership or occupation. It is almost certainly true also of water in underground aquifers in similar transboundary contexts. It is probably also true of water resources wherever located. Accordingly there is a community of interest in how all water resources are managed. It becomes a function of the law to recognise this community of interest despite the conflicts of interest that may arise. The ways in which shared water resource arrangements operate in practice are various. Many of the legal arrangements in ancient civilisations recognised this – at least implicitly if not explicitly: for example those in the early civilisations in Egypt, Mesopotamia and China. The rules of Roman law similarly recognised a community of interest in flowing water, and these rules consequently influenced the development of the civil law and of the common law. Such a community of interest was recognised notwithstanding the existence in particular sets of circumstances of private interests in water. These private interests became more significant during the period of rapid economic growth since the 17th century in parts of western Europe and in parts of north America. But in more recent years the notion of a community of interest has more explicitly returned as a basis for the way water resources are managed. This seems to have led to a reformulation of water as the common heritage of humankind. But, just as significantly, there has been a reformulation of the responsibility of humankind for the way it is managed. Hence the relevance and significance of sustainability.

(b) Formalising Coherence in International Arrangements

If sustainable development is emerging as the fundamental *Grundnorm* of a system of water resources governance, what form should it take? We need to remind ourselves that form should follow function. Let us begin with the international legal arrangements. The nature of these arrangements for the management of water resources suggests that this issue should be approached from two perspectives. The first comprises the set of norms which applies to the governance of water resources in general, and the second comprises the set of norms which governs the management of particular water resources, in particular locations and in particular circumstances. The first is a set of goals, principles and rules which constitutes the strategic framework for the management of water resources. The second is a set of specific rights and obligations which comprises the operational rules for undertaking specific activities. The first is likely to be a rule of customary international law and the second a rule contained in an interstate agreement. In any event the rules in an interstate agreement should in principle be consistent with and reflect the rules of customary international law. But the simultaneous application of the principles of sovereignty and of *pacta sunt servanda* does not require that approach. In theory the outcome reached by agreement should be the same as an outcome reached by adjudication. But in practice this is not necessarily so.

There is a further perspective to this distinction. A negotiated outcome or an adjudicated outcome comprises a set of rights and duties which apply only to the parties to the agreement or to the adjudicated settlement. On the other hand any rights and obligations acknowledged by customary international law are of general application. More particularly the obligations are not owed to any particular state or person but rather to all states and persons – *erga omnes*. In other words the framework of international norms has effect as an arrangement of strategic paralegal rules which inform the recognition or establishment of operational legal rules. Thus sustainable development may function as a strategic goal of general relevance and as an operational goal in particular sets of circumstances. The substance of the norm is the same but the function performed by it is different.

Let us take this analysis further in the context of sustainable water resources governance. The principle of reasonable and equitable utilisation of water resources indicates how water resources should be managed. It is essentially an obligation of performance rather than one of conduct: how to do rather than what to do. The question whether a use of water is reasonable and equitable may arise in two different ways: in relation to an existing use of water or in relation to a future use of water. The first is a reactive determination or adjudication about what has been happening and what is happening. The second is a prediction about what should happen in the future. This is the consequence of a planning process. The importance of the distinction is this. If reasonableness, equity or sustainability has a function to perform in the context of planning, it is either a factor in the decision-making process or the outcome of the decision making process. In the context of adjudication, it is a standard against which the lawfulness of what has happened in the past is determined. In this case it is neither a factor in planning nor an outcome of planning.

If the *Grundnorm* of international law is the principle of equitable and sustainable utilisation, or perhaps the principle of sustainable development, then it is important to recognise that the function that it performs depends upon whether decisions are being made about the lawfulness of what has happened in the past or about what should happen in the future, namely, a standard on the one hand or a factor or goal on the other hand. And in different terminology: a legal rule which is part of the microlegal system or a paralegal rule which is part of the macrolegal system. It is critical whether sustainable development is structured as an obligation of conduct, of performance, or of outcome. It may also be structured as an obligation of methodology. How it is structured and the language according to which it is structured indicate its normative status.

Let us translate this specifically to water resources governance. If sustainability is to be the *Grundnorm* of water resources governance, then this must be clear beyond doubt. It is not a factor in determining what is reasonable and equitable. It is not a factor in determining whether there has been an unreasonable interference with a water resource. It is not a factor in determining whether an ecosystem or the environment has been seriously degraded. It is not a factor in determining whether there is a need of water for basic human purposes. The principle of sustainable development – whatever its substance – transcends all of these and at the same time underpins all of these. This is so at a strategic level as well as at an operational level. However it is likely that the norm of sustainable development will be structured differently at each of these levels.

(c) A Structure for Coherence in International Arrangements

So much for generalities. How might a rule of law be structured to ensure this? The approach of international law is almost necessarily the creation of a strategic framework. It is the attainment of sustainable development that lies at the heart of this strategic framework. What is required, therefore, is not a set of arrangements that comprise an obligation of conduct; rather a set of arrangements that comprise either an obligation of performance or an obligation of methodology designed to achieve a specific outcome or a combination of both. What is being created is not a regime of liability for unlawful past conduct but a regime of decision making directed at planning and decision making for the future.

An appropriate structure might comprise three elements:

- a fundamental principle;
- the conditions according to which the principle is given effect to;
- a duty to consider the principles of sustainable development.

The first element is a statement of the relevant fundamental principle. For example:

> The fundamental principle to be applied in managing water resources is their sustainable use and development.

The expression management is wide enough to mean or be given a meaning that includes planning to undertake activities, undertaking activities, and permitting or authorising the planning or undertaking of activities. It is described as the fundamental principle to ensure that it has absolute priority even if it is suggested that other principles are relevant. It is a principle to be applied in the sense that this constitutes an obligation of methodology.

The substance of the principle is sustainable use and development. This needs to be explained. For example:

> The use of water resources is sustainable and the development of water resources is sustainable, in either case, if the social and economic needs of the community are satisfied and if the ecological needs of the environment are satisfied.

This distinguishes between use and development. A use is what happens at any moment in time and a development is what is proposed for the future. The test of sustainability is to satisfy each of the three elements of sustainable development. Each of these is a matter of substance. Moreover each of these is in the form of an obligation of outcome. Each of these outcomes must be satisfied conjunctively rather than disjunctively.

But how do you demonstrate that these conditions are satisfied? Perhaps in this way:

> In determining whether the social and economic needs of the community are satisfied and whether the ecological needs of the environment are satisfied, full and detailed consideration must be given to the principles of sustainable development recognised by international law so that these principles are given effect in making that determination.

This is, first, an obligation of procedure: the duty to consider. It is the procedural aspect of the obligation described by the fundamental principle. It is, secondly, an obligation of methodology: the duty to give effect. This explains how the decision on its substantive merits is to be made. It is the simultaneous discharge of these two obligations that is likely to ensure that sustainable development will be achieved. There is a growing consensus within the international community about the detail of these principles of sustainable development. However international law may well recognise more – or perhaps fewer – principles as knowledge and understanding change. Accordingly it would be helpful to include a statement describing the principles of sustainable development but not limited to those so included. For example, the precautionary principle, the principle of intragenerational equity, the principle of intergenerational equity, the polluter pays principle and the user pays principle.

This structure comprises a range of norms which perform different functions within this set of arrangements. What is particularly important is the relationship between all of them. One does not – indeed cannot – have effect by itself. Equally important is the absence of any reference to the persons or institutions that are subject to these obligations. This is deliberate. These norms are directed at no one in particular but everyone in general: *erga omnes*. This invokes the notion of the common responsibility of humankind for the sustainable use and development of the water resources that are the common heritage of humankind. An arrangement such as this does not deal with specific sets of circumstances. However it guides decisions in particular sets of circumstances, including those negotiated with a view to an interstate agreement. It also informs how any disputes between states and indeed others would be settled. It represents a strategic framework which constitutes the substantive standards according to which water resources are used and developed. In effect a constitution or charter to underpin sustainable water resources governance at international level.

Such a set of normative arrangements may affect not only the relationships among states themselves. It creates indirectly a set of normative arrangements according to which water resources within nation states should be managed. The principles and rules of international law mostly apply to states, but not to individuals or corporations within states. However the strategic framework described in the foregoing paragraphs is specifically structured to be of general application. If a state fails to comply with an obligation imposed by international law, it may have a consequential effect within the state's national legal system: for example, the grant by an agency of the state of an authorisation to engage in an activity relating to water resources in circumstances where that activity would be in breach of the state's international legal obligations. The breach by the state of its obligations may – depending on how the law develops – vitiate the authorisation to engage in that activity granted in accordance with the law of the state. Thus the individual or the corporation authorised to engage in an activity under the law of the state may be unable to engage in this activity because of the invalidity in international law of the state's decision. Hence the need for coherence and harmonisation, not only within international law but also between international law and the law of the states in question. Let us turn therefore to the law effective within a state.

(d) Formalising Coherence in National Arrangements

At a national level the legal arrangements need to comprise not only a strategic framework but also a set of operational rules: a microlegal system as well as a macrolegal system. A national system moreover may reflect or even incorporate local arrangements where they appear as customs or rules of the local community. This aspect is particularly important in relation to water resources in the context of the social needs of a community – in this case the individual needs of the local community. These local and national arrangements must themselves be not only consistent and coherent within themselves but also consistent and coherent with the international strategic framework. Once again – what might such a system look like?

Five principles may be identified in support of an effective national regime for sustainable water resources governance:

- the principle of comprehensive application;
- the principle of comprehensive involvement;
- the principle of total systems integration;
- the principle of information availability;
- the principle of enforceability.

The system thus applies to everyone. Everyone may participate. It applies to the totality of the hydrological cycle and the systems with which it is related. It is dependent upon comprehensive and current information. It needs to be capable of enforcement. The legal arrangements for creating such a system include:

- the formulation of a *Grundnorm* around which the whole system revolves in a set of horizontal and vertical relationships;
- strategies according to which the outcome stated in the *Grundnorm* is achieved;
- plans – based upon hydrological units – according to which strategies are implemented in particular sets of circumstances;
- the creation of regulatory instruments – such as protectable rights and enforceable duties – which facilitate and control what persons and institutions do in relation to the use and development of water resources;
- instruments which assume the form of standards of behaviour that have effect automatically and are directly enforceable through the legal system;
- instruments which assume the form of standards formulated in particular sets of circumstances by the public agency so authorised;
- instruments which assume the form of economic incentives and disincentives influencing commercial behaviour in a market context.

Protectable rights and enforceable duties lead to liabilities on the part of the persons and institutions affected. Such instruments are the traditional means for satisfying the social and economic needs of a community and of its members. These instruments do not by themselves satisfy the needs of the environment. This is simply because the environment does not exist as a juridical entity within a legal system. Protection of the environment can be achieved by the creation of protectable rights or enforceable duties. But a person or an institution must be set up to ensure protection of the right or enforcement of the duty. For example, the environment may have an entitlement to a flow of water but the entitlement itself cannot be assured in the absence of the involvement of a juridical entity.

(e) A Constitutional Framework for National Arrangements

Let us consider first of all a constitutional framework to support a system of water resources governance. These constitutional arrangements would be of general application. Once again they apply to no one in particular but to everyone in general. Increasingly the *Grundnorm* of a system has appeared as a statement of the purpose or of the object of the statutory scheme in question. While helpful, this approach does not give direct normative status to the subject matter of the statement. A statutory object is too distant in analytical terms from the decision to be made or the activity undertaken. There needs to be a link between the decision and its outcome or between the activity and its consequence. If sustainable development is to be afforded normative status, it needs to be structured as a rule that links the decision or the activity to its outcome or consequence. For example:

> The cardinal rule for the management of water resources is their sustainable use and development.

The sustainable use and development of water resources is thus a rule. It is the cardinal rule. It is not one of a number of rules. Any other rules are subsidiary to it. It applies to existing uses of water resources and it applies to the development of water resources for the future. However it applies differently to existing uses and future developments. This needs to be acknowledged.

First there is the existing use of water resources. In this case:

> The cardinal rule applies as the standard according to which the lawfulness of an existing use of a water resource is determined.

This states the question to be asked if there is a dispute about how water resources are actually being used. The question is not whether the use is reasonable or whether the use is equitable. It is whether the use is sustainable. Then there is the development of water resources. For example:

> The cardinal rule applies as the outcome to be achieved in determining whether the proposed development of a water resource is to be undertaken or to be permitted.

The question is not about what has happened in the past – although that may be relevant – but what should happen in the future. It creates an obligation of methodology in the context of the performance of a planning or regulatory function. The obligation is placed upon the person or institution undertaking the development – an entrepreneur in either the public sector or the private sector. The obligation is also placed upon the person or institution whose permission or authorisation is necessary to enable the development to take place: in this case, similarly, an institution either in the private sector or the public sector.

Further clarification is needed on what constitutes sustainable use and development. For example in relation to sustainable use:

> For the purposes of the cardinal rule an existing use is sustainable if it satisfies the social and economic needs of the community and the ecological needs of the environment and if these needs have been assessed and determined in accordance with the principles of sustainable development.

Accordingly the three substantive elements of sustainable development are included together with the principles of sustainable development. Similarly in relation to a proposed development:

> For the purpose of the cardinal rule a proposed development is sustainable if it will satisfy the social and economic needs of the community and the ecological needs of the environment and if these needs have been assessed and determined in accordance with the principles of sustainable development.

Once again, the peripatetic principles of sustainable development are included in the structure created by these rules.

The obligation in relation to the development of water resources is an obligation of outcome. It means in effect that the proposed development of the water resource in question must be sustainable. This is a prediction for the future rather than an adjudication upon past events. Whether the development is or is not sustainable will never be known until it has actually taken place. Accordingly. to facilitate the implementation of such an obligation there need to be accompanying rules to ensure a process of investigation, analysis and decision making that is more likely than not to lead to a sustainable development outcome. The process cannot guarantee that the development is sustainable. Hence the importance of including in such a process rules about public participation; the scope of enquiry; environmental, economic and social impact assessment; and decision-making methodology.

(f) The Implementation of a Constitutional Framework

While these statements assume the form of rules and contain obligations, they are not directed at anyone. In themselves they create no existing or potential liability if there is a failure to comply. In this sense they represent the normative framework of the system, in other words, a set of paralegal rules which are effective in informing and directing how the legal rules creating liability are applied in individual circumstances, but nevertheless potentially effective. The effectiveness of the framework depends upon how it is related to the responsibilities of those who are undertaking activities or making decisions about the use and development of water resources. This takes us to the second stage of a set of legal arrangements. How might the range of duties imposed upon specific persons or institutions be appropriately structured?

The most important point about the cardinal rule is that it applies automatically in all circumstances governed by these arrangements. Accordingly it controls the formulation of strategies and policies, the formulation of plans, the allocation of entitlements in relation to water, the transfer of entitlements, the exercise of rights associated with entitlements, the operation of infrastructure and all operational activities associated with the use or the development of water resources. Some of these functions are performed by the public sector and some by the private sector. The application of the cardinal rule will vary depending upon which of these functions is performed. It would accordingly be helpful to indicate how the rule is applied in these different ways.

For example in relation to water plans:

> The Minister [or the person responsible for formulating the water resources plan] must ensure that the water resources plan complies with and gives effect to the cardinal rule for the management of water resources.

Similarly in relation to entitlements:

> The Minister [or the person granting the entitlement] must ensure that the grant of the entitlement does not contravene or the activities authorised by the grant do not contravene the cardinal rule for the management of water resources.

In relation to a transfer of an entitlement:

> The Minister [or the person so authorised] must not approve a transfer of an entitlement if the transfer

> or the activities authorised as a consequence of the approval of the transfer would contravene the cardinal rule for the management of water resources.

The purpose of these provisions is to ensure that the use of water authorised by a water resources plan, an entitlement or a transfer of an entitlement is sustainable. This involves a judgement on the part of the relevant public agency but a judgement reached in accordance with the obligation of methodology prescribed by these arrangements: in other words the obligations to satisfy the conditions for sustainable use and development and to comply with the processes for reaching that conclusion: in effect this means an obligation to comply with both the substantive and the process elements of sustainable development.

So much for controlling the planning and administrative functions of the public sector. Just as important are the operational activities undertaken in accordance with these plans and these entitlements. These activities may be undertaken either by the public sector or the private sector. The public sector, for example, may be responsible for the construction and operation of a system of water supply to be used for domestic purposes or agricultural purposes such as irrigation. The activities may be undertaken by the private sector in exercise of their entitlements in relation to water: for example the extraction and use of water for a particular purpose. These activities are controlled not by obligations of methodology but by obligations of conduct, of performance or perhaps even an outcome. For example:

> A person must not use or develop water or a water resource except in accordance with the cardinal rule for the management of water resources.

Or, more generally:

> Water or water resources must not be used or developed except in accordance with cardinal rule for the management of water resources.

Once again, the purpose of these provisions is to ensure that all activities associated with the use and development of water and water resources are sustainable in the sense that they comply with the substantive and process requirements of the cardinal rule.

(g) Ensuring the Effectiveness of National Arrangements

The cardinal rule for the management of water resources applies to all water resources within the jurisdiction, to all activities associated with their use and development and to all persons and institutions in the public sector and the private sector involved in doing so: in other words the application of the principle of comprehensive application and the principle of comprehensive involvement. Compliance with the cardinal rule in these circumstances will not be easy. Further assistance perhaps needs to be provided in the legal arrangements for doing so. This may be done in a number of ways: by embedding in the legislation a set of relevant principles and by developing these principles through sets of correlated rights and duties. These principles may be included:

- the principle of community interest in shared water resources;
- the principle of common responsibility for shared water resources;
- the principle of adaptive management;

- the principle of public participation;
- the principle of public accountability;
- the principle of comprehensive availability of information;
- the principle of individual ownership of entitlements;
- the principle of open and competitive transferability of entitlements;
- the principle of enforceability of individual rights and obligations.

Let us take three of these further. Compliance with the cardinal rule for managing water resources depends in many respects upon the availability of comprehensive, accurate and current information. This applies to the formulation of policies and plans, the granting of entitlements as well as to the transfer of entitlements. Planning must be preceded by the collection and analysis of the relevant information. An open and competitive market for trading in entitlements requires detailed knowledge of and information about what is to be traded. Monitoring how water resources are used and developed requires appropriate information. Enforcement of rights and obligations requires the relevant information. Some of this information lies with the public sector and some with the private sector. The information needs to be made public to enable the proper functioning of planning, monitoring, trading and enforcing. Accordingly a range of supporting duties includes:

- a duty to collect the information;
- a duty to analyse the information;
- a duty to make the information available to the public according to appropriate accounting standards.

An information and accounting system, in other words, of the availability and use of all water within the community.

Trading in entitlements is an increasingly significant management technique in a number of jurisdictions. This may include entitlements held for environmental purposes. A regime for trading raises a number of issues:

- are all entitlements transferable;
- who may transfer an entitlement;
- is the approval of a public agency required;
- the formalities of transfers and other transactions;
- registration;
- certification.

If trading is permitted, this does not exclude the need to comply with the cardinal rule for the management of water resources. For instance, compliance with the cardinal rule may be a condition of a transfer or a condition of an approval of a transfer. Neither is an easy solution in practice. This is, once again, simply because to comply with the cardinal rule requires a prediction on the part of the transferee or on the part of the person authorising the transfer. Difficult to do but also difficult to enforce.

Then there is the principle of enforceability of rights and obligations. It is relatively easy to design a regime of criminal or civil liability for a failure to comply with a rule of law. This is much more difficult in relation to the cardinal rule for the management of water resources.

The rule is a combination of several rules; the issues reflect a variety of facts and circumstances; in some cases it involves a prediction about a future set of circumstances rather than about the consequences of past circumstances. Moreover the issues are in many respects technical, not only in economic but also in ecological terms. Economic impact analysis and environmental impact analysis are challenging enough in their own terms. But the inclusion of the social impact analysis so as to consider all three simultaneously is even more challenging.

Moreover the nature and extent of the supervisory jurisdiction of the courts vary from one system to another. In any event compliance with the cardinal rule is a responsibility of the public sector and of the private sector operating together. Accordingly there needs to be coherence between regimes of criminal and civil liability on the one hand and regimes of public accountability through judicial supervision of the public sector. Although the cardinal rule is the same in each of these contexts, the way in which it has effect differs substantially from one to the other. If the rule of law is a feature of water resources governance, then the arrangements for enforcement are just as important as the capacity for the rules themselves to be enforced. In this respect, for all of these reasons and in light of the nature of the cardinal rule, a coherent set of enforcement arrangements is an essential element of the principle of total systems integration which itself is an essential element of sustainable development. This may point in the direction of a single court or tribunal that is suitably resourced, intellectually and technically, to confront all of these practical issues associated with ensuring compliance with the cardinal rule.

The principles of coherence and integration lie at the foundations of the achievement of the sustainable use and development of water resources. This applies to all aspects of the hydrological cycle; to their relationship with all of the ecosystems of which water resources are a part; to the social, economic and ecological perspectives of these relationships; to the institutional systems associated with them; to the information systems similarly associated; to their legal relationships; and to the arrangements for enforcement and adjudication. Our immediate focus is the integration of the legal arrangements: namely, international law, constitutional values and rights, water resources plans, water entitlements, and all of the operational activities thereby authorised and undertaken. All of these must be brought within the umbrella of the cardinal rule. The arrangements for the management of water resources are a complex combination of vertical and horizontal relationships within the legal system. It is the cardinal rule that first provides coherence and then requires integration of each and every one of these various elements.

(h) Ensuring Coherence

How can this coherence and integration be achieved? Perhaps by the inclusion of a normative statement which embraces all of these structural elements. For example in relation to those undertaking activities in relation to water resources:

> It is the duty of those using or developing water resources or those using and developing water resources –
> 1) to achieve their sustainable use and development;
> 2) to comply with the cardinal rule for the management of water resources;
> 3) to comply with the rules stated in this statute;
> 4) to comply with the rules stated in the water plans made under this statute;
> 5) to comply with the conditions stated in an entitlement granted under this statute;

6) to take all reasonable and practicable measures to prevent or minimise any harm to the environment arising out of the use or development of water resources.

Then there is a duty in relation to those making decisions about the use and development of water resources, namely:

> It is the duty of those authorising, permitting or approving the use or development of water resources or the use and development of water resources:
> 1) to comply with the cardinal rule for the management of water resources;
> 2) to observe the principles for the management of water resources stated in this statute;
> 3) to ensure that the use or development authorised, permitted or approved –
>   (a) achieves the sustainable use and development of the water resource;
>   (b) complies with the rules stated in this statute;
>   (c) complies with the rules stated in the water plans made under this statute.

Then there is a statement of the persons upon whom these duties are imposed, namely:

> These duties are imposed upon all persons and institutions (including the state and its administrative, executive and judicial agencies) exercising powers, performing functions or discharging obligations in accordance with this statute, in accordance with any water plans made under this statute and in accordance with any conditions in an entitlement granted under this statute.

The purpose of this statement is to ensure that there is not only common but also individual responsibility for achieving sustainable use and development of water resources across all sectors and involving all those undertaking activities and making decisions about these activities.

Two further sets of obligations need to be put in place to assure consistency and coherence within the system overall. The first is coherence within the arrangements for water resources governance and the second is consistency with the other elements of the arrangements including the international legal arrangements. In other words, internal consistency and coherence and external consistency and coherence.

Let us consider first internal coherence. The purpose of this is, first, to ensure that the fundamental *Grundnorm* of the system – namely the cardinal rule for the management of water resources – underpins and is reflected in all the other instrumental arrangements and, secondly, to ensure that these instrumental arrangements are consistent with each other. The obligation thus relates to the instruments rather than to the persons responsible for the instruments. Perhaps something like this:

> In order to achieve the sustainable use and development of water resources –
> 1) a water entitlement must comply with and be consistent with the water plans made under this statute, the rules stated in this statute, the principles stated in this statute, and the cardinal rule for the management of water resources;
> 2) the rules stated in the water plans made under this statute must comply with and be consistent with the rules stated in this statute, the principles stated in this statute and the cardinal rule for the management of water resources.

Effectively, the obligations specified in entitlements and in water plans – because they deal with particular sets of circumstances – must be in conformity with the normative requirements of the other elements of the system. The other elements of the system – the rules in the statute, the principles stated in the statute and the cardinal rule itself – must themselves be consistent and compatible simply because they have been created as part of the one system.

Then there is the need for external coherence and consistency. This probably cannot be a substantive requirement of the system. What is proposed is that the system itself, and more particularly decisions made in accordance with it, should reflect the principles of international law and the principles and practices that are otherwise elements of the national legal system. Perhaps something like this:

> In the exercise of powers, in the performance of functions and in the discharge of obligations under this statute full and appropriate consideration must be given to:
> 1) the principles generally accepted by the international community and the strategies, policies and objectives by which they are implemented that together comprise the normative framework of international rules of law
> 2) the rules of customary international law
> 3) the obligations imposed upon the state by any international agreement to which the state is a party
> 4) the obligations imposed upon the state by any interstate agreement to which the state is a party
> 5) any relevant rules of the law of the state about the rights and duties of persons and juridical entities in their private capacities
> 6) any relevant customary rules
> 7) any relevant local practices.

While none of these is in most jurisdictions a part of the enforceable legal arrangements within a state, the way water resources are managed is influenced by these matters, and accordingly they can and indeed must be considered in the way suggested. It is a procedural norm affecting substance rather than a norm of substance in itself.

CONCLUSION

The sustainable use and development of water resources may be analysed from a multiplicity of perspectives. Our focus in this chapter has been the integration of the several elements of the legal system at both international and national levels. Some of these elements take the form of legal rules, while others are paralegal rules. The former are taken to be legally enforceable. The latter represent the normative framework within which these legal rules have effect. The expectation is that these legal rules will realise in an operational sense the goals and objectives indicated by the normative framework of paralegal rules. A system of sustainable water resources governance needs to include within it a set of legally integrated and analytically coherent arrangements which ensures – as best it can – that these goals and objectives are realised by these legal rules. It is accordingly the relationship between these paralegal and legal rules that has become critical. Such an arrangement is described in the context of the European Union as the *acquis communautaire*. This is the legal order according to which 'the formal rules of the European Community are complemented by a penumbra of non-legal rules and conventions'.[22]

It has been suggested on a number of occasions that form must follow function if the formality of the law is to achieve its objectives. Legal rules and paralegal rules perform different functions. Consequently they are structured differently and assume different forms. This chapter has suggested what these structures and what these forms might look like. The simplicity of this approach is confounded by the diversity of functions performed by these norms and by the range of persons and institutions to which they apply. The sustainable use and development of water resources involves a multiplicity of relationships. These may be formal and informal; vertical and horizontal; personal, corporate and institutional; direct and indirect; or explicit and implicit – a combination of all perhaps. For example, the relationship between the executive and the judiciary; between the executive and a citizen; between a citizen and another citizen; between a corporate citizen and an individual citizen; between agencies within the executive; even between any of these and the legislature. These relationships may be governed by the rules of the civil law, by those of administrative or public law, or by those of the constitution. They may relate to matters of procedure, process, methodology or substance. The need to ensure harmony or consistency with international norms is an additional challenge. It has been suggested that the achievement of sustainable development depends upon an integrated and coherent pattern of norms across the spectrum of the legal system: some structured as enforceable rules in the traditional way; others structured as underpinning principles; some structured as strategies and policies for their implementation; and a few designed to establish the relationships between the others. A system of legal rules and paralegal rules linked by 'interstitial normativity'. The models described in this chapter do no more than suggest how these rules may be structured so as to formalise through the law a system of sustainable water resources governance.

NOTES

1. Davis (2006, pp. 326–328).
2. Davis (2006, p. 328).
3. Davis (2006, pp. 338–351).
4. Davis (2006, p. 344).
5. Hu (2006, pp. 168–172, 192 and 193).
6. Wouters and Rieu-Clarke (2004, p. 91).
7. Rieu-Clarke (2005, pp. 159–162).
8. Rieu-Clarke (2005, p. 160).
9. Rieu-Clarke (2005, p. 160).
10. Rieu-Clarke (2005, p. 160).
11. Rieu-Clarke (2005, p. 161).
12. Rieu-Clarke (2005, p. 162).
13. Rieu-Clarke (2005, p. 162).
14. Hildering (2004, pp. 180–182).
15. Hildering (2004, pp. 171–173).
16. Hildering (2004, p. 182).
17. Hildering (2004, pp. 181 and 182).
18. Hildering (2004, p. 181).
19. Hildering (2004, p. 182).
20. Tarlock (2004, p. 120).
21. Tarlock (2004, p. 126).
22. Bronitt (1995, p. 98).

GENERAL REFERENCES

Bronitt, Simon et al. (1995), *Principles of European Community Law* (Sydney: Law Book Company).

Davis, Sandra K. (2006), 'Sustainability and freshwater in the western United States', in V.I. Grover (ed.), *Water: Global Common and Global Problems* (Enfield: Science Publishers), pp. 325–354.

Hildering, Antoinette (2004), *International Law, Sustainable Development and Water Management* (Delft, The Netherlands: Eburon).

Hu, Deshang (2006), *Water Rights: An International and Comparative Study* (London: IWA Publishing).

Rieu-Clarke, Alistair (2005), *International Law and Sustainable Development: Lessons from the Law of International Watercourses* (London: IWA Publishing).

Tarlock, A. Dan (2004), 'National water law: the foundation of sustainable water use', *Water Law* 15, 120–126.

Wouters, Patricia and Alistair Rieu-Clarke (2004), 'The role of international water law in ensuring 'good water governance': a call for renewed focus and action', *Water Law* 15, 89–92.

Bibliography

Adande, Nana et al. (2000), 'Policy foundations for sustainable development', in Jim Whitman (ed.), *The Sustainability Challenge for Southern Africa* (Basingstoke: Macmillan Press), pp. 291–295.

Ahmad, Ali (2000), 'Islamic water law as a comparative model for maintaining water quality', *The Journal of Islamic Law and Culture* 5, 159–185.

Al-Kloub, B. and T.T. Al-Shemmeri (1996), 'Application of multi-criteria decision aid to rank the Jordan–Yarmouk basin co-riparians according to the Helsinki and ILC rules', in J.A. Allan (ed.), *Water, Peace and the Middle East* (London: Tauris Academic Studies), pp. 185–207.

Allan J.A. (1995) 'Striking the "right price" for water? Achieving harmony between basic human need, available resources and commercial viability', in J.A. Allan and Chibli Mallat (eds), *Water in the Middle East: Legal, Political and Commercial Implications* (London: Tauris Academic Studies), pp. 325–346.

Anderson, Terry L. (ed.) (1983), *Water Rights: Scarce Resource Allocation, Bureaucracy and the Environment* (San Francisco, Cal.: Pacific Institute for Public Policy Research).

Anderson, Terry L. and Donald R. Leal (1991), *Free Market Environmentalism* (Oxford: Westview Press).

Arts, Bas and Henri Goverde (2006), 'The governance capacity of (new) policy arrangements: a reflective approach', in Bas Arts and Pieter Leroy (eds), *Institutional Dynamics in Environmental Governance* (Dordrecht: Springer), pp. 69–92.

Aswathanarayana, U. (2001), *Water Resources Management and the Environment* (Lisse: A.A. Balkema Publishers).

Bar, Rosmarie and Frank Kurschner-Pelkmann (2007), 'How do we guarantee the human right: sufficient water for all?', in José L. Lozán et al. (eds), *Global Change: Enough Water for All?* (Hamburg:Wissenschaftliche Auswertungen), 346–349.

Bartlett, Richard H., Alex Gardner and Bob Humphries (eds) (1996), *Water Resources Law and Management in Western Australia*, Nedlands (WA: The Centre for Commercial and Resources Law).

Beach, Heather L. et al. (eds) (2000), *Transboundary Freshwater Dispute Resolution: Theory, Practice, and Annotated References* (Tokyo: United Nations University Press).

Beintema, Nienke et al. (2006), *World Water Forum Bulletin* (New York: International Institute for Sustainable Development).

Benn, Suzanne and Dexter Dunphy (eds) (2007), *Corporate Governance and Sustainability: Challenges for Theory and Practice* (London: Routledge).

Berkes, Fikret (1996), 'Social systems, ecological systems and property rights', in Susan Hanna et al. (eds), *Rights to Nature* (Washington, DC: Island Press), pp. 87–107.

Betlem, G. (1993), *Civil Liability for Transfrontier Pollution: Dutch Environmental Tort Law in International Cases in the Light of Community Law* (London; Boston: Graham & Trotman/M. Nijhoff; Norwell, MA: Kluwer Academic Publishers).

Birnie, Patricia, Alan Boyle and Catherine Redgwell (2009), *International Law and the Environment* (3rd edn, Oxford: Oxford University Press).

Bjarup, Jes (2005) 'Social interaction: the foundation of customary law', in Peter Ørebech et al. (eds), *The Role of Customary Law in Sustainable Development* (Cambridge: Cambridge University Press), pp. 89–157.

Blackstone, William (1765–1769), *Commentaries on the Laws of England* (edn of 1809, Abingdon: Professional Texts Ltd).

Blake, Gerald H. et al. (1995), *The Peaceful Management of Transboundary Resources* (London and Boston: Graham & Trotman/Martinus Nijhoff).

Bogdanović, Slavko (2001), *International Law of Water Resources: Contribution of the International Law Association (1954–2000)* (London: Kluwer Law International).

Borkowski, Andrew and Paul du Plessis (2005), *Textbook on Roman Law* (3rd edn, Oxford: Oxford University Press).

Bosselman, Fred (2005), 'Adaptive resource management through customary law', in Peter Ørebech et al. (eds), *The Role of Customary Law in Sustainable Development* (Cambridge: Cambridge University Press), pp. 245–281.

Bosselmann, Klaus (2006), 'Ecological justice and law', in Benjamin Richardson and Stepan Wood (eds), *Environmental Law for Sustainability: a Reader* (Oxford: Hart Publishing), pp. 129–163.

Bosselmann, Klaus (2008), *The Principle of Sustainability: Ethics, Law and Governance* (Aldershot: Ashgate Publishing).

Boyle, Alan and David Freestone (1999a), 'Introduction', in Alan Boyle and David Freestone (eds), *International Law and Sustainable Development* (Oxford: Oxford University Press), pp. 1–18.

Boyle, Alan and David Freestone (eds) (1999b), *International Law and Sustainable Development: Past Achievements and Future Challenges* (Oxford: Oxford University Press).

Bradnee Chambers, W. and J.F. Green (2005a), 'Introduction: toward an effective framework for sustainable development', in W. Bradnee Chambers and J.F. Green (eds), *Reforming International Environmental Governance: From Institutional Limits to Innovative Reforms* (Tokyo: United Nations University Press), pp. 1–12.

Bradnee Chambers, W. and J.F. Green (eds) (2005b), *Reforming International Environmental Governance: From Institutional Limits to Innovative Reforms* (Tokyo: United Nations University Press).

Brans, Edward H.P. et al. (eds) (1997), *The Scarcity of Water: Emerging Legal and Policy Responses* (London; Boston: Kluwer Law International).

Bressers, Hans and Stefan Kuks (2006), 'Water basin regimes in Europe and institutional conditions for their sustainability', in V.I. Grover, (ed.), *Water: Global Common and Global Problems* (Enfield: Science Publishers), pp. 235–268.

Bronitt, Simon et al. (1995), *Principles of European Community Law* (Sydney: Law Book Company).

Brown Weiss, Edith (2005), 'Water transfers and international trade law', in E. Brown Weiss et al. (eds), *Fresh Water and International Economic Law* (Oxford and New York: Oxford University Press), pp. 35–59.

Brown Weiss, E. et al. (eds) (2005), *Fresh Water and International Economic Law* (Oxford and New York: Oxford University Press).

Bruch, Carl et al. (eds) (2005), *Public Participation in the Governance of International Freshwater Resources* (Tokyo: United Nations University Press).

Bruns, Bryan (2007), 'Community priorities for water rights: some conjectures on assumptions, principles and programmes', in B. van Koppen et al. (eds), *Community-based Water Law and Water Resource Management Reform in Developing Countries* (Wallingford: CAB International), pp. 28–45.

Burger, Joanna et al. (2001), 'Common-pool resources and commons institutions', in Joanna Burger et al. (eds), *Protecting the Commons* (Washington, DC: Island Press), pp. 1–15.

Calabresi, Guido and A.D. Melamed (1972), 'Property rules, liability rules and inalienability: one view of the cathedral', *Harvard Law Review* 85, 1089–1128.

Caponera, D.A. (1960), 'Water law principles in the Chinese legal system', *Indian Journal of International Law*, 239–275.

Caponera, D.A. (1973), *Water Laws in Moslem Countries* (Rome: Food and Agriculture Organisation of the United Nations).

Caponera, D.A. (1978), 'Water laws in hydraulic civilizations', in G.L. Ullmen (ed.), *Society and History* (The Hague: Mouton Publishers), pp. 91–106.

Caponera, D.A. (1992), *Principles of Water Law and Administration* (Rotterdam: A.A. Balkema Publishers).

Caponera, D.A. (1995) , 'Shared waters and international law', in G.H. Blake et al. (eds), *The Peaceful Management of Transboundary Resources* (London: Graham and Trotman/ Martinus Nijhoff), pp. 121–126.

Chalabi, Hasan and Tarek Majzoub (1995), 'Turkey, the waters of the Euphrates and public international law', in J.A. Allan and Chibli Mallat (eds), *Water in the Middle East* (London: Tauris Academic Studies), pp. 189–235.

Chave, Peter A. (2001), *The EU Water Framework Directive* (London: IWA Publishing).

Chimni, B.S. (2005), 'A tale of two treaties: the Ganga and Mahakali agreements and the watercourses convention', in Surya P. Subedi (ed.), *International Watercourses Law for the 21st Century* (Aldershot: Ashgate Publishing Limited), pp. 63–103.

Chomchai, Prachoom (1995), 'Management of transboundary water resources: a case study of the Mekong', in Gerald H. Blake et al. (eds), *The Peaceful Management of Transboundary Resources* (London: Graham and Trotman Limited), pp. 245–259.

Ciriacy-Wantrup, S.V. and R.C. Bishop (1975), 'Common property as a concept in natural resources policy', *Natural Resources Journal* 15, 713–729.

Clark, Sandford D. (1990), 'Tensions between water legislation and customary right', *Natural Resources Journal* 30, 503–520.

Commission of the European Communities, Directorate-General XI – Environment, Nuclear Safety and Civil Protection [European Economic Community] (1992), *European Community Environment Legislation* (Luxembourg: Office for Official Publications of the European Communities).

Connell, Daniel (2007), *Water Politics in the Murray-Darling Basin* (Sydney: Federation Press Pty Ltd).

Connor, Robin and Stephen Dovers (2004), *Institutional Change for Sustainable Development* (Cheltenham: Edward Elgar Publishing).

Cooter, R.D. and T.S. Ulen (1988), *Law and Economics* (New York: Scott Foresman and Company).

Corbridge, James N. and Teresa A. Rice (eds) (1999), *Vranesh's Colorado Water Law* (Niwot: University Press of Colorado).

Cordonier Segger, Marie-Claire (2005), 'Governing and reconciling economic, social and environmental regimes', in M.C. Cordonier Segger and C.G. Weeramantry (eds), *Sustainable Justice: Reconciling Economic, Social and Environmental Law* (Leiden: Martinus Nijhoff Publishers), pp. 561–592.

Cordonier Segger, Marie-Claire and Ashfaq Khalfan (2006), *Sustainable Development Law: Principles, Practices and Prospects* (Oxford: Oxford University Press).

Costanza, Robert and Carl Folke (1996), 'The structure and function of ecological systems in relation to property-rights regimes', in Susan Hanna et al. (eds), *Rights to Nature* (Washington, DC: Island Press), pp. 13–34.

Daibes-Murad, Fadia (2005), *A New Legal Framework for Managing the World's Shared Groundwaters : a Case Study from the Middle East* (London : IWA Publishing).

Davis, Sandra K. (2006), 'Sustainability and freshwater in the western United States', in V.I. Grover (ed.), *Water: Global Common and Global Problems* (Enfield: Science Publishers), pp. 325–354.

De Villiers, M. (2001), *Water Wars: is the World's Water Running Out*? (London: Phoenix Press).

De Waart, P.J.I.M. (1997), 'Securing access to safe drinking water through trade and international migration', in Edward H.P. Brans et al. (eds), *The Scarcity of Water: Emerging Legal and Policy Responses* (London: Kluwer Law International Ltd), pp. 100–118.

Dellapenna, Joseph W. (1994), 'Treaties as instruments for managing international shared water resources: restricted sovereignty vs community of property', *Case Western Reserve Journal of International Law* 26, 27–56.

Dellapenna, Joseph W. (1997), 'The Nile as a legal and political structure', in Edward H.P. Brans et al. (eds), *The Scarcity of Water: Emerging Legal and Policy Responses* (London and Boston: Kluwer Law International), pp. 121–134.

Devif, Michèle Zirari (1995), 'The law of small and large scale hydraulics in Morocco', in J.A. Allan and Chibli Mallat (eds), *Water in the Middle East: Legal, Political and Commercial Implications* (London: Tauris Academic Studies), pp. 139–150.

Diosdi, Gyorgy (1970), *Ownership in Ancient and Preclassical Roman Law* (Budapest: Akademici Kiado).

Dovers, Stephen and Robin Connor (2006), 'Institutional and policy change for sustainability', in Benjamin Richardson and Stepan Wood (eds), *Environmental Law for Sustainability: a Reader* (Oxford: Hart Publishing), pp. 21–60.

Dryzek, John S. (2007), 'Paradigms and discourses', in Daniel Bodansky et al. (eds), *The Oxford Handbook of International Environmental Law* (Oxford: Oxford University Press), pp. 44–62.

Du Bois, François (1995), 'Water law in the economy of nature', in J.A. Allan and Chibli Mallat (eds), *Water in the Middle East; Legal, Political and Commercial Implications* (London: Tauris Academic Studies), pp. 111–123.

Dupuy, Pierre-Marie (2007), 'Formation of customary international law and general principles', in Daniel Bodansky et al. (eds), *The Oxford Handbook of International Environmental Law* (Oxford: Oxford University Press), pp. 449–466.

Dzurik, Andrew A. (1990), *Water Resources Planning* (Savage, Mld: Rowman and Littlefield Publishers Inc).

Ebbesson, Jonas (1996), *Compatibility of International and National Environmental Law* (London: Kluwer Law International).

Economic Commission for Asia and the Far East (1968), *Water Legislation in Asia and the Far East*, Water Resources Series No. 35 (New York: United Nations).

Elhance, Arun P. (1999), *Hydropolitics in the Third World: Conflict and Cooperation in International River Basins* (Washington, DC: United States Institute of Peace Press).

European Commission (2000), *Towards Sustainable and Strategic Management of Water Resources* (Luxembourg: European Communities Official Publications).

European Environment Agency (1999a), *Sustainable Water Use in Europe: Part 1: Sectoral use of Water* (Luxembourg: European Communities Official Publications).

European Environment Agency (1999b), *Sustainable Water Use in Europe: Part 2: Demand Management* (Luxembourg: European Communities Official Publications).

Fahlbusch, Henning (2007), 'Water and its use in early history', in José L. Lozán et al. (eds), *Global Change: Enough Water for All?* (Hamburg: Wissenschaftliche Auswertungen), pp. 77–82.

Falkenmark, M. (1997), 'Water scarcity – challenges for the future', in Edward H.P. Brans et al. (eds), *The Scarcity of Water – Emerging Legal and Policy Responses* (London and Boston: Kluwer Law International), pp. 21–39.

Farrier, David et al. (2000), 'Integrated water and land management in New South Wales: plans, problems and possibilities', in D.E. Fisher and Noeleen McNamara (eds), *Integrated Water and Land Management* (Brisbane: Queensland University of Technology), pp. 291–336.

Fisher, D.E. (1987), *Natural Resources Law in Australia: a Macro-legal System in Operation* (Sydney: Law Book Company).

Fisher, D.E. (2000), *Water Law* (Sydney: LBC Information Services).

Fisher, D.E. (2003), *Australian Environmental Law* (Sydney: Lawbook Company).

Fisher, D.E. (2004a), *Trading in Water Rights – towards a National Approach* (Melbourne: Queensland University of Technology and Phillips Fox).

Fisher, D.E. (2004b), 'Rights of property in water: confusion or clarity', *Environmental Planning Law Journal* 21, 200–226.

Fisher, D.E. (2004c), 'Land, water and irrigation: hydrological and legal relationships in Australia (Part 1)', *Water Law* 15, 229–237.

Fisher, D.E. (2005), 'Land, water and irrigation: hydrological and legal relationships in Australia (Part 2)', *Water Law* 16, 14–22.

Fisher D.E. (2006a), 'Water law and policy in Australia – an overview', *Environmental Law Reporter (USA)* 36, 10264–10276.

Fisher, D.E. (2006b), 'Markets, water rights and sustainable development', *Environmental and Planning Law Journal* 23, 100–112.

Fisher D.E. (2006c), 'Water resources governance and the law', *Australasian Journal of Natural Resources Law and Policy*' 11, 1–41.

Fisher, D.E. (2007), 'Delivering the National Water Initiative: the emergence of innovative legal doctrine', in Karen Hussey and Stephen Dovers (eds), *Managing Water for Australia – the Social and Institutional Challenges* (Collingwood: CSIRO Publishing).

Fisher, D.E. (2008), 'Legal and paralegal rules for biodiversity conservation: a sequence of conceptual, linguistic and legal challenges', in Michael Jeffery et al. (eds), *Biodiversity Conservation: Law + Livelihoods* (Cambridge: Cambridge University Press), pp. 94–131.

Fisher D.E. and Noeleen McNamara (eds) (2000), *Integrated Water and Land Management* (Brisbane: Queensland University of Technology).

Fitzmaurice, Malgosia and Olufemi Elias (2004), *Watercourse Co-operation in Northern Europe: a Model for the Future* (The Hague: TMC Asser Press).
Food and Agriculture Organisation of the United Nations (1975), *Water Laws in Selected European Countries* (Rome: Food and Agriculture Organisation).
Food and Agriculture Organisation of the United Nations (1997), *Treaties Concerning the Non-navigational Uses of International Watercourses Africa* (Rome: Food and Agriculture Organisation).
Food and Agriculture Organisation of the United Nations (1998), *Sources of International Water Law* (Rome: Food and Agriculture Organisation).
Food and Agriculture Organisation of the United Nations (1973), *Water Laws in Moslem Countries* (Rome: Food and Agriculture Organisation).
French, Duncan (2005), *International Law and Policy of Sustainable Development* (Manchester: Manchester University Press).
French, Duncan (2006), 'Supporting the principle of integration in the furtherance of sustainable development: a sideways glance', *Environmental Law and Management* 18, 103–117.
Fuentes, Ximena (1998) 'Sustainable development in the equitable utilization of watercourses', *British Yearbook of International Law* 69, 119–200.
Gehring, M. and M.C. Cordonier Segger (eds) (2005), *Sustainable Development in World Trade Law* (The Hague: Kluwer Law International).
Getzler, J. (2006), *A History of Water Rights at Common Law* (Oxford: Oxford University Press).
Ghassemi, Fereidoun and Ian White (2007), *Inter-basin Water Transfer: Case Studies from Australia, United States, Canada, China and India* (Melbourne: Cambridge University Press Australia).
Ginther, Konrad and Paul de Waart (1995), 'Sustainable development as a matter of good governance: an introductory view', in Konrad Ginther et al. (eds), *Sustainable Development and Good Governance* (Dordrecht: Martinus Nijhoff Publishers), pp. 1–14.
Gischler, Christiaan E. (1979), *Water Resources in the Arab Middle East and North Africa* (Wisbech: Middle East & North Africa Studies Press Ltd).
Gleick, Peter H. (1993a), 'An introduction to global fresh water issues', in Peter H. Gleick (ed.), *Water in Crisis: a Guide to the World's Fresh Water Resources* (Oxford: Oxford University Press), pp. 3–12.
Gleick, Peter H. (1993b), 'Water in the 21st century', in Peter H. Gleick (ed.), *Water in Crisis: a Guide to the World's Fresh Water Resources* (Oxford: Oxford University Press), pp. 105–113.
Gleick, Peter H. (ed.) (1993), *Water in Crisis: a Guide to the World's Fresh Water Resources* (New York: Oxford University Press).
Gleick, Peter H. et al. (2006), *The World's Water 2006–2007: the Biennial Report on Freshwater Resources* (Washngton, DC: Island Press).
Godana, Bonaya Adhi (1985), *Africa's Shared Water Resources* (London: Francis Pinter Publishers).
Goldberg, David (1995), 'World Bank policy on projects on international waterways in the context of emerging international law and the work of the International Law Commission', in Gerald H. Blake et al. (eds), *The Peaceful Management of Transboundary Resources* (London: Graham and Trotman Limited), pp. 153–166.
Gonthier, Charles D. (2005), 'Fraternity: a global value underlying sustainable development',

in M.C. Cordonier Segger and C.G. Weeramantry (eds), *Sustainable Justice: Reconciling Economic, Social and Environmental Law* (Leiden: Martinnus Nijhoff Publishers), pp. 33–46.

Goodin, Robert (1995), *Utilitarianism as a Public Philosophy* (Cambridge: Cambridge University Press).

Grover, V.I. (2006), 'Water governance and water economics', in V.I. Grover (ed.) (2006), *Water: Global Common and Global Problems* (Enfield: Science Publishers), pp. 221–234.

Grover, V.I. (ed.) (2006), *Water: Global Common and Global Problems* (Enfield: Science Publishers).

Haas, Peter M. et al. (2004), 'Conclusion: institutional design and institutional reform for sustainable development', in Norichika Kanie and P.M. Haas (eds), *Emerging Forces in Environmental Governance* (Tokyo: United Nations University Press), pp. 263–281.

Haddadin, M.J. (1996), 'Water management: Jordanian viewpoint', in J.A. Allan (ed.), *Water, Peace and the Middle East* (London: Tauris Academic Studies), pp. 59–73.

Hanna, Susan and Svein Jentoft (1996), 'Human use of the natural environment: an overview of social and economic dimensions', in Susan Hanna et al. (eds), *Rights to Nature* (Washington: Island Press), pp. 35–55.

Hanna, Susan et al. (1996), 'Property rights and the natural environment', in Susan Hanna et al. (eds), *Rights to Nature* (Washington, DC: Island Press), pp. 1–10.

Herrera, P. et al. (2006), 'Information use and water resources laws and policies in Ecuador', in J. Wallace and P. Wouters (eds), *Hydrology and Water Law – Bridging the Gap* (London: IWA Publishing), pp. 218–246.

Hey, Ellen (1995), 'Sustainable use of shared water resources: the need for a paradigmatic shift in international watercourses law', in Gerald H. Blake et al. (eds), *The Peaceful Management of Transboundary Resources* (London: Graham and Trotman), pp. 127–152.

Hildering, Antoinette (2004), *International Law, Sustainable Development and Water Management* (Delft: Eburon Academic Publishers).

Holland, Kenneth M. et al. (eds) (1996), *Federalism and the Environment: Environmental Policy Making in Australia, Canada and the United States* (Westpoint, Conn: Greenwood Press).

Howarth, William (1992), *Wisdom's Law of Watercourses* (5th edn, Crayford: Shaw).

Howes, Michael (1985), *Whose Water? An Investigation of the Consequences of Alternative Approaches to Small Scale Irrigation in Bangladesh* (Dhaka: Bangladesh Institute of Development Studies).

Hu, Deshang (2006), *Water Rights: An International and Comparative Study* (London: IWA Publishing).

International Union for the Conservation of Nature (2004), *Draft International Covenant on Environment and Development* (3rd edn, Gland: International Union for the Conservation of Nature).

International Union for the Conservation of Nature (2008), *State of the World 2008: Water in a Sustainable Economy* (Gland: International Union for the Conservation of Nature).

International Water Tribunal (1994), *Dams* (Utrecht: International Books).

International Water Tribunal (1994), *Management* (Utrecht: International Books).

International Water Tribunal (1994), *Mining* (Utrecht: International Books).

Iqbal, Munawar (2005), 'Introduction', in Munawar Iqbal (ed.), *Islamic Perspectives on Sustainable Development* (Basingstoke: Palgrave Macmillan), pp. 1–21.

Johnson, Stanley P. (1993), *The Earth Summit: The United Nations Conference on Environment and Development (UNCED)* (London: Graham & Trotman).

Kanie, Norichika and P.M. Haas (eds) (2004), *Emerging Forces in Environmental Governance* (Tokyo: United Nations University Press).

Katō, Ichirō et al. (eds) (1983), *Water Management and Environmental Protection in Asia and the Pacific* (Tokyo: University of Tokyo Press).

Kaya, Ibrahim (2003), *Equitable Utilization: the Law of Non-navigational Uses of International Watercourses* (Aldershot: Ashgate Publishing Limited).

Kennett, Steven A. (1991), *Managing Interjurisdictional Waters in Canada: a Constitutional Analysis* (Calgary: Canadian Institute of Resources Law).

Kinzelbach, Wolfgang et al. (2007), 'Sustainable water resources management in arid and semi-arid regions', in José L. Lozán et al. (eds), *Global Change: Enough Water for All?* (Hamburg: Wissenschaftliche Auswertungen), pp. 291–295.

Lafferty, William M. (2004a), 'Introduction: form and function in governance for sustainable development', in William M. Lafferty (ed.), *Governance for Sustainable Development* (Cheltenham: Edward Elgar), pp. 1–27.

Lafferty, William M. (2004b), 'From environmental protection to sustainable development: the challenge of decoupling through sectoral integration', in William M. Lafferty (ed.), *Institutional Dynamics in Environmental Governance* (Cheltenham: Edward Elgar), pp. 191–219.

Lang, W., H. Neuhold and K. Zemanek (1991), *Environmental Protection and International Law* (London; Graham & Trotman; Norwell, MA, USA: Kluwer Academic Publishers Group).

Lang, Winfried (1995), 'How to manage sustainable development', in Konrad Ginther et al. (eds), *Sustainable Development and Good Governance* (Dordrecht: Martinus Nijhoff Publishers), pp. 90–104.

Laube, Wolfram and Nick van de Giesen (2006), 'Ghanaian water reforms: institutional and hydrological perspectives', in J. Wallace and P. Wouters (eds), *Hydrology and Water Law – Bridging the Gap* (London: IWA Publishing), pp. 157–197.

Lazonick, William and Mary O'Sullivan (eds) (2002) *Corporate Governance and Sustainable Prosperity* (Basingstoke: Palgrave).

Legge, J. (trans) (1968), *The Sacred Books of China: the Texts of Confucianism* (Delhi: Motilal Banarsidass).

Leroy, Pieter and Bas Arts (2006), 'Institutional dynamics in environmental governance', in Bas Arts and Pieter Leroy (eds), *Institutional Dynamics in Environmental Governance* (Dordrecht: Springer), pp. 1–20.

Locke, John (1690). *Two Treatises of Civil Government* (edn of 1924, London: J.M. Dent & Sons Ltd).

Louka, Elli (2006), *International Environmental Law* (Cambridge: Cambridge University Press).

Louka, Elli (2008), *Water Law and Policy: Governance without Frontiers* (Oxford: Oxford University Press).

Lowe, Vaughan (1989), 'The role of equity in international law', *Australian Yearbook of International Law* 12, 54–81.

Lowe, Vaughan (1999), 'Sustainable development and unsustainable arguments', in Alan Boyle and David Freestone (eds), *International Law and Sustainable Development* (Oxford: Oxford University Press), pp. 19–37.

Lozán, J.L. et al. (2007), 'The water problem of our Earth: from climate and the water cycle to the human right for water', in José L. Lozán et al. (eds), *Global Change: Enough Water for All?* (Hamburg: Wissen Schaftliche Auswertungen), pp. 13–18.

Lucas, Alastair R. (1990), *Security of Title in Canadian Water Rights* (Calgary, Alta: Canadian Institute of Resources Law).

Lundqvist, Lennart J. (2004), 'Management by objectives and results: a comparison of Dutch, Swedish and EU strategies for realising sustainable development', in William M. Lafferty (ed.), *Governance for Sustainable Development* (Cheltenham: Edward Elgar), pp. 95–127.

Luterbacher, Urs and Ellen Wiegandt (2005), 'Cooperation or confrontation: sustainable water use in an international context', in Edith Brown Weiss et al. (eds), *Fresh Water and International Economic Law* (Oxford: Oxford University Press), pp. 1–34.

Macpherson, C.B. (1978), 'Meaning of property', in C.B. Macpherson (ed.), *Property: Mainstream and Critical Positions* (Oxford: Basil Balckwell), pp. 1–13

Magraw, D.B. and L.D. Hawke (2007), 'Sustainable development', in Daniel Bodansky et al. (eds), The *Oxford Handbook of International Environmental Law* (Oxford: Oxford University Press), pp. 613–638.

Maktari, A.M.A. (1971), *Water Rights and Irrigation Practices in Lahj: A Study of the Application of Customary and Shari'ah Law in South-West Arabia* (Cambridge: Cambridge University Press).

Mallat, Chibli (1995), 'The quest for water use principles: reflections on Shari'a and custom in the Middle East', in J.A. Allan and Chibli Mallat (eds), *Water in the Middle East* (London: I.B. Tauris Publishers), pp. 127–137.

Mallat, Hyam (1995), 'Water laws in Lebanon', in J.A. Allan and Chibli Mallat (eds), *Water in the Middle East: Legal, Political and Commercial Implications* (London: Tauris Academic Studies), pp. 151–176.

Martin, Andrew et al. (2007), 'Towards a model of governance for sustainability', in Suzanne Benn and Dexter Dunphy (eds), *Corporate Governance and Sustainability: Challenges for Theory and Practice* (Abingdon: Routledge), pp. 94–121.

McCaffrey, Stephen C. (1997a), 'Water scarcity: institutional and legal responses', in Edward H.P. Brans et al. (eds), *The Scarcity of Water: Emerging Legal and Policy Responses* (London: Kluwer Law International), pp. 43–58.

McCaffrey, Stephen C. (1997b), 'Middle East water problems: the Jordan River', in Edward H.P. Brans et al. (eds), *The Scarcity of Water: Emerging Legal and Policy Responses* (London: Kluwer Law International), pp. 158–165.

McCaffrey, Stephen C. (2007), *The Law of International Watercourses: Non-Navigational Uses* (2nd edn, Oxford: Oxford University Press).

McCay, Bonnie J. (1996), 'Common and private concerns', in Susan Hanna et al. (eds), *Rights to Nature* (Washington, DC: Island Press), pp. 111–126.

McIntyre, Owen (2007), *Environmental Protection of International Watercourses under International Law* (Aldershot: Ashgate Publishing Ltd).

McKean, Margaret A. (1996), 'Common-property regimes as a solution to problems of scale and linkage', in Susan Hanna et al. (eds), *Rights to Nature* (Washington, DC: Island Press), pp. 223–243.

McNeill, J.R. (2000), *Something New under the Sun: an Environmental History of the Twentieth Century World* (London: W.W. Norton and Company).

Mitchell, Bruce (ed.), (1990), *Integrated Water Management: International Experience and Perspectives* (London: Belhaven Press).

Muckleston, Keith W. (1990), 'Integrated water management in the United States', in Bruce Mitchell (ed.), *Integrated Water Management: International Experiences and Perspectives* (London: Belhaven Press), pp. 22–43.

Mujwahuzi, Mark R. (2002), *Drawers of Water II: 30 Years of Change in Domestic Water Use & Enviromental Health in East Africa. Tanzania Country Study* (London: International Institute for Environment and Development).

Murray, Geoffrey and I.G. Cook (2004), *The Greening of China* (Beijing: China Intercontinental Press).

Mutembwa, Ammon (2000), 'Toward a sustainable water management strategy for Southern Africa', in Jim Whitman (ed.), *The Sustainability Challenge for Southern Africa* (Basingstoke: Macmillan Press), pp. 239–264.

Navia, Jose M. Borrero (1994), *Environmental and Water Law in the South* (Utrecht: International Books).

Nguyen, Van Duyen (2001), 'The inadequacy of environmental protection mechanisms in the Mekong River Basin Agreement', *Asia Pacific Journal of Environmental Law* 6, 349–376.

Nollkaemper, André (2005), 'The evolution of the regime for the River Rhine', in Surya P. Subedi (ed.), *International Watercourses Law for the 21st Century* (Aldershot: Ashgate Publishing Limited), pp. 151–166.

Okidi, C.O. (1997), 'International law and water scarcity in Africa', in Edward H.P. Brans et al. (eds), *The Scarcity of Water: Emerging Legal and Policy Responses* (London: Kluwer Law International), pp. 166–180.

Ørebech, Peter and Fred Bosselman (2005), 'The linkage between sustainable development and customary law', in Peter Ørebech et al. (eds), *The Role of Customary Law in Sustainable Development* (Cambridge: Cambridge University Press), pp. 12–42.

Ostrom, Elinor (2001), 'Reformulating the Commons', in Joanna Burger et al. (eds), *Protecting the Commons* (Washington, DC: Island Press), pp. 17–41.

Ostrom, Elinor and Edella Schlager (1996), 'The formation of property rights', in Susan Hanna et al. (eds), *Rights to Nature* (Washington, DC: Island Press), pp. 127–156.

Percy, David R. (1988), *The Framework of Water Rights Legislation in Canada* (Calgary, Alta: Canadian Institute of Resources Law).

Perruso, Richard (2002), 'The development of the doctrine of *res communis* in medieval and early modern Europe', *Tijdschrift voor Rechtsgeschiedenis* 70, 69–93.

Pichyakorn, Bantita (2005), 'International watercourses law: the experience of the Mekong River Basin', in Surya P. Subedi (ed.), *International Watercourses Law for the 21st Century* (Aldershot: Ashgate Publishing Limited), pp. 167–192.

Pinto, M.C.W. (1995), 'Reflections on the term sustainable development and its institutional implications', in Konrad Ginther et al. (eds), *Sustainable Development and Good Governance* (Dordrecht: Martinus Nijhoff Publishers), pp. 72–79.

Planiol, Marcel (1939), *Treatise on the Civil Law: Volume 1, Part 2* (12th edn, translated by the Louisiana State Law Institute with the authority of Librairie Générale de Droit et de Jurisprudence Paris (Louisiana State Law Institute).

Productivity Commission (2003a), *Water Rights Arrangements in Australia and Overseas: Annex I, Colorado* (Melbourne: Productivity Commission).

Productivity Commission (2003b), *Water Rights Arrangements in Australia and Overseas: Annex J, Chile* (Melbourne: Productivity Commission).

Richardson, Benjamin and Stepan Wood (eds), *Environmental Law for Sustainability: a Reader* (Oxford: Hart Publishing), pp. 1–18.

Rieu-Clarke, Alistair (2005), *International Law and Sustainable Development: Lessons from the Law of International Watercourses* (London: IWA Publishing).

Rothert, Steve (1995), *Lessons Unlearned: Damning the Mekong River* (Berkeley, CA: Internatonal Rivers Network).

Roughton, Geoffrey E. (2007), 'The ancient and the modern: environmental law and governance in Islam', *Columbia Journal of Environmental Law* 32, 99–139.

Saeijs, H.L.F. and M.J. van Berkel (1997), 'The global water crisis: the major issue of the twenty-first century: a growing and explosive problem', in Edward H.P. Brans et al. (eds), *The Scarcity of Water: Emerging Legal and Policy Responses* (London: Kluwer Law International), pp. 3–20.

Sagoff, Mark (1974), 'On preserving the natural environment', *Yale Law Journal* 84, 205–267.

Salau, Ademola T. (1990), 'Integrated water management: the Nigerian experience', in Bruce Mitchell (ed.), *Integrated Water Management: International Experiences and Perspectives* (London: Belhaven Press), pp. 188–202.

Salman, Salman M.A. and Daniel D. Bradlow (2006), *Regulatory Frameworks for Water Resources Management: A Comparative Study* (Washington, DC: World Bank).

Salman, Salman M.A. and Kishor Uprety (2002), *Conflict and Cooperation on South Asia's International Rivers: a Legal Perspective* (The Hague: Kluwer Law International).

Salman, Salman M.A. and Siobhán McInerney-Lankford (2004) *The Human Right to Water: Legal and Policy Dimensions* (Washington, DC: World Bank).

Salman, Salman M.A., *Regulatory Frameworks for Water Resources Management: a Comparative Study* (Washington, DC: World Bank).

Sands, Philippe and Paolo Galizzi (eds) (2006), *Documents in European Community Environmental Law* (2nd edn, Cambridge: Cambridge University Press).

Sands, Philippe (1994), 'International Law in the Field of Sustainable Development', *British Yearbook of International Law* 65, 303–381.

Sands, Phillipe (2004), *Principles of International Environmental Law* (2nd edn, Cambridge: Cambridge University Press).

Saunders, J. Owen (1988), *Interjurisdictional Issues in Canadian Water Management* (Calgary, Alta : Canadian Institute of Resources Law).

Scheck, Ulrich and Simone Malz (2007), 'Water demand and water use in the domestic and industrial sectors – an overview', in José L. Lozán et al. (eds), *Global Change: Enough Water for All?* (Hamburg: Wissenschaftliche Auswertangen), 95–99.

Schlager, Edella and William Blomquist (2001), 'Water resources: the southwestern United States', in Joanna Burger et al. (eds), *Protecting the Commons* (Washington, DC: Island Press), pp. 133–159.

Schlatter, R. (1951), *Private Property: The History of an Idea* (London: Allen and Unwin).

Schrijver, Nico (1999), 'The changing nature of state sovereignty', *British Yearbook of International Law* 70, 65–98.

Schrijver, Nico (2005), 'The New Delhi Declaration of principles of international law relating to sustainable development', in M.C. Cordonier Segger and C.G. Weeramantry (eds), *Sustainable Justice: Reconciling Economic, Social and Environmental Law* (Leiden: Martinus Nijhoff Publishers), pp. 549–552.

Schrijver, Nico and Friedl Weiss (eds) (2004), *International Law and Sustainable Development: Principles and Practice* (Leiden: Martinus Nijhoff).

Scott, Anthony and Georgina Coustalin (1995), 'The evolution of water rights', *Natural Resources Journal* 35, 821–979.

Sherk, George William (2000), *Dividing the Waters: the Resolution of Interstate Water Conflicts in the United States* (London and Boston: Kluwer Law International).

Shiklomanov, Igor A. (1993), 'Global fresh water resources', in Peter H. Gleick (ed.), *Water in Crisis: a Guide to the World's Fresh Water Resources* (Oxford: Oxford University Press), 13–24.

Shrubsole, D.A. (1990), 'Integrated water management strategies in Canada', in Bruce Mitchell (ed.), *Integrated Water Management: International Experiences and Perspectives* (London: Belhaven Press), pp. 88–118.

Siebert, Stefan and Petra Döll (2007), 'Irrigation water use – a global perspective', in José L. Lozán et al. (eds), *Global Change: Enough Water for All?* (Hamburg: Wissenschaftliche Auswertungen), pp. 104–107.

Simonis, Udo E. (2007), 'Water related conflicts – and the need for a global water strategy', in José L. Lozán et al. (eds), *Global Change: Enough Water for All?* (Hamburg: Wissen Schaftliche Auswertungen), pp. 329–333.

Singh, Chhatrapati (1991), *Water Rights and the Principles of Water Resource Management* (Bombay: N. M. Tripathi Pty Ltd).

Singh, Chhatrapati (2000), 'Water rights and the principles of water management in India', in D.E. Fisher and Noeleen McNamara (eds), *Integrated Water and Land Management* (Brisbane: Queensland University of Technology), pp. 49–79.

Sklar, Leonard et al. (1994), *Damning the Rivers: the World Bank's Lending for Large Dams* (Berkeley, CA: Internatonal Rivers Network).

Storer, David (1995), 'The potential role of privatization in the management of water resources in the Middle East', in J.A. Allan and Chibli Mallat (eds), *Water in the Middle East: Legal, Political and Commercial Implications* (London: Tauris Academic Studies), pp. 261–269.

Subedi, Surya P. (2005), 'Regulation of shared water resources in international law: the challenge of balancing competing demands', in Surya P. Subedi (ed.), *International Watercourses Law for the 21st Century* (Aldershot: Ashgate Publishing Ltd), pp. 7–18.

Tamrat, Imeru (1995), 'Constraints and opportunities for basin-wide cooperation in the Nile: a legal perspective', in J.A. Allan and Chibli Mallat (eds), *Water in the Middle East: Legal, Political and Commercial Implications* (London: Tauris Academic Studies), pp. 177–188.

Tarlock, A. Dan (1997), 'Current trends in United States water law and policy: private property rights, public interest limitations and the creation of markets', in Edward H.P. Brans et al. (eds), *The Scarcity of Water: Emerging Legal and Policy Responses* (London: Kluwer Law International Ltd), pp. 183–195.

Tarlock, A. Dan (2000), 'Putting rivers back into the landscape: the revival of watershed management in the United States', in D.E. Fisher and Noeleen McNamara (eds), *Integrated Water and Land Management* (Brisbane: Queensland University of Technology), pp. 195–247.

Tarlock, A. Dan (2004), 'National water law: the foundation of sustainable water use', *Water Law* 15, 120–126.

Tarlock, A. Dan (2005), 'Water transfers: a means to achieve sustainable water use', in Edith Brown Weiss et al. (eds), *Fresh Water and International Economic Law* (Oxford: Oxford University Press), pp. 35–59.

Tarlock, A. Dan et al. (2002), *Water Resource Management: A Casebook in Law and Public Policy* (5th edn, New York: Foundation Press).

Tolentino, Amado S. (2000), 'Integrated water and land management in the Philippines', in D.E. Fisher and Noeleen McNamara (eds), *Integrated Water and Land Management* (Brisbane: Queensland University of Technology), pp. 81–109.

Tumwine, James K. (2002), *Drawers of Water II: 30 Years of Change in Domestic Use & Enviromental Health in East Africa. Uganda Country Study* (London: International Institute for Environment and Development).

Upadhyay, Surya Nath (2005), 'The legal regime of water resources in Nepal', in Surya P. Subedi (ed.), *International Watercourses Law for the 21st Century* (Aldershot: Ashgate Publishing Limited), pp. 195–209.

Van Dunné, Jan M. (1997), 'Liability in torts for the detrimental use of fresh water resources under Dutch law in domestic and international cases', in Edward H.P. Brans et al. (eds), *The Scarcity of Water: Emerging Legal and Policy Responses* (London: Kluwer Law International Ltd), pp. 196–211.

Van Koppen, B. et al. (eds) (2007), *Community-Based Water Law and Water Resource Management Reform in Developing Countries* (Wallingford: CAB International).

von Benda-Beckmann, K. et al. (1997), 'Disputing water rights: scarcity of water in Nepal hill irrigation', in Edward H.P. Brans et al. (eds), *The Scarcity of Water: Emerging Legal and Policy Responses* (London: Kluwer Law International Ltd), pp. 224–242.

Wallace, J. and P. Wouters (eds) (2006), *Hydrology and Water Law: Bridging the Gap* (London : IWA Publishing).

Ware, Eugene F. (1905), *Roman Water Law – Translated from the Pandects of Justinian* (HeinOnline).

Weeramantry, C.G. (1998), 'Separate opinion in the case concerning the Gabcikovo–Nagymaros Project', *International Legal Materials* 37, 204–217.

White, Gilbert F. et al. (1972), *Drawers of Water: Domestic Water Use in East Africa* (Chicago: University of Chicago Press).

Wieringa, Keimpe (1995), *Environment in the European Union 1995: Report for the Review of the Fifth Envirornmental Action Programme* (Luxembourg: European Environment Agency).

Wilkinson, David (2002), *Environment and Law* (London: Routledge).

Winpenny, James (1994), *Managing Water as an Economic Resource* (London: Routledge).

Woolrych, Humphry W. (2005), *A Treatise of the Law of Waters: including the Law relating to Rights in the Sea, and Rights concerning Rivers, Canals, Dock Companies, Fisheries, Mills, Watercourses, etc., with a note concerning the Rights of the Crown to the Land between High and Low Water Mark* (1st Am. edn, Clark, N.J.: Lawbook Exchange).

World Commission on Environment and Development (1987), *Our Common Future* (The Brundtland Report) (Oxford: Oxford University Press).

World Health Organisation (2003), *The Right to Water* (Geneva: World Health Organisation).

Wouters, Patricia (ed.) (1997), *International Water Law: Selected Writings of Professor Charles B. Bourne* (London: Kluwer Law International).

Wouters, Patricia and Alistair Rieu-Clarke (2004), 'The role of international water law in ensuring "good water governance": a call for renewed focus and action', *Water Law* 15, 89–92.

Xi, Wang et al. (2002), 'Managing water resources for a sustainable future: law, policy and methodology of China', *Asia Pacific Journal of Environmental Law*, 7, 5–41.

Yandle, Bruce (1983), 'Resource economics: a property rights perspective', *Journal of Energy Law and Policy* 5, 1–19.

Yousri, Abd al-Rahman (2005), 'Sustainable development: an evaluation of conventional and Islamic perspectives', in Munawar Iqbal (ed.), *Islamic Perspectives on Sustainable Development* (Basingstoke: Palgrave Macmillan), pp. 22–57.

Zhang, Jiebin (2006), 'Water management issues and legal framework development of the Tarim River Basin', in J. Wallace and P. Wouters (eds), *Hydrology and Water Law – Bridging the Gap* (London: IWA Publishing), pp. 108–142.

Index